The Professional Pool
Maintenance Manual

The Professional Pool Maintenance Manual

Terry Tamminen

Illustrated by Owen W. Smith
Professor, California State University Northridge

McGraw-Hill, Inc.

New York San Francisco Washington, D.C. Auckland Bogotá
Caracas Lisbon London Madrid Mexico City Milan
Montreal New Delhi San Juan Singapore
Sydney Tokyo Toronto

Library of Congress Cataloging-in-Publication Data

Tamminen, Terry.
    The professional pool maintenance manual / by Terry Tamminen.
      p.   cm.
    Includes index.
    ISBN 0-07-061407-5
    1. Swimming pools—Water-supply. 2. Hot tubs—Water-supply.
    3. Water—Purification. I. Title.
    TH4763.T36 1994
    690'.89—dc20   94-12061
                                                    CIP

Acquisitions editor: April Nolan
Editorial team: Laura J. Bader, Editor
                Susan W. Kagey, Managing Editor
                Joanne Slike, Executive Editor
                Joann Woy, Indexer
Production team: Katherine G. Brown, Director
                Lisa M. Mellott, Coding
                Jan Fisher, Desktop Operator
                Linda L. King, Proofreading
Design team: Jaclyn J. Boone, Designer   TAB1
              Brian Allison, Associate Designer   0614075

To Norman James Ackerberg, a man who knows all about hot water and business . . . and who got me into both! Thanks for the love and friendship.

Contents

8 Water Chemistry   245

9   Cleaning and Servicing   281

10   Special Procedures   299

Acknowledgments

There can be only one person who "taught me all I know." That person for me was my partner Ritchie Creevy.

My special thanks to Paul Van Bloem for all the help in getting me started. Thanks for various assistance and technical support to Ruby Raitt, Dee Dee Dalton, Ross McCulloch, and Edward deMahy. My thanks also to Rick and Langley Luskin, John Cronin, Robb Rice, and especially Mr. and Mrs. Peter Just for the use of their pools in testing my theories.

Many manufacturers of pool and spa equipment and supplies helped me with this book. I have tried to take my examples from industry leaders—products you will encounter most often in the field. Excluding a good product from this company or that one is not a suggestion that one product is not as good as another. In some cases, I used manufacturers' drawings that best illustrated the point I was trying to make in the text. Virtually every manufacturer and supplier in the industry has been cooperative and supportive of this book; however, some were able to provide the needed illustrations before my publisher's deadline!

In short, I owe thanks to just about every company in the water technology industry for their training, product innovations, product literature, and advice. The following list reflects my special gratitude to those who made specific contributions to the book.

Aladdin Equipment Company, Inc., Sarasota, Florida, Lindy Smith
American Leak Detection, San Jose, California, Paul Routon and Pat Serra
Anthony Pools, Montebello, California, Scot Buxbaum
Aqua-Flo, Inc., Chino, California, Mark Coito
Aquaswim 'n' Spa, Rio Plastics, Inc., Brownsville, Texas, Carol Meyer
Arneson Products, Inc., Fairfield, California, Dan Rothschild
Blue White Industries, Westminster, California, Selma Ponsell
Brett Aqualine, Inc., Huntington Beach, California, Jim Brett
Caldera Spas and Baths, El Cajon, California, Marcy Francis and Brian Roble
California Cooperage and Hot Shells, Santa Monica, California, Tom Healy
California Pools & Spas, Ontario, California, W. Doug Steimle
Cantar Corporation, Youngstown, Ohio, Brian Frost
CEnTech Corporation, Puyallup, Washington, Gary L. Perry and Roger Perry
Conger Manufacturing, Paso Robles, California, Janis Conger
Cover Pools, Inc., Salt Lake City, Utah, Elayna Dikeon
Creative Homeowner Press, Upper Saddle River, New Jersey, Ted Marcus
Franklin Electric, Bluffton, Indiana, Mel Haag
Harmsco, Inc., North Palm Beach, Florida, Hiedy L. Hansel
Hayward Pool Products, Inc., Pomona, California, William G. Treene and Terry Rake

Heat Siphon ® Swimming Pool Heat Pumps, Latrobe, Pennsylvania, Bill Bernardi

Intermatic, Inc., Spring Grove, Illinois, Karen Hahn and Eugene D. Banathy, P.E.

Jacuzzi, Inc., Little Rock, Arkansas, Lesa Poe

Kreepy Krauly USA, Inc., Sunrise, Florida, Leon Rawitz

Lass Enterprises, Inc., Altamonte Springs, Florida, Alan D. Vanderboegh

Len Gordon Company, Las Vegas, Nevada, Brent Blanchard and Tim Ware

Loop-Loc Ltd., Hauppauge, New York, Marilyn DiDomenico

Los Angeles County Health Department, Gil Rodriguez

MagneTek, St. Louis, Missouri, Renee Deering

Melnor, Inc., Moonachie, New Jersey, Deborah J. Wright

MP Industries, Garden Grove, California, Dan Morrison

Odyssey Reel Systems, Ltd., San Clemente, California

Pac-Fab, Inc., El Monte, California, Michael Tacconi and Becky Martin

Polaris Pool Systems, Inc., San Marcos, California, Robert Stuart and Barbara Michaels

Pool and Spa News, Los Angeles, California, Jules Field and Rodney Collins

Pool Plans, Inc., Garden Grove, California, B.C. Adams and Ginette Adams

Pool Tool Company, Ventura, California, Herb Tilsner

Pool Water Products, Canoga Park, California, David Winge and his staff, especially Don Cole, Steve McCool, and Bill Mace

Premier Spring Water, Inc., Pacoima, California, Paul R. Blum

Purex Pool Systems, Inc., City of Industry, California, Tom Phillips

Questar Pools and Spas, Escondido, California, Skip Phillips and photographer Marv Sloban (cover photo)

Rainbow Lifeguard Products, El Monte, California, Richard Joseph

Raypak, Inc., Westlake Village, California, Kathy Etidi, Tony Ots, and Tom Apt

Sentry Pool, Inc., Moline, Illinois, Bob Gallagher

Service Industry News, Torrance, California, Dave Dickman

Spelvin Industries, George Spelvin

Sta-Rite Industries, Inc., Delevan, Wisconsin, Pam Spotz and Karen E Spors

Sundance Spas, Chino, California, Dix Henneke

Sunstar Enterprises, San Marcos, California, Laurie J. Batter

Swimrite Company, New York

Teledyne Laars, Moorpark, California, Pat Coblegh, Paula Dolan, and Bob Hamos

Tropicana Pools and Spas, Lagos, Nigeria, Wanle Akinboboge

United Leisure Industries, St. Petersburg, Florida, Ray Westlund, Jr.

Val-Pak Products, Canyon Country, California, Ben Solakian

Finally, but perhaps most importantly, thanks to Art Arndt and Iris Langdon, without whom this book would not exist.

Foreword

When Terry Tamminen first contacted me about this book, I had my concerns. Is this going to be another do-it-yourself book that does a disservice to the client and the professionals? It didn't take long for me to understand that Terry's approach was one of increasing the level of professionalism within our industry and, most particularly, of the pool service technician.

Pool builders have traditionally completed their projects and left the care and maintenance up to the homeowner and the professional service technician. The builder is always pleased when the homeowner chooses the professional. It usually makes everybody's job easier and helps ensure the proper use and ultimate enjoyment of the pool or spa.

Unfortunately, however, at times the pool and spa technician does not work closely with the pool builder. Where there ought to be a strong liaison, due to misinformed or self-serving interest, there is just the opposite. The service technician pretends to be more knowledgeable than he or she is, taking the position of (and is viewed by the client as) an expert, offering suggestions, criticisms, or recommendations to the client that destroy the bond that should exist between technician and builder. The builder is often guilty of the same. Rather than building up the credibility of the professional technician, he or she responds, "Oh, he's just another guy with a pick-up truck, brush, and skimmer net—anyone can do that." Builders would do well to learn more about the sophisticated process of properly diagnosing and servicing the product we manufacture.

If you are reading this book, you are obviously the type of professional who intends to serve his or her clients well and make sure that the "dream" of owning a pool and spa is our client's reality. It's a major investment on their part and one of the best they can make. This book, and Terry's ardent desire to have us all be professional, helps ensure the satisfaction of our clients and our industry.

W. Douglas Steimle, President
California Pools and Spas
Ontario, California

Introduction

Dustin Hoffman, Madonna, Stacy Keach, Dick Clark, Barbra Streisand, Charles Bronson, David Letterman, Rich Little, Carroll O'Connor, Lou Gossett, Dyan Cannon, Kareem Abdul Jabbar, Walt Disney, Martin Sheen, Sting, Goldie Hawn, Olivia Newton-John, Roy Orbison, Diana Ross, George C. Scott, Dick Van Dyke, Bruce Willis.

That is a partial list of the celebrities for whom I have done pool or spa work in the past 17 years (actually, the list of agents, producers, artists, and authors is even more impressive, but you might not recognize the names). These celebrities are, after all, just people who want a clean pool, spa, or water feature like anybody else.

Yes, I am the Malibu Pool Man to the Stars. Big deal! My more anonymous clients are just as important, because this is a business based on word-of-mouth referrals. But more on that later.

For now . . . Welcome to the world of water maintenance. Yes, water maintenance, not pool or spa maintenance. A pool or spa is merely the chosen vessel, but the water is the real product. It might be flowing in a pool, spa, fountain, or commercial application, but the point is that you are dealing with water and its effects on the vessel, plumbing, and related equipment. By accepting this basic premise, you will approach every aspect of this book in a way that will give you understanding of the *why* and *how* and not just procedures. In short, you will achieve better results in all aspects of pool, spa, and fountain *water* maintenance.

When I first began in the water business in 1976, I couldn't believe that there was no authoritative text (or any text) on the subject of pool, spa, and water maintenance. There were licensing courses in various states and municipalities (all widely varied in subject matter and emphasis), but none were taught with a text book. Information was passed down from one old pro to another, complete with his or her preferences, prejudices, and mistakes. (By the way, in the interest of space and style that is the last time I will use the his/hers format. When I use the masculine gender in this book, it is for simplicity of writing rather than gender discrimination. Some of the most competent technicians I know are female; however, because there are more males in the trade, I will use the masculine in my writing and ask that you "fill in the blanks." Thanks.).

This is how I was forced to learn, the famous OJT (on-the-job training), often at my expense, doing jobs three or four times to get them right while learning from my mistakes. I would have given anything for a reference guide on the chemical, mechanical, and business aspects of the water trade. There is still no such book available, so after years of experience and gathering technical information from factory training, health department licensing courses, and OJT, here it is.

After all this time, I am still amazed that our profession is still not a "profession." Unlike most other building and maintenance trades such as plumbers, electricians, carpenters, masons, engineers, and architects, the pool and spa technician is still regarded by many as a janitor, when in fact they must be expert in aspects of each of those trades mentioned. (As an example, the State of California Workman's Compensation Fund classifies pool and spa technicians under janitorial.)

In part, this perception is caused by incompetent practitioners—the typical pool man who cleans a few pools, giving equal space in his truck to pool equipment and a surfboard, working out of his (or his parents') garage, and never making any attempt to understand the industry in which he works or the chemistry and equipment involved.

Oh sure, there is the NSPI (National Spa and Pool Institute) and various independent pool and spa service technician groups, all of which try to promulgate standards. But there is a shocking lack of uniformity in standards and practices from group to group, city to city, and state to state. I believe this is due in part to the lack of accurate, current information in a readily accessible format about water maintenance and the equipment used to clean, move, filter, and heat water. I hope this book will help to address at least that part of the problem.

In fact, the typical pool and spa maintenance technician is an amazing jack-of-all-trades. He is required to be an expert chemist, plumber, electrician, furnace operator, mason, and driver—in short, he deals with many trades which when combined deal with serious matters of health and safety. Consider, for example, the potentially lethal combinations of water, electricity, natural or propane gas, and caustic chemicals, yet people are not surprised to pay a plumber $50 an hour, but balk at paying a competent, trained pool and spa technician $25 an hour. Why do I mention all this? Because I believe the competent, trained individuals will eventually run the fly-by-nights out of business, and with the spiraling increase across the country in pools and spas, more states and municipalities will require increased levels of professionalism for water technicians.

In short, if you treat this as a professional trade, you will be ahead of your competition and be prepared for the future. Also, you will be able to justify being paid as a professional.

This book is designed for the pool and spa professional, but because levels of understanding in this industry vary so greatly, it assumes little knowledge of each topic and presents material from the most basic conceptual discussion, graduating to the most complex. I hope this book will aid those just starting in the field and the do-it-yourself homeowner. The glossary at the end of the book will assist you in gaining a thorough understanding of water maintenance.

It will be helpful to master the information in this book and then stay on top of this rapidly changing field by attending factory seminars, where available, and reading professional journals and publications to remain current on the latest products and techniques. Some of these publications are listed in the reference list at the end of the book. To help the technician or homeowner budget repairs or maintenance, I quote prices and costs based on national averages at the date of publication. You will need to adjust these according to local prices, shipping (if you are not in a major metropolitan area), and inflation. Special tools, some of which you can make from simple materials, are detailed in each chapter.

Finally, what makes me the "water pro" who should write "The Book?" I enjoy the field and approach it as a great profession; my experience is what makes this book work. Everything discussed in this book I have done, hands on, countless times, and continue to do in my business in Malibu, California (I have also done pool work in Florida, Africa, Europe, and South America). I have participated daily in every aspect of the water business from construction to cleaning, repairs to marketing. I hold a Los Angeles County Health Department license and factory authorized training from countless manufacturers and tradespeople. In my years in the water trade, both in California and Florida, I have also made many contacts and friends in the field, related trades, and manufacturing.

As you can see from the acknowledgments, their assistance, input, and in some cases actual review of the text, has been essential to making this book not only accurate but practical. Lastly, no one else has taken the time to do this, so until someone comes out with a better book, this is it.

Good luck.

1
The pool and spa

As I said in the Introduction, this is a book about water before it is a book about pools, spas, fountains, or other water containers. If you wanted to be a banker, it would be nice to understand something about the bank, vault, and cash drawers, but the real business is the money and how it is used. Similarly, this book will contain appropriate information about the "containers", but the fact is, most of us will never build a pool, spa, fountain, hot tub, or other such container. Our business is water and the related products and equipment that move or change it.

OVERVIEW

I will begin with a basic overview of the typical container and water system. Figure 1-1 shows a typical pool and spa and its related equipment. A hot tub or spa alone is plumbed and serviced in a similar manner. Just as in banking where you follow the money if you want to learn that business, in this business you must follow the water. That is also how this chapter (and the entire book) is outlined—in the logical pattern that the water travels from pool through plumbing to the pump/motor, filter, heater, and back to the pool.

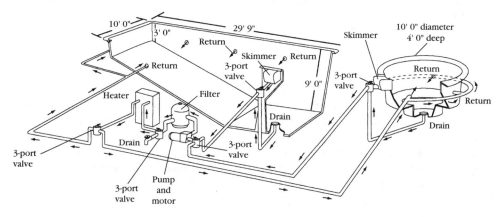

1-1 Typical pool and spa with equipment.

Follow the arrows in Fig. 1-1 to follow the path of the water. The water enters the plumbing through a main drain and/or a surface skimmer (components of either a pool or a spa). It does this thanks to suction created by a pump and motor. After passing through the pump, the water is cleansed by a filter, warmed by a heater, and returned to the pool or spa through return outlets.

CAPACITY CALCULATIONS

Because many of the calculations in this book depend on knowing the quantity of water involved, here is how to calculate the volume of your pool or spa.

Square or rectangular

The formula is simple:

$$Length \times width \times average\ depth \times 7.5 = volume\ (in\ gallons)$$

Let's first examine the parts of the formula. Length times width gives the surface area of the pool. Multiplying that by the average depth gives the volume in cubic feet. Since there are 7.5 gallons in each cubic foot, you multiply the cubic feet of the pool by 7.5 to arrive at the volume of the pool (expressed in gallons).

The formula is simple and so is the procedure. Measure the length, width, and average depth of the pool, rounding each measurement off to the nearest foot or percentage of one foot. If math was not your strong suit in school, remember one inch equals 0.0833 feet. Therefore, multiply the number of inches in your measurements by 0.0833 to get the appropriate percentage of one foot.

$$Example:\ 29\ ft,\ 9\ in. = 29\ ft + (9\ in. \times 0.0833)$$
$$= 29 + 0.75$$
$$= 29.75\ ft$$

Average depth will be only an estimate, but obviously if the shallow end is 3 feet and the deep end is 9 feet, and assuming the slope of the pool bottom is gradual and even, then the average depth is 6 feet. If most of the pool is only 3 or 4 feet and then a small area drops off suddenly to 10 feet, you will have a different average depth. In such a case, you might want to treat the pool as two parts. Measure the length, width, and average depth of the shallow section, then take the same measurements for the deeper section. Calculate the volume of the shallow section and add that to the volume you calculate for the deeper section.

In either case, be sure to use the actual water depth in your calculations, not the depth of the container. For example, the hot tub depicted in Fig. 1-2 is 4 feet deep, but the water is only filled to about 3 feet. Using 4 feet in this calculation will result in a volume 33 percent greater than the actual amount of water. This could mean serious errors when adding chemicals for example, which are administered based on the volume of water in question. There might be a time when you want to know the potential volume, if filled to the brim. Then, of course, you would use the actual depth (or average depth) measurement. In the example, that was 4 feet.

Try to calculate the volume of the pool in Fig. 1-1:

$$Length \times width \times average\ depth \times 7.5 = volume\ (in\ gallons)$$
$$29.75\ ft \times 10\ ft \times 6\ ft \times 7.5 = 13,387.5\ gallons$$

Circular

The formula:

$$3.14 \times \text{radius squared} \times \text{average depth} \times 7.5 = \text{volume (in gallons)}$$

The first part, 3.14, refers to pi, which is a mathematical constant. It doesn't matter why it is 3.14 (actually the exact value of pi cannot be calculated but who cares?). For our purposes, we need only accept this as fact.

The radius is one-half the diameter, so measure the distance across the broadest part of the circle and divide it in half to arrive at the radius. Squared means multiplied by itself, so multiply the radius by itself. For example, if you measure the radius as 5 feet, multiply 5 feet by 5 feet to arrive at 25 feet. The rest of the equation was explained in the square or rectangular calculation.

Use the hot tub in Fig. 1-2 to calculate the volume of a round container. Let's do the tricky part first. The diameter of the tub is 10 feet. Half of that is 5 feet. Squared (multiplied by itself) means 5 feet times 5 feet equals 25 square feet. Knowing this, you can return to the formula:

$$3.14 \times \text{radius squared} \times \text{average depth} \times 7.5 = \text{volume (in gallons)}$$
$$3.14 \times 25 \text{ ft} \times 3 \text{ ft} \times 7.5 = 1766.25 \text{ gallons}$$

Note that in measuring the capacity of a circular spa, you might need to calculate two or three areas within the spa and add them together to arrive at a total volume. An empty circular spa looks like an upside-down wedding cake, because of the seats, as in Fig. 1-3. Therefore, you might want to treat it as two separate volumes—the volume above the seat line and the volume below. In the wooden hot tub depicted in Fig. 1-2, where there is actually water above and below the seats, the tub can be measured as if there are no seats because this difference is negligible.

1-2 Cross-section of a typical hot tub.

1-3 Cross-section of a typical spa.

Kidney or irregular shapes

There are two methods used to calculate the capacity of irregular shapes. First, in Fig. 1-4, you can imagine the pool or spa as a combination of smaller, regular shapes. Measure these various areas and use the calculations described previously for each square or rectangular area and for each circular area. Add these volumes together to determine the total capacity. Figure 1-4 contains one rectangle and one circle (shown in two halves).

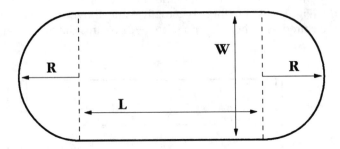

1-4 Volume of irregular shapes.

The second method is as diagrammed in Fig. 1-5:

$$0.45 \times (A+B) \times \text{length} \times \text{average depth} \times 7.5 = \text{volume (in gallons)}$$

The total of measurement A plus measurement B multiplied by 0.45 multiplied by the length gives you the surface area of the kidney shape. (A + B = 18 feet). The rest of the calculations you are now familiar with. Try this volume calculation:

$$0.45 \times (A+B) \times \text{length} \times \text{average depth} \times 7.5 = \text{volume (in gallons)}$$
$$0.45 \times 18 \text{ ft} \times 25 \text{ ft} \times 5 \text{ ft} \times 7.5 = 7593.75 \text{ gallons}$$

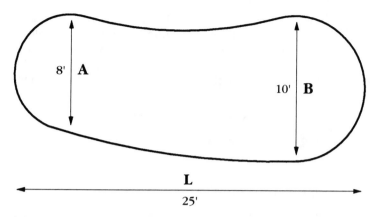

1-5 Volume of kidney shapes.

Parts per million (ppm)

One other important calculation you will use is parts per million (ppm). The amount of solids and liquids in the water is measured in parts per million, as in three parts of chlorine in every one million parts of water (or 3 ppm). However, one gallon of chlorine, for example, poured into one million gallons of water does not equal 1 ppm. That is because the two liquids are not of equal density. This becomes obvious when you discover that a gallon of water weighs 8.3 pounds but a gallon of chlorine weighs 10 pounds (in a 15 percent solution, as described later). The chlorine is a more dense liquid—there's more of it than an equal volume of water.

To calculate parts per million, use the following example:

1 gallon of chlorine in 25,000 gallons of water = 10 pounds of chlorine in 207,500 pounds of water

Now dividing each by 10 gives you:

1 pound of chlorine in 20,750 pounds of water

So you see that 1 part of chlorine is in each 20,750 parts of water. But how does that translate to parts of chlorine per one million parts of water? To learn that, you must find out how many 20,750s there are in a million.

$$1,000,000 \div 20,750 = 48.19$$
$$48.19 \times 1 \text{ part of chlorine} = 48.19$$

There are 48.19 parts of chlorine in each million parts of water, expressed as 48.19 ppm.

Using the same formula without first translating the two liquids into pounds would give an answer of 40 ppm. Obviously this great discrepancy can result in substantial errors in treating water chemistry problems. But we're not through just yet.

If chlorine were 100 percent strength as it comes out of the bottle, that would be all there is to this calculation. As you will see in later chapters, that is not the case.

In fact, liquid chlorine is produced in 10 to 15 percent solution, meaning 10 to 15 percent of what comes out of the bottle is chlorine and the rest is filler. Therefore, to really know how many parts of chlorine are in each million parts of water, you must adjust your result for the real amount of chlorine. Usually liquid chlorine is 15 percent strength (common laundry bleach is the same product, but around 3 percent strength), so:

$$48.19 \times 0.15 = 7.23 \text{ ppm}$$

Therefore, 7.23 ppm is our true chlorine strength in the example of 25,000 gallons of water.

TYPES OF POOLS AND SPAS

Ever since the invention of the creekside swimming hole, complete with swinging rope or tire, pool builders have invented new and creative ways to capture water in our backyards. Today, because of modern materials, engineering, and building techniques, there are countless types of pools, spas, fountains, and ponds. Here are some of the most common. You will find that others are variations of these basic types.

Concrete and plaster

Concrete and plaster pools are the most typical "hole-in-the-ground" pool, using steel-reinforced concrete to form a shell (Fig. 1-6). Because concrete is porous, the shell is coated with plaster to hold water and for cosmetic purposes.

Questar Pools and Spas, Escondido, Ca.

1-6 Reinforcing steel bars (rebar) laid out for pool construction.

Concrete can be sprayed over a latticework (Fig. 1-7) of reinforcing steel bars (rebar) or forms, or it can be poured from a mixing truck. The sprayed types are the most common because the material is easier to work to create free-form shapes.

1-7 Gunite sprayed onto rebar.

Sprayed concrete types include gunite (an almost dry mix of sand and cement) or shotcrete (a wetter version of gunite). Poured concrete requires forms into which the wet concrete is poured.

Sometimes, the pool walls are made of hollow block that is reinforced with rebar and filled with poured concrete, but that is not common. They are more common in countries where modern equipment is scarce. Figure 1-8 shows a pool I built in Nigeria. It was dug completely by hand and the forms were built of wood into which concrete was poured to form the walls.

In any case, a cutaway of these types of construction is shown in Fig. 1-9.

Vinyl-lined

Vinyl-lined pools or spas are built as metal or plastic frames (less frequently masonry blocks or pressure-treated wood is used) above the ground or set into a hole in the ground. Prefabricated panels of plastic, aluminum, or (rarely) wood are joined to the frame making a form that is then lined with heavy vinyl to create the actual pool shell (Fig. 1-10). These pools require somewhat different treatment and maintenance methods than a concrete pool. Prefabrication and easier assembly make these pools less costly than concrete styles. Some small units are made as do-it-yourself backyard kits consisting of self-supporting aluminum or steel frames with a vinyl liner, sometimes with stairs, decks, and equipment all packaged together.

Laurie and Ted Latour

Laurie and Ted Latour

1-8
Pool construction methods in Nigeria.

1-8 Continued.

Concrete
pool deck

Expansion joint

Coping

Grout or caulk

Tile

Sand or
clean fill

Mortar bed

Cap leveling mud

Bond beam

Rebar

Dirt
base

Brown coat

Thin plaster coating

Wall

Concrete
(gunite or shotcrete mix)

Dirt
base

1-9 Cross-section of a typical gunite pool wall.

Fiberglass

Fiberglass pools or spas consist of a fiberglass shell resting in a hole in the ground. Aboveground models are framed with metal or plastic like vinyl-lined pools. Some concrete pools are lined with fiberglass instead of plaster and, in fact, this process is becoming more popular because fiberglass requires less maintenance than plaster.

Some fiberglass pools or spas are assembled on-site from panels, some are an entire molded shell, and some are fiberglass walls mounted on a concrete pool bottom.

Spas, if not built with a pool or as an architectural feature in the home, are almost all some form of fiberglass shell (Fig. 1-11). These can range from a shell set directly into the ground to small portable units (Figs. 1-12 and 1-13) that are self-contained wooden frames with wooden skirting containing the fiberglass spa and all the equipment in one package.

Plumbing and installation of spas is discussed in a later chapter.

A

Coping

Sand or
clean fill

Vinyl liner

Wall panel
(plastic, fiberglass,
or metal)

Braces
(metal or plastic)

Concrete

Dirt
base

Anchor pin

B

I-10 Construction (A) and cross-section (B) of a typical vinyl-lined pool.

Types of pools and spas   **11**

1-11 Typical acrylic-over-fiberglass preplumbed spa.

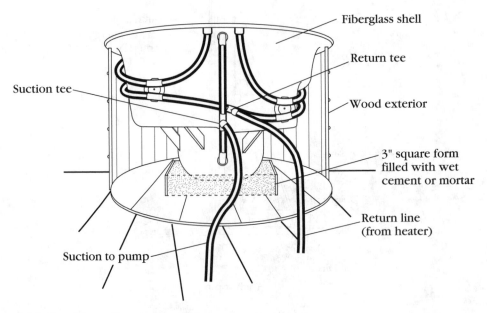

Fiberglass shell

Return tee

Suction tee

Wood exterior

3" square form
filled with wet
cement or mortar

Return line
(from heater)

Suction to pump

1-12 Acrylic-over-fiberglass spa built into wooden hot tub. Hot Shell-California Cooperage, Santa Monica, Ca. (A Woodward-Healy concept)

1-13 Self-contained spa.

Inflatable

Zodiac, makers of the famous inflatable boat, and other manufacturers are now applying PVC technology to pools and spas. No longer just the "kiddie" pool, inflatables are an inexpensive and portable solution to the swimming or spa needs of many.

Wood

Hot tubs are made from a variety of woods, most commonly redwood (but cedar, teak, mahogany, and more exotic woods are also used). In fact, they can be made from any wood, but those mentioned are most resistant to rot. The key to hot tub installations is a well-drained area to keep moisture (which sooner or later will rot any wood) away from the tub (Fig. 1-14). Hot tub assembly and installation are discussed in a later chapter.

1-14
Typical wooden hot tub (profile).

POOL AND SPA DESIGN AND CONSTRUCTION

This book is not designed to make you a pool or spa builder or design expert. In the reference sources at the end of the book you will find a list of excellent resouces to help you do those things, taking into consideration your ground plan, slope and geology, wind and sun conditions, landscaping, home value, use and need, and a host of other planning and building aspects. Ask four pool builders their opinion of this process and you will get at least five answers and discover that no two are alike in approach or priority. An entire book could be written on that subject alone, but unless you are planning to study for a pool contractor's license or are a homeowner with a self-destructive mentality, you will not be building or designing pools. Generally you will be dealing with pools or spas that are already in use.

Still it is worth understanding how that pool or spa came into being. As a pool technician or someone buying a home with a pool, you will want to recognize good (or bad) construction. This knowledge will help you estimate the maintenance costs, future potential repair expenses, or the problems involved in upgrading the pool or spa in question.

The construction of a pool or spa is basically approached in the following way (usually in this order).

Plans and permits

The architect or pool builder supplies plans (Fig. 1-15) and hires an engineer to provide the steel (rebar) schedule based on calculations of soil stability, geology, etc. The schedule includes the specifications of thickness, tensile strength, and how close each bar will be to the next based on how strong the final product needs to be. This will differ from one part of the pool to another. For example, the bottom needs to support more weight than the sides.

A local building permit is issued when plans and steel schedules are approved. Careful planning should include consideration of backyard access. Be sure heavy equipment will not be running through fences or over fragile planting areas, septic tanks, pipes not deeply buried, etc. Also make sure the area to be excavated (including pipe trenches) does not traverse underground gas pipes, water pipes, phone lines, or electrical conduit. If so, make plans to reroute these and excavate them carefully.

Excavation

The excavator lays out the pool diagram, based on the plans, using bender board or wooden planks (Fig. 1-16). He digs according to the plan *plus* one foot to allow for the thickness of material and plumbing. He digs straight down approximately 3 feet, then slopes the remainder to the bottom. The actual slope and contour of the pool is determined by the rebar sculpting. The excavator also cuts out a 2 cubic foot area for each skimmer and for each light to be installed. He also trenches from the poolside to the equipment area for laying pipe. It should also be the responsibility of the excavator to remove the dirt from the job site, as well as tamp down, compact, and add a layer of gravel to the finished hole. The gravel aids drainage if groundwater is present from below or if leaks occur in the structure.

SLOPE GRADE TO DRAIN AWAY FROM POOL

CAP

GUNITE

2 #3a TOP

VENEER

MORTAR/GROUT

#3a 12" O.C. B.W. (BOTH WAYS.)

W.L.

HAND GRIP (OPTIONAL)

A

2'-0" MAX.

"H"

3" 3"

3" 3"

6" MIN.

4 #3 HORIZ.

SEE SCHEDULES

VERT. STEEL

GUNITE POOL SHELL

NOTCH ② RAISED

"H"

STUCCO SCREED

EXP. JOINT @ HOUSE P.A.

DECK

B

TYP. SEPARATION & EXPANSION JOINT @ POOL @ POOL DECK & @ STEPS.

JOINT @ STEP ⑪

1-15 Typical pool and spa plans. A-D: Pool Plans, Inc., Garden Grove, Ca.
E and F: Questar Pools and Spas, Escondido, Ca.

Pool and spa design and construction **15**

TYPICAL POOL SECTION ⑰

C

1-15 Continued.

D

○ APPROVED ANTI-VORTEX
SUCTION COVERS
○ SPA FLOOR IN SOIL 3" TO POOL

SPA SECTION ⑫

E

1-15 Continued.

POOL GENERAL SPECIFICATIONS

SIZE 31½x21 IRREGULAR DEPTH 3½/6½
SHAPE CUSTOM
LINER PEBBLETECH
POOL CAPACITY 20,000 GALS.
PUMP HAYWARD
MOTOR H.P. 2 H.P.
FILTER C-4000 SQ. FT.
VACUUM LINE & SKIMMER 3"
RETURN LINE 3" (4 RET.)
MAIN DRAIN 1½
SKIMMER MODEL U-3/2
BACKWASH LINE N/A
32' OF ½" FILL LINE AUTOFILL
ANTI SIPHON VALVE STANDARD
HEATER HAY. SIZE 406 BTU
GASLINE BY GAS, VENTED BY: "
LIGHT 4 POOL | SPA / OPTICS
CLOCK TM1
ELECTRIC BY QUESTAR
ELECTRICAL BONDING BY: 11
POOL CLEANER PLUMBED
CHLORINATOR —
BOARD – SIZE —
BOARD SUPPORTS —
LADDER – MODEL —
SLIDE # ___ Color ___ Water Hookup ___
GRADING 2 HRS.
STUB PLUMB ☐ YES ☒ NO
DECK BY: OWNER

NOTES
SCALE 1/8" = 1'0"

SPA HAS GRANITE
SPILLWAY UNDER
16" REEF STONE
BBB ON SPA
& POOL ELEVATION
IN GLASS MOSAIC

DWN BY ___ DATE ___
CK'D BY ___ DATE ___

NOT TO SCALE
6½' DEEP END 3½
SHALLOW END
UNLESS OTHERWISE SPECIFIED:
POOL IS __ SHALLOW TO __ DEEP
I HAVE RECEIVED A COPY OF
THIS PLAN AND HEREBY
APPROVE POOL AND
EQUIPMENT LOCATION

CUSTOMER'S SIGNATURE   DATE

SPA GENERAL SPECIFICATIONS

SPA TYPE GUN. MDL # ___
DIMENSION: 8½ x 8
DEPTH: ___
COLOR — TOTAL GALLONS 850
SPA JETS 7 TILE
HEATER: —
PUMP & MOTOR: —
AIR BLOWER: —
GAS LINE: —
PLUMBING FOR SPA 3"
ELECTRICAL: — CLOCK: —
EXCAVATION: —
DECKING —
MISCELLANEOUS: —

SOLAR GENERAL SPECIFICATIONS

SQ. FT. POOL ___ SQ. FT. PANEL ___
PANEL TYPE ___ PANEL SIZE ___
NUMBER PANELS ___ PLUMB RUN ___
AUTOMATIC ___ MANUAL ___
THERMOMETERS ___ BOOSTER PUMP ___
SINGLE ☐ DOUBLE ☐ ELECTRIC BY: ___

JOB NO. ___
MAP BOOK NO. ___
LEGAL DESCRIPTION ___
___
___
___
LOT NO. ___
TRACT NO. ___
BOOK ___ PAGE ___ BLOCK ___
ESCROW CLOSE ___
TENTATIVE DIG DATE ___
PERMIT OFFICE COUNTY
MGR. ___
SALESMAN ___

OWNER:
TO DETERMINE
APPROXIMATE ELEVATION OF
POOL ON DAY OF EXCAVATION

POOL AREA TO BE FENCED,
BY OWNER PER COUNTY OR CITY
ORDINANCE. GATES TO BE SELF
CLOSING AND SELF LATCHING.

DO NOT TURN ON POOL LIGHT
WHEN POOL IS EMPTY

POOL OR SPA
NAME ___
ADDRESS ___

CROSS STREETS ___
RES. PHONE ___
BUS. PHONE ___

F

I-15 Continued.

Anthony Pools, Montebello, Ca.

Anthony Pools, Montebello, Ca.

I-16 Typical phases of pool construction.

Anthony Pools, Montebello, Ca.

Anthony Pools, Montebello, Ca.

1-16 Continued.

1-16 Continued.

Anthony Pools, Montebello, Ca.

Anthony Pools, Montebello, Ca.

1-16 Continued.

Plumbing

Sometimes the steel rebar is laid first, with the steel man laying in the main drain and leaving room for the plumber to work around the steel, but often the plumber lays in his plumbing first. Skimmers are laid in where the prevailing wind in that area will push debris toward them (a change that should be made on the job if it was not figured into the original plan). Plumbing must be at least 18 inches below ground and gas lines 12 inches below (unless PVC gas line is used which must be 18 inches below as well).

The plumbing includes the main drain and skimmers, pipes to the equipment area and back to the pool, return outlets, automatic cleaner piping, waterfall lines, a water filler line, or other design requirements. All plumbing is then sealed off and tested under pressure to test for leaks. Some large pools or commercial installations might have water sent back to the bottom of the pool for even distribution of filtered water, chemicals, and heat.

Steel

The rebar is laid in a crisscross pattern and formed or sculpted to the final contour and design of the pool. Generally, rebar is centered 12 inches apart on pool walls and 6 inches apart on the bottom and stress points (it does no good to locate steel any closer than this because there will not be enough room for the concrete mix, which needs space to build up strength). Wherever the rebar crosses, it is tied together with heavy wire to create a large, single, mesh bowl. The steel man must be sure to closely cut the ends after making these ties. Long, loose tails that stick up will rust and later show through the gunite and plaster layers.

Heavier steel is often used for the top 12 inches and over the edge, called the *bond beam*. The bond beam supports the coping and sometimes the edge of the deck so it must be extra strong (often engineered to support up to 1000 pounds per linear foot). Sometimes the bond beam extends up several feet above the waterline for waterfalls or tile areas. If the job is a vinyl-lined pool, this steel process will instead be the layout of the support structure.

Electrical

The electrician grounds or bonds the steel (and any other metal within 5 feet of the water's edge). He also adds light fixtures, which must usually be at least 24 inches below the water surface. Leaks in pools often occur at the light fixture cutout area or *niche*. The electrician is concerned that his wiring is completed to local code and connected properly, but often cares little about waterproofing. That is not his job! Therefore, builders often end up filling conduit and the joints where the light niche and pool wall meet with silicone sealant.

Gunite

The exact gunite or shotcrete mix is specified by the engineer based on strength and weight-bearing needs (as you have seen, water weighs a lot!). Often the mix will be five parts sand or gravel to one part cement.

The mixture is shot under pressure with a hose and nozzle from the mixing truck in and around the rebar to create the pool or spa shell, usually 4 to 5 inches thick for walls, 6 inches on the deep end floor, and 11 inches for the bond beam. Like spraying water from a garden hose, some waste, splashing, and overspray occurs.

Some of the gunite does not adhere to the rest of the mix and falls away from the surface being sprayed. This waste is called *rebound* and should be cleaned up and thrown away. Some builders use it to fill in step, love-seat, or other contour areas. This is a poor practice because rebound hardens quickly and when it is used as filler it creates air pockets, which later settle and cause leaks. This is why a large percentage of cracks and leaks are found on or near pool steps. In a vinyl-lined pool, this gunite process will instead be the installation of the panels that form the shell.

Tile

Tile is added, usually at the waterline (Fig. 1-17) to create an aesthetically pleasing finish and to provide a material at the surface (where oil and scum accumulate) that is not porous and is therefore easier to clean. Sometimes decorative tile patterns or racing lanes are installed to match the tile surface line.

Of course, the entire pool can be tiled. I have serviced several such pools, and although the original cost is high, I can enthusiastically recommend this design. An all-tile pool is strikingly beautiful, holds a pH better than plaster pools, never needs refinishing, and does not stain or etch.

Rock, brick, or stone

Rock, brick, or stone brought to the edge of the pool, or in some cases over the edge and below the waterline, create unique designs and natural pond looks. Until a few years ago, rocks were trucked in and cemented in place around the pool. Unfortunately such shipping and installation was very expensive and the bond beam needed extra, costly reinforcement to support the weight.

Since the late 1980s, however, *rocks* are formed with light rebar or chicken wire sculpting, then covered with a special plaster, concrete, and sand mix colored to look like natural rock (Fig. 1-18). Made on-site, such artificial rock is made to conform exactly to design specifications and is far cheaper and lighter than real rock. Rock, slate, and waterfalls are added using Thoroseal (a waterproof concrete sealer) to set them in place.

The only drawback to these and natural rock is that they are porous and you will soon see unsightly white scale forming at the waterline. Even if you maintain perfect chemistry in the pool, natural evaporation leaves behind any mineral present in the water as *scale* (Fig. 1-19), which appears as white scum, mostly calcium, around the waterline. Although this can be scrubbed off of nonporous tile, it must be sandblasted off rocks and will reappear in short order. One solution is to keep a constant water level, replacing water as it evaporates so the scale line is hidden under the waterline. Of course, maintaining proper water chemistry balance also reduces scale. These aspects are discussed in later chapters.

1-17 Tile at the waterline (A) and an all-tile pool (B).

1-18 Artificial rock pools, spas, and waterfalls.

1-19 Scale deposits on tile.

Coping, decks, and expansion joints

Coping (Fig. 1-20) is the finish work done to the top of the pool wall, usually attached directly to the bond beam. Coping stones are often precast and made of porous material to provide better traction for the wet feet and hands of swimmers entering or leaving the pool.

Questar Pools and Spas, Escondido, Ca.

1-20 Typical pool coping, deck, and expansion joints.

After the coping is laid on, the deck is poured (or deck brick or stone work is done) up to the edge of the coping, leaving a ¼- to ½-inch gap. The gap, which allows expansion or contraction of the deck and coping materials in hot and cold temperatures, is filled with silicone caulking to keep out water. All of this work should be done before plastering or finish work, because it is usually the messiest procedure. I have seen builders forced to drain and replaster newly finished pools because they completed the pool before the deck work was done, only to find sloppy deck workers scatter the fresh, soft pool plaster with cement, gravel, and stone chips.

Equipment set

The pump(s), filter, heater, and ancillary equipment is set on its concrete pad and connected to the plumbing (Fig. 1-21). The electrician and plumber finish these hookups.

1-21 Typical pool equipment set.

Cleanup

A smart builder cleans the area of loose debris, rebound, concrete dust, other debris—
in short, anything that might end up in the pool after the water goes in or which might
stain the new plaster. Even dirt taken from the excavation will stain fresh plaster.

Plaster

The fine white or colored plaster, also called marcite in some regions, is now trow-
eled over the gunite, about ½ inch thick. Plaster can now be mixed to just about any
color you can imagine. Rails, ladders, drain covers, rope hooks, and so on are added
and plastered in place. Water is added immediately because the plaster hardens and
cures underwater. If the plaster is allowed to dry out before water is added, the
weight of the water can create stress and cracks.

I deal at greater length with plaster and related problems in the chemistry and
replastering chapters; however, it is worth pointing out here that if five plaster experts were asked a question about their field, you would have at least six opinions.
Pool men, plasterers, and builders disagree about every aspect of plaster application
and maintenance.

If this installation is a vinyl-lined pool, the laying out of the liner takes the place
of plastering.

Start-up

Plaster requires a break-in to prevent staining. The water also must be treated before swimming. This process is treated in detail in a later chapter.

2

Basic plumbing systems

To understand a pool or spa, you must follow the path of the water. This and all following chapters have been arranged in this manner and you will find it is easy to troubleshoot a pool or spa problem by following this path.

As can be seen in Fig. 2-1, water from the pool or spa (not both at the same time) enters the equipment system through a main drain on the floor, through a surface skimmer, or through a combination of both main drain and skimmer. It travels to a three-port valve (if there is no spa, there will be no such valve) and into the pump, which is driven by the attached motor. From the pump, the water travels through a filter, up to solar panels (if part of the installation), then to the heater, and back through three-port valves to the pool or spa return lines.

SKIMMERS

Some pools have more than one skimmer. The purpose of the skimmer, as the name implies, is to pull water into the system from the surface with a skimming action, pulling in leaves, oil, and dirt before they can sink to the bottom of the pool. It also provides a conveniently located suction line for vacuuming the pool. Most skimmers today are molded, one-piece plastic units. Older pools have built-in-place concrete skimmers.

In either case, the skimmer is accessed through a cover on top that sits on the deck at the edge of the pool (the cover will be plastic or concrete) or by reaching into the skimmer through the opening that faces the pool itself. As shown in Fig. 2-2, most skimmers are built into the deck, connecting to the pool out of sight. Some, as with portable or aboveground pools, are separate units that hang on the edge of the pool (in the water or outside of it). Redwood hot tubs use a flat, vertical skimmer that has no basket but skims the surface and pulls any floating debris to a plastic screen. Some portable spa skimmers have a cartridge filter built in. Some pool skimmers include automatic water level controls and automatic chlorinators.

2-1 Typical pool and spa plumbing layout.

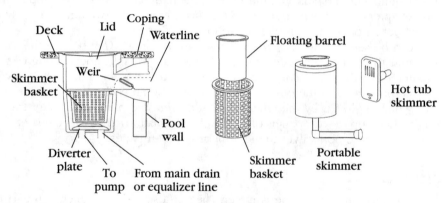

2-2 Typical skimmers.

The water pours over a floating weir (Fig. 2-2) that allows debris to enter, but when the pump is shut off and the suction stops, the weir floats into a vertical position, preventing debris from floating back into the pool. Some skimmers have no such weir (although spring-loaded weirs are available that can be fitted into any skimmer mouth) and use a floating barrel as part of the skimmer basket. The purpose of the basket is to collect leaves and large debris so they can then be easily removed.

The disadvantage of both types of weirs is that leaves can cause them to jam in a fixed position, thus preventing water from flowing into the skimmer. When this happens, the pump will lose prime (water flow) and run dry, causing damage to its components. Therefore, during windy periods it might be better to remove the weir from the skimmer to prevent such problems.

By the way, you should exercise care when working around the skimmer when the pump is on. I have nearly had fingers broken when placing my hand over a skimmer suction opening and have lost various pieces of equipment, T-shirts, bolts, and plastic parts, which invariably end up clogged in the pipe at some turning point where leaves, hair, and debris later catch and close off the pipe completely. Keep small objects away from the skimmer opening when the basket is removed and especially keep your hands from covering that suction hole.

There are two types of skimmer plumbing. The first one has a single visible suction port (opening). Actually the pipe from the main drain and the pipe from the skimmer connect just below the visible opening and a combination diverter is inserted to regulate the amount of suction from one or the other (Fig. 2-3). A neck on the diverter extends up from the skimmer bottom for attaching your vacuum hose when cleaning the pool. The value of this system is that when vacuuming the pool, you can divert all of the pump suction to the skimmer bottom, in effect, shutting off the main drain.

2-3 Single-port skimmer with diverter unit.

The diverter also has a nipple aligned horizontally to the opening. Usually when the nipple faces away from the pool, the flare on the bottom of the diverter closes the main drain pipe and all of the suction from the pump is now at the skimmer. When the nipple faces toward the pool, the body of the diverter closes the pipe from the skimmer and all of the suction is now at the main drain. Obviously various degrees between these two settings will divide suction between the skimmer and main drain.

Each pool needs its own setting to compensate for various factors such as wind conditions, equipment efficiency, and type of cleaning conditions. For example, if the pool gets more leaves than dirt, the diverter should be set to make the suction in the skimmer stronger than in the main drain. That will help the skimmer pull the leaves into the skimmer basket. If the pool tends to get more dirt or sand than leaves, the diverter should be set to strengthen the main drain suction. When dirt falls to the bottom of the pool, the strong suction from the main drain will pull the dirt toward it. This is also helpful when brushing the pool bottom, because suspended dirt will be pulled into the main drain.

Diverter units are made of plastic or bronze. I always carry a bronze unit with me because the plastic ones tend to come loose and float out during vacuuming if the suction from the pump is not strong. They also tend to rotate in the skimmer as you work, changing the amount of suction in the skimmer from what you have set. The bronze diverter, obviously heavier, solves these problems.

The other type of skimmer plumbing (see Fig. 2-2) has two separate ports—one is a pipe that goes directly to the main drain or to an equalizer line, while the other goes directly to the pump. In this type of skimmer, a diverter plate regulates the suction between the main drain and the skimmer. Usually the port farthest from the pool edge is the pipe that goes to the pump, and the port closest to the pool goes to the main drain or equalizer line. An *equalizer line* is simply a pipe that extends from the skimmer bottom down 18 to 24 inches and through the pool wall just below the skimmer.

In both styles of skimmer, the idea is that if the pool runs low on water, the pump can pull water from the bottom of the pool via the main drain instead of the empty skimmer (or from the side of the pool below the skimmer in the case of the equalizer line) so the pump will not run dry.

Some older (often concrete) skimmers have odd-sized ports that can't accommodate your vacuum hose. In these cases, a special cover plate (Fig. 2-4) can create a generic adaptor.

2-4
Skimmer vacuum adapter plate.
Jacuzzi, Inc., Little Rock, Ark.

MAIN DRAINS

The main drain has one or more plumbing ports. One port feeds a pipe to the pump. In a spa, there might be several ports for several pipes leading to different pumps (for jet action).

Another port is a one-way valve (check valve) that allows water that might collect under the pool to enter the pool, but no water can flow out. Figure 2-5 depicts a main drain with a *hydrostatic valve*.

Water collecting under the pool creates extreme upward pressure that can crack the pool. This pressure, called *hydrostatic pressure*, is relieved by this valve. To illustrate this yourself, take an empty bucket and, holding it upright, try to press the bottom of the bucket into a tub of water. The hydrostatic pressure makes it nearly impossible. Now try the same experiment with a full bucket. It sinks easily into the larger container of water. The water inside compensates for the hydrostatic pressure on the outside. You can try a smaller version of this experiment by pressing an empty glass down into a bowl of water, followed by a full glass.

Pool floor

Main drain

To suction

Float

Dirt base

Groundwater pressure

2-5 Main drain with a hydrostatic valve.

Hydrostatic pressure is an important consideration when planning to drain a pool for any reason. Obviously it is not wise to drain a pool completely during the rainy season or if there is any other suspicion of groundwater.

In some spas, there might be more than one main drain so that if one becomes covered with a foot or hand, water is pulled from the other, avoiding injury to the bather. These drains are usually located at least 12 inches apart. Obviously in a pool where the main drain is very deep, this is not a concern, so safety suction lines are not added. Also, the suction in a pool is usually divided between the main drain and the skimmer, so one is not dangerously stronger than the other.

Because spas are relatively shallow, strong suction can create a whirlpool effect. To prevent this, many spa main drains are fitted with antivortex drain covers which are slightly dome-shaped with the openings located around the sides of the dome (Fig. 2-6). Pool main drain covers are flat with the openings on top.

In any case, the drain area is covered by a grate, usually 8 to 12 inches in diameter that screws or twist-locks into a ring that has been plastered into the pool bottom.

GENERAL PLUMBING GUIDELINES

Before proceeding to specific instructions on working with PVC plastic, galvanized, or copper plumbing, here are a few general guidelines that I think are important regardless of the material you are using.

2-6 Main drain antivortex cover and standard cover.

Measure the pipe run carefully, particularly if you are repairing a section between plumbing that is already in place. In measuring, remember to include the amount of pipe that fits inside the connection fitting, usually about 1½ inches at each joint (Fig. 2-7).

2-7 Pipe measuring and fitting.

When working on in-place plumbing, support your work by building up wood or bricks under the pipe on each side of your work area. This prevents vibration as you cut, which can damage pipes or joints further down the line. Also, unsupported pipe sags and binds when you cut it. That is, as you cut, it pinches the saw blade making cutting difficult, and straight, clean cuts impossible!

Threaded fittings are obvious and simple; however, leaks occur most often in these connections. The key is to carefully cover the male threads with Teflon tape and to tighten the fitting as far as possible without cracking.

Teflon tape fills the gaps between the threads to prevent leaking. Apply the tape over each thread twice, pulling it tight as you go so you can see the threads. Apply the tape clockwise (Fig. 2-8) as you face the open end of the male threaded fitting. If you apply the tape backwards, when you screw on the female fitting, the tape will skid off the joint. Try it both ways to see what I mean and you will only make that mistake once.

Another method of sealing threads is to apply joint stick or pipe dope. These are odd names for useful products that are applied in similar ways (Fig. 2-9). Joint stick

2-8
Correct application of
Teflon tape.

2-9 Pipe dope and joint stick.

is a crayon-type stick of a gum-like substance that works like Teflon tape. Rub the joint stick over the threads so that the gum fills the threads. Apply pipe dope the same way. The only difference is that dope comes in a can with a brush and is slightly more fluid than joint stick. The key to success with joint stick or pipe dope is to apply it liberally and around all sides of the male threaded fitting, so that you have even coverage when you finally screw the fittings together. Some product will ooze out as you tighten the fittings together, but that proves that you have applied enough.

I use Teflon tape because I know upon application that it is an even and complete coverage of the threads. Pipe dope or joint stick might not apply evenly or can bunch up when threading the joint together. If you use dope or stick, be sure it is a nonpetroleum-based material (such as silicone). Petroleum-based products will dissolve plastic over time, creating leaks.

When working with PVC pipe and fittings, tighten threaded fittings with channel lock pliers of adequate size to grip the pipe. Using pipe wrenches usually results in application of too much force and cracking of the fittings. Save the pipe wrenches for copper or galvanized plumbing. If you don't have pliers large enough for the work and must use a pipe wrench, tighten the work slowly and gently—not bad advice with copper either, because copper is soft and will bend or crimp if too much force is applied.

To avoid slipping off the work and damaging fittings, always tighten with channel locks or wrenches into the jaw, not away from it (Fig. 2-10). Another way to think of it is into their base, not their head.

2-10
Correct wrench use.

Now a word about pipe cutters (Fig. 2-11). Made for PVC or metal pipe, these are adjustable wrench-like devices that have cutting wheels. You lock the device around the pipe and rotate it, constantly tightening it as you go, until the pipe is cut. They provide the straightest, cleanest cut of all. However, in pool work you will be dealing mostly with 1½- to 2-inch diameter pipe in close quarters. You rarely have the luxury of enough space to get around the entire pipe, and these cutters take far

2-11 Pipe cutters.

longer than a good, fresh hacksaw blade. Use them if you like, but I think you will soon abandon them in favor of simple old Mr. Hacksaw.

The most important advice I can give you on odd-job or tight-quarters plumbing, indeed on any of this plumbing, is to ask questions. There are so many different, unique fittings and fixtures for cutting, joining, and repairing plumbing that they can fill a book of their own, and it would still be out of date because of constant revision and new products. So if you run across a tough connection of odd pipe or different materials, ask questions at your local plumbing supply house. Most of the counter help is knowledgeable and willing to advise you because their advice sells their products. A good idea is to take Polaroid snapshots of the job (or bring the materials in with you if you must cut them out anyway) so they will thoroughly understand your specific needs.

PVC PLUMBING

If you played with Tinker Toys, Lego blocks or Lincoln Logs as a kid, you will find working with PVC plumbing literally, well, a snap.

Pool plumbing is prepared with plastic or metal lengths of pipe and connection fittings that join those lengths together. The pipe acts as the *male* which fits and is glued into the *female* openings of these connection fittings. Alternatively, connection is made by each side having threads, joined by screwing them together. The plastic pipe used is PVC (polyvinyl chloride) and it is manufactured in a variety of different strengths depending on the intended use.

To help identify the relative strength of PVC, it is labeled by a *schedule* number; the higher the number, the heavier and stronger the pipe. Pool plumbing is done with PVC schedule 40. Some gas lines are plumbed with PVC schedule 80.

PVC is designed to carry unheated water (under 100°F). CPVC is formulated to withstand higher temperatures for connection close to (or in some cases directly to) a pool or spa heater.

Ultraviolet (UV) light from the sun causes PVC to become brittle over time, losing its strength under pressure and creating cracks. Chemical inhibitors are added to some PVC to prevent this, the most common and cheapest being simple carbon black (which is why plastic pumps and other pool and spa equipment is often black). Another common preventive measure is to simply paint any pipe exposed to sunlight.

PVC pipe is manufactured in a flexible version, making plumbing easier in tight spaces and for spas and jetted tubs. Flex PVC is available in colors for cosmetic purposes and has the same characteristics and specifications as rigid PVC of the same schedule and size.

All pipe is measured by its diameter, expressed in inches. Typically pool plumbing is done with 1½- or 2-inch pipe, referring to the interior diameter (the diameter of the pipe that is in actual contact with the water). The exterior diameter of the pipe differs depending on the material. For example, the exterior diameter of 2-inch PVC pipe is greater than that of 2-inch copper pipe because the PVC pipe walls are thicker than those of copper.

All pipe is connected with fittings (Fig. 2-12). Fittings allow connection of pipe along a straight run (called couplings), right angles (called 90-degree couplings or elbows), 45-degree angles, T fittings, and a variety of other formats. In the case of PVC, such fittings are most often smooth-fitted and glued together (called *slip* fittings).

Some fittings are threaded (called threaded fittings) with a standard plumbing thread size so they can be screwed into comparable connecting fittings in pumps or other plumbing parts. National Pipe Thread (NPT) standards are used in the United States so different products of various materials by different manufacturers will all work together. The NPT standard includes a slight tapering between the male and female connections. The importance of this is that because of this taper, it is easy to overtighten plastic threaded fittings and crack them. Great Britain, Europe, and Asia not only operate on metric measurements, but also have their own unique thread standards. Fittings with male (external) threads are called *mip* and fittings with female (internal) threads are called *fip*. If one side of the fitting is mip and one side is slip, you order it as *mip by slip*, and so on.

In most cases with pool and spa plumbing, the long runs of pipe will be underground. Sometimes, however, horizontal runs will be under a house or deck or over a slope where support is needed. In this case, pipe should be supported every 6 to 8 feet, hung with plumber's tape to joists or supported with wooden bracing. PVC does not require support on vertical runs because of its stiffness, but common sense and local building codes might require strapping it to walls or vertical beams to keep it from shifting or falling over. Remember, the pipe becomes considerably heavier when it is filled with water and might vibrate along with pump vibration.

2-12 Pipe connection fittings. Top row (left to right): straight slip coupling, T coupling, 45-degree slip coupling, 90-degree slip coupling; second row: MIP, FIP, plumbers strap tape, reducer bushing, 90-degree slip street coupling; third row: close nipple, compression coupling; bottom row: male threaded plug, male slip plug, female slip cap.

Supplies

The supplies and tools you need for PVC plumbing are
- Hacksaw with spare blades (coarse: 12–18 teeth per inch)
- PVC glue and primer
- Cleanup rags
- Fine sandpaper
- Teflon tape or joint stick

Plumbing methods

The concept of joining PVC pipe involves welding the material together by using glue that actually melts the plastic parts to each other. In truth, each joint will have an area that is slightly tighter than the rest. In the tightest parts, this welding actually occurs. In the remainder, the glue bonds to each surface and itself becomes the bonding agent. Obviously the strongest part of each joint is the welded portion; but in either case, the key is to use enough glue to ensure total coverage of the surfaces to be joined.

Following is the correct procedure for plumbing with PVC (Fig. 2-13):

2-13 Step-by-step PVC plumbing.

1. Cut and dry fit all joints and plumbing planned. It is easy to make mistakes in measuring or cutting and sometimes fittings are not uniform so they don't fit well. Dry fitting ensures the job is right before gluing. If you need the fitting and pipe to line up exactly for alignment with other parts, make a line on the fitting and pipe (Fig. 2-13A) with a marker when dry fitting so you have a reference when you glue them together.
2. Lightly sand the pipe (Fig. 2-13B) and inside the fittings so they are free of burrs. The slightly rough surface will also help the glue adhere better.
3. You might need to apply a preparation material, called *primer,* to the areas to be joined before gluing (Fig. 2-13C). Some PVC glues are solvent/glue combinations and no primer is required. In some states, however, use of primer might be required by building code, so check that before selecting an all-in-one product. If you are using primer, apply it with the swab provided to both the pipe and the inside of the fitting. Read and follow the directions on the can.
4. Before gluing, be ready to fit the components together quickly because PVC glue sets up in 5 to 10 seconds. Apply glue to the pipe and inside of the fitting (Fig. 2-13D).
5. Fit the pipe and fitting together, duplicating your dry fit, and twist about a half turn to help distribute the glue evenly, realigning the lines drawn on the pipe and on the fitting. If using flexible PVC, because it is made by coiling a thin piece of material and bonding it together, do not twist it clockwise. This can make the material swell and push the pipe out of the

fitting. Get in the habit of twisting all pipe counterclockwise (even though it makes no difference with rigid PVC) and you will never make that mistake.

6. With rigid PVC, hold the joint together about a minute to ensure a tight fit; about two minutes with flex PVC. Although the joint will hold the required working pressure in a few minutes (and long before the glue is totally dry), allow overnight drying before running water through the pipe to be sure. I have seen demonstrations with some products (notably Pool-Tite solvent/glue) where the gluing was done underwater, put immediately under pressure, and it held just fine. I don't, however, recommend this procedure as I have gone back on too many plumbing jobs to fix leaks a few weeks later because I hastily fired up the system after allowing only a few minutes drying time.

Tips

1. Make all threaded connections first, so if you crack one while tightening it can be easily removed. Then glue the remaining joints to the threaded work.

2. When cutting PVC pipe, hacksaw blades of 12 teeth per inch are best, particularly if the pipe is wet (as when making an on-site repair). Finer blades will clog with soggy, plastic particles and stop cutting. In all cases, the key is a fresh, sharp blade. For the few pennies involved, change blades in your saw frequently rather than hacking away with dull blades—you'll notice the difference immediately.

3. No matter how careful you are, you will drip some glue on the area or yourself. That's why I always carry a supply of dry, clean rags to keep myself, the work area, and the customer's equipment clean of glue.

4. Try to make as many *free* joints as possible first. By that I mean the joints that do not require an exact angle or which are not attached to equipment or existing plumbing. The free joints are those that you can easily redo if you make a mistake. Do the hard ones last—those that commit your work to the equipment or existing plumbing and cannot be undone without cutting out the entire thing and starting over.

5. Use as much glue as you need to be sure there is enough in the joint. It's easier to wipe off excess glue than to discover that a small portion of the joint has no glue and leaks.

6. Practice. PVC pipe and fittings are relatively cheap, so make several practice joints and test them for leaks in the shop before working on someone's equipment in tight quarters in the field.

7. Flexible PVC is the same as rigid, but when you insert the pipe into a fitting, hold it in place for a minute or longer because flex PVC has a habit of backing out somewhat, causing leaks.

8. In cold weather, more time is required to obtain a pressure-tight joint, so be patient and hold each joint together longer before going on to the next.

9. Bring extra fittings and pipe to each job site. Bring extras of the types you expect to use, as well as types you don't expect to use, because you just might need them. Nothing is worse than completing a difficult plumbing job and being short just one fitting, or needing to cut out some of your work and not having the fittings or a few feet of pipe to replace them. It is often

several miles back to the office or the nearest hardware store to grab that extra fitting that should have been in your truck in the first place. Bring extra glue, sandpaper, and rags too.

COPPER PLUMBING

Copper plumbing is quickly disappearing from the pool and spa scene for a number of reasons. Unlike PVC, metal plumbing such as copper will corrode, especially in the presence of caustic pool and spa chemicals moving at high speed and under pressure through the pipes. Copper plumbing is also more difficult to install and repair, and it has become extremely expensive in recent years.

Copper was more recently still in use where dissipation of heat was important, such as in plumbing directly connected to the heater. Stainless steel heat risers and threaded galvanized plumbing have replaced that function, making copper plumbing for pools and spas obsolete.

In fact, copper pipes are something of a hazard, because even the most careful pool technician can make chemistry mistakes. When water becomes acidic, copper oxidizes and deposits greenish black metals on porous plaster surfaces. Not only is this unsightly and difficult to remove, but the copper pipe becomes increasingly thin every time such oxidization occurs until finally, leaks begin.

Having said all this, copper is still widely present in various older installations. Understanding copper and its repair techniques is therefore an important part of a complete education in this field. Copper pipe and fittings look similar to those made of PVC and come in similar sizes and fitting types.

Copper pipe is made in three thicknesses, designated by the letters K (thick wall), L (medium), and M (thin). I know, M should have been medium, but I didn't invent this system. Most pool plumbing is done with L (medium thickness) material.

Supplies

The supplies and tools you need for copper plumbing are
- Hacksaw with spare blades (fine: 32 teeth per inch)
- 50/50 solid wire solder
- Flux with application brush
- Emery cloth or fine steel wool
- Butane torch (self-igniting, with lighter or propane torch)

Plumbing methods

Be sure to read the general plumbing guidelines section and the one on PVC plumbing because some of the commonsense methods apply to copper work as well.

Following is the correct procedure for plumbing with copper (Fig. 2-14):
1. The secret to successful copper soldering (called *sweating*) is clean pipe and fittings. Start by cutting and dry fitting all intended connections, then clean the ends of the pipe (Fig. 2-14A) and inside the fittings with the emery cloth (or fine steel wool), sanding until the copper is shiny and bright.

   Apply a thin coating of flux to the pipe (Fig. 2-14B) and inside the fitting with a small, stiff brush, coating the entire area where you want solder to make a connection.

2. Fit the parts together and twist a half turn to evenly and thoroughly distribute the flux. Heat the joint with your torch (Fig. 2-14C), moving the flame back and forth and around the joint area to distribute the heat evenly until the flux starts to bubble. Be sure the entire joint is hot. A good way to know is to keep touching the solder to the joint until it melts. When it does, you have the right temperature.

   Here is where three plumbers will give you four opinions. Some say when that temperature is reached, remove the torch and solder the connection quickly. Others say move the heat source an inch behind the joint to keep some heat but do not overheat. The problem is that if the joint cools, the solder will not melt, but if the joint gets too hot, the flux will burn off (and the solder only goes where the flux is present). The correct way? Practice it yourself on scrap material and develop your own approach. Whatever you are comfortable with is the correct way.

3. Touch the solder to the joint (Fig. 2-14D). Solder with lead melts and flows more easily. Recently many states have restricted the use of lead in solder, at least when soldering plumbing that is used for drinking water, so it might not apply to pool installations in your state. I find 50/50 (50 percent lead, 50 percent amalgams) works best, but if no-lead solder is used, even more heat must be applied and maintained during the soldering operation to ensure melting and even distribution.

   My method is to apply solder to the fitting while continuing to heat the joint, about an inch behind the joint. Solder is drawn toward the heat. The solder melts and enters the joint wherever there is flux, drawn in by capillary action, forming a seal. Work around the joint, making sure solder is drawn into the entire joint.

4. Clean excess solder away with a damp cloth (Fig. 2-14E) before it cools; but be careful, the heat turns the moisture on the rag to steam which burns you worse than grabbing the hot pipe directly. Allow the work to cool.

Two last items relating to copper plumbing. First, threaded fittings are handled like PVC threaded fittings (see previous section) using pipe dope or Teflon tape to ensure leak-free connections.

2-14 Step-by-step copper plumbing.

Second, another type of copper connection fitting is the compression fitting (Fig. 2-15). This fitting is used in small diameter (¼- to ½-inch) pipe, often inside the heater (see the pressure switch section in the heater chapter). The compression nut is placed on the pipe followed by the compression ring. The pipe is placed inside the opposing fitting and when the nut is screwed onto that fitting, the compression ring tightens around the pipe and seals it.

Copper pipe

Broad-shouldered compression nut (threaded inside)

Threaded to fit compression nut

Standard NPT thread

Compression ring

Flared to fit compression ring

2-15 Copper compression fittings.

Tips

1. Copper pipe is cut with a hacksaw like PVC. I prefer a fine blade for copper (at least 24 teeth per inch or preferably 32) as it cuts faster with each stroke.
2. When making several connections with various fittings, I like to sweat as many joints as possible while they are not connected to the equipment (just like the free joints I mentioned in the PVC section). This way, I can test each joint as it is completed. Hopefully only one or two joints will then have to be soldered in-place and trusted to the luck and skill of the work.

    To check if the joint is leak free, hold the work with one hand over one end of the pipe or fitting and blow hard through the other open end. No air should escape through the joint—if air doesn't leak, neither will water. When sweating a joint attached to equipment or other in-place plumbing, the only leak testing will be when the job is complete and the system is started up with water.
3. Solder will not seal if there is any moisture. If you must solder where some water is weeping from a pipe connected to equipment, stuff the line with bread to absorb the water. When the system is turned on again, the force of the water will break down the bread and allow it to be filtered or removed.
4. Most leaks in sweating are caused by moisture, overheating the joint, and most of all, unclean pipes or fittings.
5. Unlike PVC, sweating can be undone if you need to repair a bad job or take old work apart to perform new installations or add-ons. Apply heat to the joint, just as in sweating, until the solder melts and pull the joint apart. If you intend to reuse the fittings or that part of the pipe, carefully clean and sand the copper to a good shine before reuse.

6. Practice makes perfect. Although copper is expensive, it is more costly to make errors in the field. Take some pipe and a fitting and try this process several times at your shop. After sweating and testing your work for leaks, heat the joint until the solder melts and take it apart, clean it, and do it again. Because of tight quarters and odd angles, it only gets tougher in the field, so if you can't do it in the shop, you won't be successful on the job.

7. Read the directions on the torch, the can of flux, and the spool of solder. Sometimes reading another person's directions for performing the same task will make more sense than mine. Don't worry, I won't be offended, as long as your work doesn't leak. Also, labels can provide helpful hints that make for better, quicker jobs.

8. When supporting copper pipe to joists for long horizontal runs, use plumber's tape every 6 to 8 feet as with PVC, but wrap the pipe in that spot with insulating tape first. If you fail to do this, the different metals (copper pipe and galvanized plumber's tape) will cause electrolysis, corrosion, and leaks.

GALVANIZED PLUMBING

Galvanized means iron pipe that has been coated with zinc or other alloys to prevent corrosion. You will encounter galvanized plumbing in some installations. It is inferior to PVC and copper because it is heavier and more difficult to use. Because the interior surface of the pipe is rougher, it takes, for example, a ¾-inch galvanized pipe to carry as much water (at the same pressure) as a ½-inch copper or PVC pipe. The rough surface also promotes calcium buildup inside the pipe.

Galvanized plumbing can only be joined by threading the pipe ends and fittings and screwing them together. There is no simple sweating or gluing method as with copper or PVC, so exact lengths and threading must be accomplished.

Having said that however, it does make an easy heat dispersal pipe (called a *heat sink* or *heat riser*) for plumbing directly to the heater (Fig. 2-16). As shown, you can use a threaded galvanized nipple to thread directly into the heater and then continue with PVC beyond that, also threaded directly to the galvanized nipple. As with threaded PVC or copper, use plenty of Teflon tape or pipe dope and tighten the connections adequately. If connecting threaded PVC to galvanized, apply Teflon tape and join as discussed previously.

If connecting galvanized to copper, you must use a special union called a *dielectric fitting*. Because of the different properties of the two metals, if they are connected directly, they will corrode each other through electrolysis. Because this book is not meant as a science lesson, I won't go into an explanation of electrolysis, but take my word for it (if you slept through that part of high school). The dielectric union connects the two metals while shielding each from the other. You probably won't use many of these, so if you need one, ask at your plumbing supply house for more details. At this point, you should just be aware of the problem and the solution.

MISCELLANEOUS PLUMBING

Sometimes in tight quarters or for temporary connections you can use rubber connection fittings called mission clamps, balloon fittings, or no-hub connectors (Fig. 2-17).

Note: Metal pipe to act as heat sink. When joining metal and PVC, use metal male/PVC female connection.

8" galvanized nipple

10" galvanized nipple

Rigid PVC

Manual bypass valve

Rigid or flex PVC

2-16 Galvanized plumbing used as a heat sink.
Teledyne Laars, Moorpark, Ca., modified by author

2-17
Mission clamps, balloon fittings, and no-hub connectors.

These fittings are handy for connecting pipes of different sizes or types, clamping directly onto the pipe or fitting without gluing, threading, or sweating. The hazard is that these can leak, wear out, or fail under extreme pressure as when there is a restriction in the system from debris or a dirty filter.

I don't recommend these fittings, but I carry several different sizes anyway in case I need to make a quick repair (usually at 4:30 P.M. on a Friday afternoon) that will be improved later.

SIZING OF PLUMBING

Most building codes restrict the speed of flow through pipes to prevent stripping, breakdown, and erosion of the pipe material. Typically water may not move faster than 8 feet per second through copper pipe, and 10 feet per second through PVC. Suction pipes of any type are typically restricted to 8 feet per second.

Obviously the larger the pipe, the better. There is less restriction and therefore less strain on all equipment and plumbing. Use the largest diameter pipe and fittings you can for the job.

Typically pool and spa equipment is already built and plumbed for 1½- or 2-inch plumbing, and while you can adapt 2-inch pipe to a pump that is designed for 1½-inch fittings, you don't want to use the reverse. In that case, the pump will be trying to push the proverbial 10 pounds of potatoes into a 5-pound bag.

I discuss sizing in more detail in the chapter on pumps, which largely dictates the plumbing sizing; however, the considerations in all cases are
- Desired flow rate of water (measured in gallons per minute),
- Length of plumbing runs,
- Number and angles of connection fittings,
- Pump efficiency and capacity, and
- Equipment and restrictions after the pump.

3
Advanced plumbing systems

Perhaps the fastest growing segment of the pool and spa industry is in advanced plumbing systems, including automated valves, reverse-flow heating, solar heating, and the use of space-age materials. In this chapter, I discuss some of the more common applications of advanced plumbing and the maintenance of these systems.

MANUAL THREE-PORT VALVES

The design of three-port valves takes water flow from one direction and divides it into a choice of two other directions. Picture a Y, for example, with the water coming up the stem, then a diverter allows a choice between one of two directions (or a combination thereof). Conversely, the water flow might be coming from the top of the Y, from two different sources, and the diverter decides which source will continue down the stem or mixes some from each together.

Figure 3-1 shows a typical Y or three-port valve. This is a Hydrotech Ortega valve style, which is very common and easy to use and maintain. Other manufacturers make similar units based on the same concept.

Operation

Several manufacturers make three-port valves, in Noryl plastic (a type of PVC) or brass, but the concept is the same with them all. A housing (Fig. 3-1, item 1), built to 1½- or 2-inch plumbing size, houses a diverter (Fig. 3-1, item 3 or 4) that is moved by a handle (Fig. 3-1, item 10) on top of the unit. Typically these valves are used when a pool and spa are both operated from the same pump, filter, and heater equipment.

The suction line from the pool enters one arm of the valve body; suction from the spa enters the other. The diverter between the two arms determines which line is connected with the stem, from which the water continues to the pump.

1 Valve body
2 Handle screw
3 Diverter (for separate shaft)
4 Diverter (with shaft built in)
5 Diverter seal gasket
6 Shaft
7 Shaft O-ring
8 Cover O-ring
9 Cover
10 Handle
11 Cover screw
12 Diverter stop screw

3-1 Construction of a typical three-port valve. Purex Pool Systems, Inc., City of Industry, Ca.

Conversely, when the water leaves the equipment, it passes through another three-port valve. The water this time passes up the stem and the diverter determines if the water flows to the arm plumbed into the pool return or the one plumbed into the spa return. By setting the diverter equally between the two, water from each side is mixed. Sometimes this creates a pool or spa draining problem that must be corrected. These repairs are dealt with in following sections.

Construction

The diverter is surrounded by a custom-made gasket (Fig. 3-1, item 5) so that no water can bypass the intended direction. A valve with this type of diverter and gasket is called a *positive seal* valve.

Some valves, for use where such water bypass is not considered a problem, have no such gaskets and, in fact, the diverter is designed more like a shovel head than a barrel. These divert most of the water in the desired direction with a lesser amount going in the other direction. These are called *nonpositive* valves.

The diverter is held in the valve housing by a cover (Fig. 3-1, item 9) that screws to the housing with sheet metal screws (Fig. 3-1, item 11), and is sealed with an O-ring (Fig. 3-1, item 8) to make it watertight.

Notice that besides the four screw holes in the cover and housing, there is a fifth hole in the cover that corresponds to a post on the housing. This is to ensure that the cover lines up properly with the housing, because on the underside of the cover are specially molded stops. A small screw (Fig. 3-1, item 12) on top of the diverter hits these stops molded into the underside of the cover. This allows the diverter to be turned only 180 degrees (one-half turn), in either direction, ensuring that the diverter stops turning when facing precisely one side or the other.

This screw is removed when the valve is motorized because the motors only rotate in one direction and are already precise in stopping every half-turn. Small machine screws (Fig. 3-1, item 2) hold the handle on the shaft. A hole in the center of the cover allows a shaft from the diverter to attach to the handle for manual operation of the valve. To make the shaft hole in the cover watertight, two small O-rings (Fig. 3-1, item 7) slide on the shaft in a groove under the cover.

Maintenance and repair

As the simple parts suggest, there's not too much that can go wrong with manual three-port valves.

Installation

Noryl three-port valves are glued directly to PVC pipe using regular PVC cement like any other PVC fitting. Care should be exercised not to use too much glue, as excess glue can spill onto the diverter and cement it to the housing. Excess glue can also dry hard and sharp, cutting into the gasket each time the diverter is turned, creating leaks from one side to the other.

Brass valves are sweated onto copper pipe like any other fitting. Be sure to remove the diverter when sweating so the heat doesn't melt the gasket.

Lubrication

For smooth operation, the gasket must be lubricated with pure silicone lubricant. Vaseline-like in consistency, this lubricant cannot be substituted. Most other lubricants are petroleum-based which will dissolve the gasket material and cause leaks. Lubrication should be done every six months or when operation feels stiff. This preventive maintenance is particularly important with motorized valves because the motor will continue to fight against the sticky valve until either the diverter and shaft break apart or, more often, an expensive valve motor burns out.

Lubrication is the most important maintenance item with any three-port valve. When the valve becomes stiff to turn it places stress on the shaft. On older models, the shaft is a separate piece that bolts onto the diverter (Fig 3-1, items 2, 3, and 6). Particularly on these models, but actually on any model, the stress of forcing a sticky valve will separate the stem from the diverter. If this happens, turning the handle and stem does not affect the diverter. To repair this, remove the handle and cover, pull out the diverter, and replace it with a one-piece unit (Fig. 3-1, item 4). If the gasket looks worn, replace it and lube it generously before reassembly.

Repairs

Few things go wrong with these valves, but the breakdowns that do occur are annoying and recurrent. Before disassembling any valve, check its location in relation to the pool or spa water level. If it is below the water level, opening the valve will flood the area. You must first shut off both the suction and return lines. When installations are made below water level, they are usually equipped with shutoff valves to isolate the equipment for just such repair or maintenance work. If the valves are above the water level, you will need to reprime the system after making repairs (see the section on priming).

Leaks are the most common failure in these valves. The valve will sometimes leak from under the cover. Either the cover gasket is too compressed and needs replacement or the cover is loose. The cover is attached to the Noryl valve housing with sheet metal screws. If tightened too much, the screw strips out the hole and you will be unable to tighten it. The only remedy is to use a slightly larger or longer screw to get a new grip on the plastic material of the housing. Be sure to use stainless steel screws or the screw will rust and break down, causing a new leak. If new screws have already been used and there is not enough material left in the housing for the screw to grip, you must replace the housing. I have managed to fill the enlarged hole with super-type glue or PVC glue and, after it dries, replace the screw. These repairs usually leak and are only temporary measures. You can also fill the hole with fiberglass resin, which usually lasts longer.

Leaks also occur where the shaft comes through the cover. Remove the handle and cover and replace the two small O-rings. Apply some silicone lubricant to the shaft before reassembly. This lubricates the operation of the valve, decreasing friction that can wear out the O-rings. The lubricant also acts as a sealant.

Leaks can occur where the pipes join to the housing ports. In this case, the only solution is replacing the housing. I have tried to reglue the leaking area by removing the diverter and gluing from both inside and outside of the joint. This has never worked! Try if you will, but I think you'll be wasting your time.

Finally, leaks occur inside the valve with no visible external evidence. By this I mean that water is not completely diverted in the intended direction, but slips past the diverter seal to the closed side of the valve. The symptom will be a spa that drains for no reason (or overflows for no reason). The cause might be a diverter that is not aligned precisely toward the intended port. Remove the diverter and make sure the shaft has not separated or become loose from the diverter.

With a motorized unit, be sure the motor is clean, free of rust, and able to turn its precise one-half turn each time. The other and most usual cause, however, is that the diverter gasket has worn out or become too compressed to stop all water from getting past. You might visually inspect the gasket and find that it looks good. Replace it anyway. It takes very little deterioration or compression to cause these bypass-type leaks. Again, lubricate the gasket well before reassembly for smooth operation and because the lube acts as a sealant.

If a new gasket doesn't stop the water bypass, you might find the diverter itself has shrunk or warped slightly. This doesn't occur often, but particularly with hot spa water or if the system has been allowed to run dry and heat up, you might be looking at a diverter that is not large enough to contain the water flow to one side only. It is difficult to see because such shrinkage is minimal, but it only takes a little to allow the bypass problem. Replace the diverter and see if this solves the leakage problem.

In even rarer circumstances, I have seen valve housings that have expanded or warped from overheating, usually when the system has been allowed to run dry and extreme heat builds up in the plumbing. If the new diverter and gasket seem loose, this might be the problem.

Such overheating and warping is a more common problem in systems with solar heating. Water in solar panels often exceeds 200°F, so if the system backflows or any of this superheated water is allowed to sit in the valves, they will warp in a short time. Even a slight warp is enough to allow leaks to occur.

MOTORIZED/AUTOMATED THREE-PORT VALVE SYSTEMS

The three-port valves just described are manually operated. These same valves can have small motors mounted in place of the manual handle for automatic or remote operation (Fig. 3-2).

3-2
Typical motorized three-port valve.

3-2 Continued Purex Pool
Systems, Inc., City of Industry, Ca.

Operation

The value of motorization is that the pool or spa equipment is usually located away from the pool and spa, making manual operation inconvenient. Some builders place the manual valves near the spa rather than in the equipment area; however, a small switch that operates the valve motors is often preferred. A variation of that concept is to locate the motor switch with the equipment and operate it with a remote control unit. The remote might also operate switches for lights, spa booster motors and blowers, or other optional accessories.

Construction

To motorize an existing valve, the handle and the screws holding the cover in place are removed. A mounting bracket is set on top of the cover and slightly longer sheet metal screws (to allow for the added thickness of the bracket metal) are used to re-fasten the bracket and cover to the housing. These longer screws are provided with the motor bracket kit. The machine screws that normally hold the handle in place are left on the shaft. The motor unit mounts on the bracket, held in place by two screws, and the motor shaft slips over the diverter's shaft. The two shafts are secured together by tightening the screws of the diverter shaft.

Wiring diagrams are provided with each type of valve motor and they are designed to operate on standard 110 volt, 220 volt, or from an automated system that has been transformed to 12 or 24 volts. Remote and automated systems are dealt with in more detail in a later chapter.

Maintenance and repair

Few problems occur with motorized valves (beyond those discussed in the section on manual valves). As mentioned, if the valves are not properly lubricated or become jammed with debris, the motor will continue to try to rotate the valve, finally burning itself out.

To determine if the motor has burned out, using your electrical multimeter, verify that current is getting to the motor. Obviously if there is no current, the problem is in the switch or power supply and probably not the motor. If you are not familiar with basic electricity, call an electrician to help you or study the basic electricity chapter later in this book.

If current is present, remove the motor unit from the valve and try to operate the system. If the motor rotates the shaft normally, then the problem is a stuck valve and not a burnt motor. Tear down and repair the valve as described in the previous section.

If the motor is burnt out, it can easily be replaced without replacing the entire unit or valve. Although slightly different with each manufacturer, the process is usually no more than four screws and three wires and will be obvious when opening the motor housing.

Another problem that can occur with motorized valves is that if the mounting bracket or screws holding the unit together become loose, the unit will not align correctly with the valve. The motor will then rotate its 180 degrees, but it will not fully rotate the valve diverter to match. The solution is to tighten all hardware and replace any rusted screws.

Finally, a less frequent problem can be caused by electrolysis or simply a leaking valve. The motor shaft is usually made of galvanized metal or aluminum. If the valve is leaking or the motor housing is not watertight and a combination of moisture and electricity is present as a result, electrolysis will disintegrate the soft metal of the motor shaft. The motor might continue to operate, but as the shaft dissolves, it will not turn (or not completely turn) the valve diverter.

I have also seen salt in the moist air near the ocean eat away the motor shaft. The only solution in this case is to replace the motorized unit and be sure all leaks are sealed. Near the ocean or in damp climates, enclosing motorized units is a good idea. If it is not practical to enclose the unit, cover it with plastic and sealing tape. Obviously don't tape anything to the shaft itself that will bind up the unit.

REVERSE FLOW AND HEATER PLUMBING

Motorized valves are often installed in plumbing systems designed with reverse flow. During normal circulation of the pool, water is taken from the skimmer and main drain and returned to outlets located about 18 inches below the water surface. The concept of reverse flow is that when the heater is turned on to heat the water, motorized valves reverse the flow. Water is taken through these shallow outlets and returned through the main drain or specially installed return outlets in the floor of the pool (Fig. 3-3). The thought is that because hot water (like air) rises through cooler water, the heated water will rise through the pool and heat the pool more uniformly. When the warm water is returned in a normal system to the shallow outlets, only the top 2 or 3 feet of the pool water is heated.

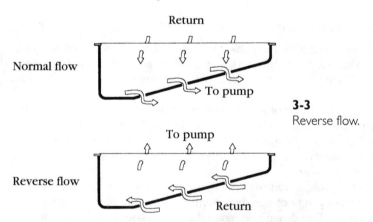

3-3
Reverse flow.

Okay, I'll grant the genius who thought this up the fact that more uniform heating is achieved. But who cares? Everyone I know with a pool only swims in the top 2 or 3 feet of the pool, so why spend the money and time to heat the bottom 6 feet of water? An engineer from the gas company or a motorized valve manufacturer must have designed this incredibly stupid, unnecessarily complex, and expensive system. They are the only ones who profit. Think about it. I have convinced my clients who have this system to disable it and allow the normal one-way circulation to operate. At least they will always have a good inventory of spare parts for the motorized valves they have in their pool or spa system.

If you see a system with far more valves and plumbing than you can justify, a reverse flow system is probably the reason. Observe the normal operation and mark the pipes and valves with a marker. When you are sure of the correct flow pattern—in from the pool or spa, through the equipment, and back to the pool or spa—then turn on the heater and observe it again. If it is a reverse flow system, do yourself and your customer a favor and return the system to normal circulation, then remove the excess valve motors to disable the system (with permission from the owner, of course).

Don't try to leave the units intact and simply disconnect the wiring—you might also disconnect the heater on/off or pump switches that are associated with the remote control system (if part of the installation). It is easier (and as mentioned, you'll create spare parts) to manually remove the motors while leaving the valves in a normal circulation position, clip off and cap the wiring to those motors, and leave the rest alone.

UNIONS

An improvement for installing equipment is the plumbing union. When you need to repair or replace a pump, filter, or other equipment that is plumbed into the system, you must cut out the plumbing and replumb upon reinstallation. The concept of the union is that when you remove a particular piece of equipment, you need only unscrew the plumbing and reinstall it later the same, simple way.

Although unions add a few dollars to your initial installation, they allow you to easily remove and replace equipment without doing any new plumbing. Unions, like other plumbing, are made of plastic or metal in standard diameters and are adapted to plumbing like any other component (gluing, threading, or sweating).

Figure 3-4 shows a typical plumbing union. A nut is placed over the end of one pipe, then male and female fittings (called *shoulders*) are plumbed onto each end of the pipes to be joined. As can be seen, the joint is made by screwing the nut down on the male fitting. Teflon tape or other sealants are not needed as the design of the union prevents leaking (either by the lip design as shown or by use of an O-ring seated between the shoulders).

3-4
Plumbing union.

Unions are made for direct adaptation to pool and spa equipment, where the pipe with the nut and female shoulder is male-threaded at its other end for direct attachment to the pump, filter, or any other female threaded equipment. Then, only the male shoulder need be added to the next pipe and the piece of equipment can be screwed into place.

GATE AND BALL VALVES

Gate and ball valves are designed to shut off the flow of water in a pipe and are used to isolate equipment or regulate water flow. You might see these in systems where the equipment is installed below the water level of the pool. Without them, when you open or remove a piece of equipment or plumbing you will flood out the neighborhood.

Some fountains where water pressure and flow must be precisely regulated use shutoff valves to adjust the water flow in the plumbing. Finally, on older plumbing systems, before the development of three-port valves, shutoff valves were used on each pipe to manually determine water flow from and to pools and spas.

There are basically two types of shutoff valves. Figure 3-5B shows the *gate valve*. As the name implies, it has a disc-shaped gate inside a housing that screws into place across the diameter of the pipe, shutting off water flow. A variation of this is the *slide valve* (Fig. 3-5A), where a simple guillotine-like plate is pushed into place across the diameter of the pipe.

The other design is the *ball valve* (Fig. 3-5A), where the valve housing contains a ball with a hole in it of similar diameter as the pipe. A handle on the valve turns the ball so the hole aligns with the pipe, allowing water flow, or aligns across the pipe, blocking flow. In each of these designs, flow can be controlled by degree as well as total on or total off.

The gate valve is operated by a handle that drives a worm screw-style shaft inside a threaded gate. If the gate sticks from obstruction or rust and too much force is applied to the handle, the screw threads will strip out, making the valve useless. The valve cap (also called the bonnet) can be removed (unscrewed) and the drive gear and gate can be removed and repaired or replaced without removing the entire valve housing. Most plumbing supply houses sell these replacement guts, but the parts from one manufacturer are not interchangeable with another.

Also notice the packing gland (Fig. 3-5C) that prevents leaks where the shaft enters the valve body. If leaks occur here, the packing material can be replaced by unscrewing the cap nut, removing the old twine (specially treated graphite-impregnated twine), and rewinding new twine. Sometimes just tightening the cap nut will stop the leak, but it also tightens the packing material on the shaft, making it more difficult to turn. Plastic gate and ball valves use O-rings to prevent leaks in this location.

CHECK VALVES

The purpose of the check valve is to check the water flow—to allow it to go only in one direction. The uses are many and will be noted in each equipment chapter where they are employed; however, four common uses are

- In heater plumbing (to keep hot water from flowing back into the filter)
- In spa air blower plumbing (to make sure air is blown into the pipe but water cannot flow back up the pipe and into the blower machinery),

Handle
Packing nut
Stem
Packing
Bonnet
Body

3-5 Ball (A left), slide (A right), and gate (B), valves and valve packing (C).

- With chlorinators to keep the flow of caustic chemicals moving in the desired direction
- In front of the pump when it is located above the pool water level (to keep water from flowing back into the pool when the pump is turned off).

There are two types of check valves (Fig. 3-6). One is a flapper gate (also called a swing gate valve) and the other is a spring-loaded gate. The flapper style opens or closes with water flow, while the spring-loaded style can be designed to respond to certain water pressure. Depending on the strength of the spring, it might require one, two, or more pounds of pressure before the spring-loaded gate opens. As with other valves, check valves are made of plastic or metal in standard plumbing sizes and are plumbed in place with standard glue, thread, or sweat methods.

3-6 In-line swing gate, in-line spring-loaded gate, and 90-degree spring-loaded check valves.

The flapper-style valve is simple and little can go wrong with it. It must be installed with the hinge of the flapper on top. If it is installed on the bottom, gravity will pull the flapper open all the time. Sometimes with metal flapper check valves you will hear them *chatter* as the flapper opens and closes, particularly if there is air in the system. This is not a malfunction or a problem of the valve (see the section on priming).

A problem I have encountered with swing gate-type check valves, especially metal ones, is that the gate comes off the hinge from rust or obstruction damage. In this case, the valve must be replaced. The spring-loaded valve rarely breaks.

The only real weakness of all check valves is that they clog easily with debris, remaining permanently open or permanently closed. Because of the extra parts inside a spring-loaded check valve, they are more prone to failure from any debris allowed in the line. If the valve is threaded or installed with unions, it is easy to remove it, clear the obstruction, and reinstall it.

Another solution is to use the 90-degree check valve. This valve allows you to unscrew the cap, remove the spring and gate, remove any obstruction, and reassemble. Be careful not to overtighten the cap—they crack easily on older models; newer models are made with beefier caps to prevent this problem. These units have an O-ring in the cap to prevent leaks. It is wise to clean these out every few months (or more frequently, depending on how dirty the pool or spa normally gets) and lubricate the gate (using silicone lube only).

These valves are great because of their ease of cleaning and repair, but I don't recommend using them unless you need to make the 90-degree turn in your plumbing anyway. Otherwise you are adding more angles to your plumbing, which restricts water flow unnecessarily. Maybe someday they'll invent a nonangled unit that can be cleaned out as easily. Why don't you come up with such a design and retire on the profits?

Some check valves are made of clear PVC, which allows you to see if they are operating normally.

SOLAR HEATING SYSTEMS

Solar heating systems are discussed here in the advanced plumbing chapter because it is the plumbing of these systems that most concerns the pool and spa technician.

Certainly an entire book could be written about solar heating and installations. This is particularly true in today's market where new technologies, alloys, and plastics are being used to manufacture solar panels. For example, modern solar panels with sensors track the sun and actually rotate around one or more axes to receive maximum sun exposure. Such systems are fairly costly and complex. Therefore, you might be advised to leave solar heating system installation and repair to the experts—call in a subcontractor and earn a referral fee.

Having said that, many repairs and some installations are simple and profitable, so if you need the work, don't be shy. As mentioned, the plumbing is the same as regular pool and spa plumbing, so leak repair is easy, although some of it might be on your customer's roof. In any case, here are some guidelines for approaching solar heating systems.

Types of solar heating systems

The function of the solar panel is to absorb heat from the sun which is transferred to the liquid as it passes through. Designs and materials are hotly debated (pun intended) between various manufacturers, but efficiency of a solar heating system is less a factor of panel design than a factor of system setup. Exposure to direct sunlight, hours of sunlight, and amount of wind, clouds, or fog are all important factors that will impact on efficiency when setting up a system.

Solar panels are made from plastic or metal and are then glazed (covered in glass) or left unglazed. Obviously the glazed panel is heavier and more expensive; however, they absorb and retain more solar heat and are therefore more efficient (fewer panels are required to accomplish the same amount of solar heating).

Open loop systems
The first system, shown in Fig. 3-7, is called an *open loop* system, meaning it is open to the pool water. This is the most common type you will encounter. A variation on this is a system that is not connected to the pool or spa equipment, but rather which has its own plumbing from the pool and back, along with its own circulating pump.

Closed loop systems
The other type of system is called a *closed loop*, where a separate pump circulates antifreeze through the solar panels. This liquid is heated in the panels and sent to

3-7 Typical solar plumbing and heating systems.

coils inside a heat exchanger. The pool water is circulated through the heat exchanger, flowing around these coils so the heat from the coils is transferred to the water. These systems are used in cold climates or in desert climates, where it is hot by day but very cold by night, where water in an open loop system might freeze, expand, and crack the panels. Another advantage of the closed loop is that harsh chemicals in the pool water or hard, scaling water are not circulated into the panels with the potential to clog or corrode them, creating the need for expensive repairs.

Plumbing

Plumbing for solar heating is no different from other pool and spa plumbing. It is located between the filter and the heater (Fig. 3-7) so water going to the solar panels is free of debris and is available for free solar heating before costly gas or electric heating by the system's mechanical heater.

A thermostat on the solar panel determines the water temperature and if it is warmer than the water coming out of the filter, a three-port motorized valve (called an *automatic bypass valve* in solar installations—it's the same motorized valve described earlier in this chapter) sends the water to the solar panels for heating and returns it to the plumbing that enters the heater. The heater thermostat senses the temperature of this solar heated water and if it is still not as hot as desired, the heater will come on to heat it further before returning it to the pool.

Therefore a main component of solar heat plumbing is the three-port valve that either sends the water from the filter directly to the heater or sends it first to the solar panels and then the heater. A check valve is installed on the pipe that returns water from the solar panel to the heater to prevent water from entering this return line when the solar panels are not in use. This might instead be another three-port valve that performs the same function as the check valve but also ensures that when not in use, the solar panels will not drain out. This might not be important where solar panels are installed at or below the water level of the equipment and pool, but most installations of panels are on rooftops, high above the water level.

To solar or not to solar?

To effectively heat with solar, regardless of the type of panel, the general rule of thumb is 75 percent of the surface area of the pool is the surface area of panels needed. For example, if the pool is 20 feet by 40 feet, the surface area is 20 × 40 = 800 square feet × 0.75 = 600 square feet of panels needed. Because panels are generally 4-by-8 feet (32 square feet) or 4-by-10 feet (40 square feet), our example pool would need 19 of the small panels (600 square feet ÷ 32 square feet per panel) or 15 of the larger panels.

Some say as little as 60 percent of the pool surface area can be used for these calculations, but I have found it is better to have a few more square feet of panels because you can't really have too many—but you can certainly have too few. Added panels will heat the water faster or, at least, more effectively on cloudy or windy days. For the few extra dollars, the customer will be happier in the long run. The manufacturer can tell you the weight of each panel with water, so you can determine if the customer's roof has the space and weight-bearing capacity for the installation.

Next, a location must be found where the panels can face the sun. It might differ in your area, but generally the best position is facing south to obtain the most hours of sun per year—winter and summer. Another factor is prevailing wind. High winds can tear panels from the roof or create so much cooling that the panels will not be effective. Such concerns will also dictate how many panels you need to install.

Knowing only these facts, you can help your customer determine if solar is likely to be practical. The cost of the system will be paid off with energy savings and tax benefits depending on how much gas or electricity would otherwise be used, but your customer might be more influenced by ecological concerns, added value to the home, or other personal concerns.

Installation

If you decide to proceed, you might want to purchase a solar heating package from your pool supply house that includes panels, plumbing, controls, and instructions. You might want to hire a licensed carpenter (let him get the building permits and take the liability) to help with the installation and support of the panels while you complete the plumbing into the system.

The existing pump will probably be adequate for circulation when adding solar equipment, because the gravity and siphon effect balances the additional pressure the pump experiences trying to push the water up to the panels. Therefore, the height of the panels above the water has no effect on the pump's ability to perform.

You do, however, have to calculate the effect of the length of pipe and fittings as with any plumbing installation (see the section on hydraulics).

As mentioned, there are numerous manufacturers and styles of panels and controls, far more than can be outlined here. Do some homework on what is available in your area. Just so you know what to look for, here are a few types:

- Plastic panels (glazed or unglazed)
- Metal panels (glazed or unglazed)
- Thin, lightweight aluminum panels
- Rubber panels (like doormats) that nail directly to the roof
- Flexible plastic or metal hose that is coiled on the roof or built into a concrete deck (the sun heating the deck in turn heats the solar coils)

Whatever the style, remember when planning an installation that the pipes to and from the panels should be insulated so heat is not lost along the way. A good idea is to attend the next pool and spa industry convention in your area and check out the wide variety of makes and models. You will receive literature and even small panel samples that can help you and your customers make decisions. Much like portable spa manufacturers, however, solar panel makers have swiftly come and gone out of business.

Although the panels themselves are fairly breakproof, choose a simple style that doesn't require replacement parts from manufacturers who might not be in business next year when you need to make repairs. Better yet, choose a manufacturer who has been around awhile.

Maintenance and repair

Once installed, most homeowners and pool technicians tend to forget about the solar heating system. Inspection every two or three months should be made to check for leaks. Leaks can easily occur because of the extremes of hot and cold temperatures that cause the panel materials to expand and contract. Leak repair depends on the type of material in the panel or plumbing, and each manufacturer makes leak repair kits with instructions. The plumbing to and from the panels can be repaired as needed using the techniques outlined in the chapter on basic plumbing.

The second common problem is dirty panels. Dirt prevents the panels from absorbing heat and can cut efficiency by as much as 50 percent. A pool technician can make good profits by charging customers for regular solar panel cleaning, requiring no more than soap and water.

Finally, panels can become clogged with scale from hard pool water and chemicals. Poor circulation is the tip-off, and the solution is to disassemble the panels from each end, exposing the pipes of the panel that actually carry the water, and reaming these out with special brushes attached to your power drill. Again, how you make this repair depends on the maker of the panel and its style. The maker should provide instructions and special tools for this procedure. As with leak repair or cleaning of solar panels, reaming is simple to perform using techniques and skills learned elsewhere in this book.

WATER LEVEL CONTROLS

The most failureproof (read idiotproof) method of replacing evaporated water in the pool or spa is to turn on the hose. Unfortunately the pool technician doesn't have the hour or two to stand around waiting for the level to come back up, and your customer will most likely forget to turn the hose off even though you tell them to do it in an hour or two.

Alternatively there are two kinds of automated water fill systems and variations on those themes. If the pool or spa was built with a water fill line plumbed in place, the on/off antisiphon valve (Fig. 3-8C) can be replaced with a mechanically timed valve. In this way, you can set the water to run 1 to 60 minutes.

Instructions for removing the manual valve and installing this timer valve are in the package with the valve and are simple to follow. It does mean you must turn off the household main water supply, usually at the meter at the street, to make the swap. This is probably the only pool or spa repair requiring shutoff of the household water supply.

To perform this task, locate the supply meter, usually in front of or alongside the house, in the ground, mounted in a concrete box. Inside the box you will see the meter with a gate valve (Fig. 3-9) on the outflow side of the meter. Turn this off. If it is stuck or rusted, there is a shutoff valve on the inflow side of the meter, but not with a standard handle. This valve can be turned off with a channel lock pliers or a small pipe wrench by gripping the post of the valve and turning it so it lines up across the pipe (in line with the pipe means it is open).

These fill units are reliable, but as they age they tend to stick in the open position and not shut off. George C. Scott and his lovely wife, actress Trish Van Devere, were greatly upset one night when one of these valves stuck open and flooded their backyard. They thought I had left the water running and left some hot messages on my answering machine that would have made General Patton blush. In the morning when I arrived and replaced the timer, George was gracious in transferring the blame from me to the faulty device, but I learned to replace these devices every two years for all my customers that have them as a preventive measure.

A variation on the timer valve is a similar unit calibrated by gallons rather than time (Fig. 3-8B). It screws onto a hose bib, the hose is screwed onto the timer unit, the hose bib is turned on, and the dial is set for the number of gallons you need. You determine the gallons needed by calculating how many inches of water are needed, and how many gallons are in those inches (see chapter 1).

The idea of these units is terrific because the addition of water by gallons is more precise than unmeasured gallons by minutes. Unfortunately these units sometimes fail to shut off, and I have stopped using them because even the newer models break down too often. Perhaps in future models, the problems will be eliminated.

There are, however, some models available in gardening supply shops that include solid-state components for regularly scheduled timed water flow or preset volume flow (Fig. 3-8A). These require batteries and careful setting and work well if you carefully check and recheck your settings and change the batteries frequently so the system doesn't fail. Because of these limitations, they are more suited to a homeowner who is there every day to keep an eye on the system, rather than a pool technician who depends on it to do the job while being checked only once a week.

The other type of water level control works much like the float valve in your toilet. A float (Fig. 3-10), which opens and closes a valve attached to a water supply line, is located at the water level desired. When the level drops, the float drops, opening the valve. As the level rises, the opposite happens. The float and plumbing can be located in the skimmer, but is more often located in a small separate tank

3-8 Water level control devices: by volume or by time. B: Melnor, Inc., Moonachie, NJ.

3-9 Typical household water main meter and shutoff.

(called a reservoir), perhaps not even near the pool. The tank must be set at the same level as the pool, so the water in it imitates the water level of the pool. A pipe connects this tank with the pool so the actual water and its level are the same in each. Also, as the level drops, the water fills the pool through this common pipe.

These units are reliable and can be adjusted for water level by bending the arm on the float to the desired level or by setting the elbow in the float arm accordingly. A setscrew loosens the elbow to allow adjustment. The small valve is threaded, so if it rusts, clogs, or fails, it can be unscrewed and easily replaced.

In recent years, several manufacturers have developed systems based on the float concept. Many of these involve costly, complicated, or unsightly additional plumbing, new skimmers, or elaborate mechanical installations. I have yet to see one of these I can recommend, so I suggest the simple units described previously or a garden hose.

Water level controls serve another valuable purpose. As water evaporates, it leaves minerals (mostly calcium) behind as scale that builds up on tiles and artificial rocks. In later chapters I review prevention and removal techniques, but the simplest method is to keep a constant water level in the pool. As water evaporates and leaves scale, fresh water is introduced to refill the pool to cover the scale line. Simple and effective!

Adjusting screw
for water level

Valve

Float

3/8" threaded fitting

1/2" pipe fitting for
water supply

3³/8"

10" Dia.

Autofill unit must be
installed flush with the
deck. To adjust for water
level, loosen adjusting
screw indicated above and
move float up or down.
Then tighten screw.

9³/8"

12³/8"

1" slip

3-10 Float water level control. MP Industries, Garden Grove, Ca.

4
Pumps and motors

Let's begin by eliminating a common error in terminology. The pump and motor are two different elements of the water circulation system, not the same thing, not interchangeable. The *motor* is the device that converts electrical energy into mechanical energy. It powers the *pump*, which is the device that actually causes the water to move. One is not much use without the other.

While built of various metals or plastics, all pump and motor combinations are composed of essentially the same components. If you understand the basic concept and components, you can find your way around almost any pump or motor. Before discussing the components of a pump and motor, let's understand the concept of what they do and how they work.

OVERVIEW

Pool and spa pumps are classified as *centrifugal* pumps. That is, they accomplish their task of moving water thanks to the principal of centrifugal force. To imagine this concept, hold a bucket with some water in it at the end of your arm and spin it around in a big sweeping circle (Fig. 4-1).

4-1
Centrifugal force. Sta-Rite Industries, Delevan, Wisc.

Centrifugal force is what keeps the water in the bucket as you spin it. If you poke a hole in the bucket and spin it again, that same force, pushing the water to the bottom of the bucket, sends it shooting out the hole. If you spin the bucket faster, the water shoots out with more force. Obviously, if you make a larger hole, more water will shoot out as you spin it around.

The pump operates the same way (Fig. 4-2). The impeller in the pump spins, shooting water out of it. As the water escapes, a vacuum is created that demands more water to equalize this force. Water is pulled from the pool or spa and sent on its way through the circulation plumbing. Just as various designs in your swinging bucket and it's hole determine the amount of water and how fast it escapes, so too the various designs of impellers, diffusers, and volutes determine the same features in a pool pump. This is discussed in more detail in later sections.

4-2 Typical pump and motor combination. Sta-Rite Industries, Delevan, Wisc.

Pumps used for pools are *self-priming*; that is, they expel the air inside upon start-up, creating a vacuum that starts suction. Once water is flowing through the pump, if you close a valve on the outflow side of the pump, restricting all flow, maximum possible pressure is created. However, unlike a piston or gear pump, there is no destructive force created—the impeller simply spins in the liquid indefinitely.

Let's examine the components of the pump and motor partnership (Figs. 4-3 and 4-4).

STRAINER POT AND BASKET

The plumbing from the pool or spa main drain and skimmer runs to the inlet port of the pump, which is usually female threaded for easy plumbing, although some designs are male and female threaded.

Water flows into a chamber, called the *strainer pot* or *hair and lint trap*, which holds a basket (generally 4 to 6 inches in diameter and 5 to 9 inches deep) of plastic mesh that permits passage of water but traps small debris. Some baskets simply rest in the pot, others twist-lock in place.

Most have handles to make them easier to remove, although I have yet to see a design where the handle is not so flimsy that it doesn't break off the second or third time you use it. The strainer basket is similar to the skimmer basket which traps larger debris.

The strainer pot is a separate component in some pumps that bolts to the volute with a gasket or O-ring in between to prevent leaks. Sometimes the pot includes a male threaded port that screws into a female threaded port in the volute. In some pumps, it is a component molded together with the volute as one piece (Fig. 4-2). In bathtub spas or booster pumps, where debris is not a problem, there is no strainer pot and basket at all.

To clean out the strainer basket, an access is provided. The strainer cover is often made of clear plastic so you can see if the basket needs emptying. It is held in place by two bolts that have a T top (Fig. 4-3) or plastic handle (Fig. 4-4) for easy gripping and turning. Some pumps have a metal clamp that fits around the edge of the cover and strainer pot. These are tightened with a bolt and nut combination. On others the cover is male threaded (Fig. 4-4) and screws into the female threads of the pot.

In all styles of pot, the strainer cover has an O-ring that seats between it and the lip of the strainer pot, preventing suction leaks. If this O-ring fails, the pump sucks air through this leaking area instead of pulling water from the pool or spa.

Notice in Figs. 4-3 and 4-4 that the pot has a small threaded plug that screws into the bottom. This plug is designed to allow complete drainage of the pot when winterizing the pump. It is made of a weaker material than the pot (on metal pots, for example, the plug is made of plastic, soft lead, or brass). If the water in the pot freezes, this sacrificial plug pops out as the freezing water expands, relieving the pressure in the pot. Otherwise, of course, the pot itself will crack.

VOLUTE

The *volute* is the chamber in which the impeller spins. Combined with the impeller, the volute forces water out of the pump and into the plumbing that takes the water to the filter (or directly back to the pool, spa, or fountain if the system is not filtered or heated). The outlet port is usually female threaded for easy plumbing. When the impeller (Fig. 4-5) moves water, it sucks it from the strainer pot.

The resulting vacuum in the pot is compensated for by water filling the void. The rushing water is contained by the volute which directs it out of the pump. Therefore the pot can be considered a vacuum chamber and the volute a pressure chamber.

1 Motor (C frame)	12 Hex bolt
2 Shaft extender	13 Lock washer
3 Allen setscrew	14 Drain plug
4 Bracket	15 Trap gasket
5/6 Hex bolt	16 Strainer pot
7 Volute gasket	17 Strainer basket
8 Shaft seal	18 Lid O-ring
9 Impeller	19 Strainer pot lid
10 Volute	20 T-handle
10A Alternate volute	21 Gasket
(for use without strainer pot)	22 Suction flange
11 Base	(for use without strainer pot)
	23 Assembly screw

4-3 Typical bronze pump and motor, exploded view. Aqua-Flo, Inc., Chino, Ca.

The impeller by itself will move the water, but it cannot create a strong vacuum by itself to make the water flow begin. The area immediately around the impeller must be limited to eliminate air and help start the water flow. A diffuser (Fig. 4-4) and/or closed-face impeller help this process, but in many pump designs, the volute serves this purpose.

Figure 4-2 shows how the volute closely encircles the impeller. Figure 4-4 shows a design with a separate diffuser that houses the impeller. In some designs, the inside of the volute is ribbed to improve the flow efficiency.

1 Motor (square flange)	14 Drain plug
2 Bonding screw	15 Strainer pot gasket
3 Bonding lug	16 Lock washer
4 Water slinger	17 Hex bolt
5 Seal plate	18 Strainer pot drain plug
6 Seal plate O-ring	19 Strainer pot
7 Seal insert	20 Strainer basket
8 Shaft seal	21 Lid O-ring
9 Impeller	22 Strainer pot lid
10 Diffuser	23 Star washer
11 Diffuser O-ring	24 Assembly machine screw
12 Clamp	25 Assembly allen-head bolt
13 Volute	26 Motor mount pad
	27 Motor mount

4-4 Typical Noryl plastic pump, exploded view. Sta-Rite Industries, Delevan, Wisc.

IMPELLER

The impeller is the ribbed disk (the curved ribs are called *vanes* and the disk is called a *shroud*) that spins inside the volute. As water enters the center or eye of the impeller (Fig. 4-5), it is forced by the vanes to the outside edge of the disk, just like our spinning bucket example. As the water is moved to the edge, there is a resulting drop in pressure at the eye, creating a vacuum that is the suction of the pump. The amount of suction is determined by the design of the impeller and pump components and the strength of the motor that spins the impeller.

There are essentially two types of impellers: closed-face (Fig. 4-5) and semiopen-face (Fig. 4-6). Some publications call the semiopen-face impeller an open face. This is not accurate, because for the pump to be self-priming, which most pool and spa pumps are, it needs a disk (shroud) on the front face as well. Therefore, although a particular impeller itself has no front shroud and might be called open when it stands alone, it does in fact make use of some sort of front shroud, either a diffuser located closely around the impeller or the interior surface of the volute.

4-5
Water's-eye view of the
impeller. Sta-Rite Industries, Delevan,
Wisc.

Impeller eye — Water flow

WATER

Impeller

Discharge

Volute

Rotation
of impeller

4-6
Front and side view of the
impeller inside the volute. Sta-Rite
Industries, Delevan, Wisc.

Impeller

Flow

Set screws

Clearance

In Fig. 4-6, you get another water's-eye view of a volute and impeller. The side view shows how close the open-face impeller is to the interior side of the volute, effectively forming a front shroud. The clearance between the volute interior and impeller face is critical. Too far away and there will be insufficient pressure created in the volute, causing weak or no suction. Too close and the impeller might rub against the volute or jam if small debris lodges between the two. As discussed later in the section on the shaft, the semiopen impeller pump can be adjusted for optimum efficiency of the impeller.

In the closed-face impeller, as the name implies, the vanes of the impeller are covered in both front and back. Water flows into the hole in the center and is forced out at the end of each vane along the edge of the impeller. This type of impeller is very efficient in moving water—so efficient that a diffuser must be added to the design (Fig. 4-4, item 10) to slow the speed of the water before it leaves the volute.

If the closed-face impeller is so much more efficient and requires no shaft extender or adjustment, why have the semiopen-face impeller designs survived? Because the downside of the closed face is that small stones, pine needles, and other fine debris can get past both the skimmer and strainer baskets and clog the closed vanes, slowing or completely shutting off water movement. The semiopen-face design allows this small debris to pass (actually it is usually pulverized) to the filter, although it is not impossible for a heavy volume of small debris to clog the open vanes as well.

One of the chief culprits in clogging of both semiopen and closed impeller designs is DE (diatomaceous earth). As is described in the chapter on filters, this white, powdery material is used to precoat some designs of filter grids. If introduced too quickly, DE will clog any restricted area—plumbing elbows, strainer baskets, and impeller vanes.

Most impellers on pool and spa pumps have a female threaded hole on the center back side (the side facing away from the water source) that screws onto the male threaded end of the motor shaft (or shaft extender). The rotation of the shaft is just like a bolt being threaded into a nut. As the shaft turns, it tightens the impeller on itself. Others, like the old Purex AH-8 models, are fitted with setscrews that clamp the impeller onto the end of the motor shaft.

Impellers are rated by horsepower to match the motor horsepower that is used. This, in turn, determines the horsepower rating of the pump or pump and motor you have. What if you used a 2-hp motor and a 1-hp impeller? The pump would still only pump the volume of the 1-hp impeller, but the motor would not have to work as hard as it was designed to work. No problem.

But what about the reverse? A 2-hp impeller will move its rated volume and speed of water even with a 1-hp motor, but the motor works harder than it is designed to and soon overheats and burns out. Big problem! Moral of the story—if you must assemble miscellaneous unmatched parts, always make the impeller rating equal to or less than the rating of the motor.

Let's say you have a box of spare impellers and don't know what their rated horsepower is or what pump they came from. Start by asking at your pump rebuilding shop or supply house. Another way is to examine the impeller for codes or markings. On many bronze semiopen-face impellers, a 0.5, 1, 2, etc., is engraved on the inside of one of the vanes, telling you the horsepower. On Sta-Rite plastic closed-face impellers, a code is used, such as 137-PD. By checking your supply house catalog for Sta-Rite pumps, you will see the model code 137-PD refers to a 2-hp pump.

Failing any identifying marks such as these, don't try to guess at the rating. It is cheaper to buy a new impeller for a particular pump than to install one of greater horsepower than the motor and, ultimately, damage the motor.

SEAL PLATE AND ADAPTER BRACKET

The volute is the pressure chamber in which the impeller spins to create suction. If this were all one piece, there would be no way to remove the impeller or to access the seal. Therefore this chamber is divided into two sections. The actual curved housing is called the *volute*, while its other half is called the *seal plate* or *adapter bracket*.

The seal plate (Fig. 4-4, item 5) is joined to the volute with a clamp. An O-ring between them makes this joint watertight. The motor is bolted directly onto this type of seal plate. In other designs, the seal plate is molded together with an adapter bracket that supports the motor (Fig. 4-3, item 4) and bolts to the volute, with a paper or rubber gasket between them to create a watertight joint. In yet another style, Jacuzzi brand units for example, the pump sections are joined with a threaded union type of clamp, like the lid of a jar. This allows disassembly by hand.

Note in Fig. 4-4, the clamp (item 12) that joins the volute and seal plate is made tight by a bolt and nut. For some inexplicable reason, Sta-Rite, the maker of this model of pump, assembles the pump with this bolt underneath the pump. When the pump is bolted to the deck, it makes it a knuckle-busting, cussword of a job to remove the clamp. Moreover, if the pump leaks at all, the bolt gets and stays wet, causing it to rust, making unscrewing it even tougher, or the bolt breaks altogether. Fortunately Sta-Rite no longer assembles pumps this way, but if you come across one, move the clamp bolt to the top of the pump before installation.

In both cases, the shaft of the motor passes through a hole in the center of the seal plate and the impeller is attached, threaded onto the shaft. The bracket allows access to the shaft extender (see following section) for adjusting the clearance between the impeller and volute (Fig. 4-6). The pump design shown in Fig. 4-4 is a closed face and needs no such adjustment, so the shaft need not be exposed.

SHAFT AND SHAFT EXTENDER

The shaft of the motor is the part that turns the impeller, creating water flow. Figure 4-3 shows a motor with shaft. In this design of pump, the impeller needs to be adjusted in relation to the volute, so a shaft extender has been created. The extender, made of brass or bronze, slides over the motor shaft and is secured by three allen-head setscrews (Fig. 4-6). The male threaded end of the shaft extender then fits through the seal plate and the impeller is screwed into place.

Note that the extender is round, but a flat area has been created on two sides. In this way, a ¾-inch box wrench can be used to prevent the extender from spinning when performing maintenance. Note also that some designs require an O-ring near the threads of the extender to ensure a watertight seal. Other designs rely only on the pump seal. Once assembled, the clearance between the impeller and volute can be adjusted as shown in Fig. 4-6 (side view).

The shaft of the motor in this example (Fig. 4-3) is called a *keyed shaft*. This means the cylindrical shaft has a groove running the length of the shaft to accept the setscrews. In this way, when the setscrews are in place, they prevent the motor shaft from slipping or skidding inside the extender, thus failing to turn the impeller. Figure 4-4 shows a pump style that requires no shaft extension. The motor shaft, already engineered to the exact length required, has a threaded end to accept the impeller.

The shaft should never be in contact with electric current, but water is a great conductor and wet conditions around pool equipment can circumvent the best of designs. Because of this, most motor shafts today are designed with a special internal sleeve to insulate the electricity in the motor from the water in the pump.

SEAL

Obviously if the shaft passed through the large hole of the seal plate without some kind of sealing, the pump would leak water profusely. If the hole was made small and tight, perhaps of tight-fitting rubber, the high-speed spinning of the shaft would create friction and burn up the components or the shaft would bind up and not turn at all. Some clever engineer devised a solution to this problem called a *seal*.

The seal allows the shaft to turn freely while keeping the water from leaking out of the pump. In Fig. 4-3, the seal (item 8) is in two parts. The right half of the seal is composed of a rubber gasket or O-ring around a ceramic ring. This assembly fits into a groove in the back of the impeller. The ceramic ring can withstand the heat created by friction. The left half fits into a groove in the seal plate and is composed of a metal bushing containing a spring. A heat-resistant graphite facing material is added to the end of the spring that faces the ceramic ring in the other half.

The tight fit of the seal halves prevents water from leaking out of the pump (item 7). The spring puts pressure on the two halves to prevent them from leaking. As the shaft turns, these two halves spin against each other but do not burn up because their materials are heat-resistant and the entire seal is cooled by the water around it. Therefore, if the pump is allowed to run dry, the seal is the first component to overheat and fail.

The pump design in Fig. 4-4 includes an additional seal housing or insert. If the pump runs dry, the heat build-up not only melts the seal, but also the plastic seal plate in which it is mounted. The inset helps isolate the seal plate from this heat and the cone-shaped unit diffuses heat.

Pump makers are always improving the heat dissipation (heat sink) capabilities of their pumps so that dry operation will result in little or no damage to the seal or pump components. Still, pumps are not designed to run without water for more than a few minutes while priming.

MOTOR

Before reading this section, a basic knowledge of electricity is helpful, so you might want to review the chapter on basic electricity. After all, the motor is the device that converts electricity into mechanical power.

Motors, like the pumps they drive, are rated by horsepower, usually in pool and spa work as ½, ¾, 1.0, 1.5, and 2.0 horsepower. Commercial installations might use higher rated systems, but these are the most common.

As shown in Fig. 4-7, electricity flows through the motor windings, which are thin strands of coiled copper or aluminum wire. The windings magnetize the iron stator. If you paid attention in your first grade science class, you recall that opposite poles of a magnet attract each other, but like poles repel. Using this concept, the rotor spins, turning the shaft. Some designs employ one set of windings for the start-

4-7
Typical pool and spa motor.
Franklin Electric, Bluffton, Ind.

up phase where greater turning power (torque) is needed and another set for normal running.

The shaft rides on ball bearings at each end. A built-in fan cools the windings because some of the electrical energy is lost as heat. The caps on each end of the motor housing are called *end bells*. A starting switch is mounted on one end with a small removable panel for connection and maintenance access.

This is where you will also find the thermal overload protector. This heat-sensitive switch is like a small circuit breaker. If the internal temperature gets too hot, it shuts off the flow of electricity to the motor to prevent greater damage. As this protector cools, it automatically restarts the motor, but if the unit overheats again, it will continue to cycle on and off until the problem is solved or the protector burns out.

It takes a great deal of electricity to start a motor but far less to keep it going (in fact, about five to six times as much). The capacitor, as the name implies, has a capacity to store an electrical charge. The capacitor is discharged to give the motor enough of a jolt to start, then it is able to run on a lower amount of electricity. Without the capacitor, the motor would need to be served by very heavy wiring and high-amp circuit breakers to carry the starting amps. The start-up amperage of a motor is about twice that of its running amperage. The capacitor is located in a separate housing mounted atop the motor housing (as in Fig. 4-7) or inside the front end bell.

Some motors are designed to operate at two speeds. For example, some spas operate at high speed for jet action, but lower speed for circulation and heating. In pool and spa work, the normal rotation speed is 3450 revolutions per minute (rpm) and the low speed is 1750 rpm.

Types

Now that you understand these basic concepts, I will discuss the three main types of motors you will find in water work.

Split phase
These motors are ¾-hp or less, and are found in small fountain applications where start-up power requirements are very low—there is no capacitor.

Capacitor start, induction run (CSI)
The most typical motor used in the pool business. This motor uses a capacitor and starting windings to start up, then these are shut down and a running winding takes

over. As noted previously, the capacitor and start-up windings allow faster, stronger torque to overcome the initial resistance of the impeller against standing water, then when the water is moving and less power is needed to keep it moving, the system shuts off and the lighter running winding takes over.

Capacitor start, capacitor run (CSR)

This is the concept of the energy efficient motor. The difference between a CSR and CSI motor is that the CSR motor employs a capacitor on the running windings as well. This smoothes out the variations in the alternating current (ac) power that helps reduce heat loss in the winding (remember, heat loss is electricity wasted). In short, CSR motors are more efficient but cost more because of the added parts. These motors are also called switchless, because on some designs the run capacitor makes a start switch unnecessary. This is a good thing because start switches get dirty and fail and tend to be fragile parts that readily break when brought into sharp contact with a screwdriver.

Energy efficient

Energy efficient motors are CSR motors that have heavier wire in the windings to lower the electricity wasted from heat loss.

A good way to compare energy efficiency between two motors is to compare the gallons pumped to kilowatts used. Let's say one pump produces a flow rate of 50 gallons per minute, which is 3000 gallons per hour. Divide that by the kilowatts used per hour. The higher the resulting number, the more efficient is the pump and motor.

By the way, kilowattage is determined by multiplying amps by voltage. Lets say the unit runs at 9 amps at 220 volts—9 × 220 = 1980 watts. Kilowatts (meaning 1000 watts) would then be 1980 divided by 1000. So this unit uses 1.98 kilowatts each hour. The 3000 gallons divided by 1.98 equals a rating of 1515 gallons pumped for every kilowatt used. You can now make similar calculations for other pump and motor units for comparison.

Voltage

I have been discussing typical 110/220-volt motors. In fact, most motors are designed to be connected to either power source. By changing a wire or two internally, you determine which voltage is used. The instructions for this conversion are printed on the motor housing or inside the access cover.

If your motor is wired for 220 volts and you feed it 110 volts, it will run slowly or not start. If your motor is wired for 110 volts and you feed it 220 volts, the thermal overload protector should overheat and cut the circuit. Higher horsepower motors might run on three-phase current. You don't want to fool with that. Call an electrician.

Housing design

The housing of the motor is designed to adapt to various pump designs. Figure 4-3 shows a motor called a *C frame*, because the face of the motor resembles a C. All this means is that it will fit certain kinds of pumps. Figure 4-4 shows a motor called a *square flange*, for equally obvious reasons. There are other types, such as the 48, uniseal flange, and those designed for automatic pool cleaner booster pumps. When replacing a motor, you need to buy the proper housing type.

Ratings

The *service factor* of a motor is a fascinating piece of marketing sleight of hand. Many moons ago, some engineer figured out that his company might sell more motors than the other guy if the buyer thought his motors were better, even if they're exactly alike. Although nominally rated for a certain horsepower, motors are generally made beefier than the rating to be sure they produce the promised horsepower over their lifetime. They also need more strength (horsepower) to start up than to run, so again they need to be made heftier than the actual rating.

So some genius came up with the service factor. This is a multiplier, a number. When you multiply the service factor number by the rated horsepower number, you get the real horsepower at which the motor is designed to operate on a continuous basis. As an example, a motor rated at 1 hp with a service factor of 1.5 can actually safely run a 1.5-hp pump (1.0 × 1.5 = 1.5 hp).

The service factor has no real value. So why learn about it? Well, you don't want to look stupid if your kids ask you someday when they're reading old motor nameplates. Then, as I discovered on the job, there might be the rare occasion when the information is useful.

The motor on the fish pond at comedian Rich Little's house burned out one day. It was a 2-hp motor. It was too late in the day to get to the supply house and buy a new one, and if the pump didn't run, the very large and very expensive koi would end up in goldfish heaven by morning. All I had in my shop was a 1.5-hp motor, but it was rated with a service factor of 1.5, meaning 1.5 × 1.5 equals 2.25 hp. I could safely use this motor on the existing pump and expect it to do the job. It worked!

Nameplate

All of what you need to know about a motor is printed on the nameplate (Fig. 4-8), a sticker applied to the housing. Here's what you can learn from the nameplate.

Electrical specifications

A diagram shows how to wire the starting switch plate for 110- or 220-volt supply. If this diagram is not on the outside of the motor, remove the small access door in the end bell and it should be printed on a sticker in there. These stickers frequently come off as the motor gets older, so if no diagram is available, refer to the manufacturer's guidebook available at your supply house.

The nameplate also tells you the amperage. It might say "10.5/5.2." This means the start-up draw is 10.5 amps, and the normal running draw is 5.2 amps. Make sure you are reading this information from the area that says "Maximum Load" or "Maximum Amps." For some silly reason, some makers publish the amps required to power the nominal horsepower as well. These are lower than the maximum and should not be used when sizing wire or circuit breakers. (They should not even appear on the nameplate.)

The nameplate also lists the electrical phasing (single phase or 1) and cycle frequency (called *hertz*). Alternating current in the United States runs on 60 hertz. In

Current nameplate

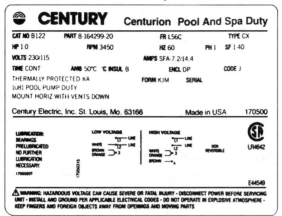

4-8
Typical motor nameplate.
MagneTek, St. Louis, Mo.

Nameplate prior to August 1985.

Europe it is 50 hertz, and that's why you can't use some appliances from one country in another, even with a voltage converter, because things like VCRs and TVs rely on the cycles as well as the voltage. "Stupid" appliances like toasters or shavers only care about voltage. Why? Go read the toaster and shaver books. It only hertz once. (Sorry, I couldn't resist that one.)

Manufacturer and date
The manufacturer's name appears on the rating plate, along with the model and serial numbers, and includes the day, month, and year the motor was built. On Century motors, the serial number gives you a code which in turn tells you the year, if you have the code list (see Fig. A-1). Why does it have to be so complicated? It doesn't. By the way, most motor rebuilders agree that pool and spa motors are designed to run about 7 years—if the unit is over 10 years old, junk it.

Horsepower
The relative strength of the motor is expressed in horsepower and corresponds to the specifications of the pump that is to be driven by the motor.

Installation
As noted previously, a diagram of the wiring connections is printed on the nameplate. There will also be a few words about mounting, such as "mount horizontally" or "mount with vents down."

Duty rating

Pool and spa motors are designed for continuous duty, meaning they can run 24 hours a day for their entire service life without stopping. The nameplate shows this by the rating "Continuous Duty." The horsepower, service factor, rpm, and frame style of the housing are listed. If the motor has a thermal overload protector, the nameplate will indicate it.

Other essentially useless information is also shown on the nameplate, such as the starting method (C means capacitor for example), the insulation category, a rating of UL (Underwriter's Laboratory) or CSA (Canadian Standards Association) approval, and ambient (surrounding) temperature requirements.

HORSEPOWER AND HYDRAULICS EQUALS SIZING

So now you know what a pump and motor are and, if you are asked to replace either, in most cases you will assume the original designer or builder used the correct size pump and motor for the job and make your replacement with the same size. Or will you?

What if the original equipment was too small or too large? What if the plumbing has been repaired (which might have added or deleted pipe and fittings) or equipment has been added or deleted, thus changing the system and requiring the pump to work more (or less)? What if the identifying rating plates have been removed or are so weatherworn that you can't tell what size the existing equipment is? Finally, what if it is a brand new installation? How do you decide what is the right pump and motor for the job?

Well I'm glad you asked all of those intelligent questions. The answer is that you need to know a little about the pool or spa system's needs and hydraulics.

Hydraulics

Hydraulics, the study of water flow and the factors affecting that flow, is important to understand because its principles affect plumbing and equipment sizing choices. Understanding hydraulics as it applies here is actually quite simple and the math involved is very basic. It only looks tough because there are so many factors to consider. Before starting on this complex series of calculations, think about this. Unless you are building a pool or spa with clean components able to operate as the manufacturer recommends and you have blueprints to know what plumbing components have been included underground, then all of this section is theoretical. In fact, most of the time you won't know what plumbing exists out of sight or how much resistance to water flow is being created by old filters, heaters, solar panels, and so on.

So why learn hydraulics? First, because understanding it helps you estimate the right pump for a replacement or when water circulation is poor on a particular job. It helps you avoid adding plumbing or other components that might aggravate an already bad situation. Second, in some cases, you might be the designer or installer of a pond, spa, or pool, so this information is essential.

But most of all, you need to learn this section because it is probably required knowledge to pass the contractor's license or board of health certification in your area. Having said that however, the following information is generally interesting theory only because it is not something you will use in everyday water maintenance and repair.

Terms

First, a few terms used in this section must be explained.

Head and flow rate *Head* is the resistance of water flow through plumbing and equipment expressed in feet. (The lower, the better.) *Flow rate* is the volume of water moved in a given period of time. Here's an example.

Let's say you have a pump and motor with a 1 foot pipe attached to the outflow port sticking straight up in the air. For the moment ignore the source of the water on the suction side. You turn on the pump and water flows out at a rate of 10 gallons per minute (oh yes, you had a 10-gallon bucket and stopwatch nearby).

The 1-foot vertical distance is the head and the 10 gallons per minute (gpm) is the flow rate. The pump is rated at 10 gpm at 1 foot of head (or 1 foot of resistance). But what if that resistance is increased? What will happen to the flow rate?

Suppose you are moving your furniture around the living room for fun and profit. You can raise the 100-pound couch over your head by 2 feet. You ask your friend to get on the couch, making it now weigh 200 pounds. Now you are able to lift it only 1 foot over your head.

This rather silly example tells you what to expect when adding resistance to the pump. Let's go back to our pump with the pipe sticking straight up and make the outflow pipe 8 feet, then measure the flow rate. It is down to 5 gpm. The additional resistance (head) of that added pipe means that the pump cannot push the water as fast. This loss of flow, as head increases, is called *head loss*, a somewhat deceptive term because it is actually flow that is lost. The term actually means flow loss caused by head increase, but just to confuse us, engineers call this process head loss. Got it?

If you continue to increase the head (resistance) by adding more vertical pipe, the flow rate will continue to decrease until at last, no water comes out at all. In the example, let's say that happens when you add 10 feet of pipe. This pump can now be charted on a graph (Fig. 4-9), allowing you to study its performance characteristics. With such a graph, you can choose any flow rate for your hypothetical pump and learn what the maximum amount of resistance can be if you are to maintain that flow rate. Conversely, you can choose any amount of head you think a certain plumbing system might create, then learn the flow rate expected out of that system.

Figure 4-9 shows the graph for this hypothetical (and rather small) pump. On the left side of the graph is the possible feet of head, from 0 to 10 feet. On the bottom is the possible flow rate, from 0 to 10 gpm. By finding where the head and flow rates intersect, you can create a pump curve for your pump.

Because you generally have no way to measure these factors, the manufacturer provides a pump curve for each pump. If you know the amount of head (resistance) in your pool or spa system and you know the desired flow rate, then you can determine which pump will satisfy those needs by referring to the manufacturer's pump curves.

Pumps are designated low, medium, high, or ultra-high head. The higher the head designation, the less strain is placed on the pump and motor components:

- Low head—suck well; push poorly.
- Medium head—suck well; push well.
- High head—suck poorly; push well (most common in pools and spas).
- Ultra-high head—suck poorly; push well (pool sweeps).

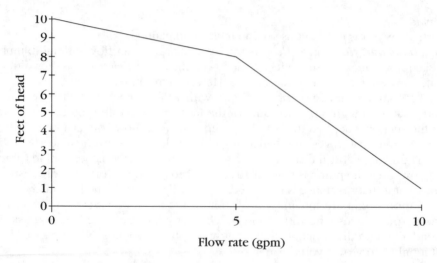

4-9 Sample pump performance curve. Ross McCulloch

One factor affecting which type a particular pump will be is its impeller. Thin vents on the face (closed or semiopen) result in greater push but poor suck; in other words, poor self-priming capabilities but good circulating flows.

Suction head So far I have only discussed the head created by adding resistance to the outflow side of the pump. By restricting the intake or requiring the pump to lift water from a source below it, you also create head. This is called *suction head*, sometimes called vertical feet of water. Don't be confused, it's the same thing.

Each foot on the suction side equals a similar foot on the discharge side, called *discharge head*. The only thing to remember here is that head (resistance) is created on both sides and must be calculated when determining pump size.

Dynamic and static head Up to now I have described *static head*—the head created by the weight of standing (static) water. This is only a small portion of the total head in the system. The rest is created by the friction of water flowing through the entire system, called *dynamic* (moving) *head*. The diameter of the pipe and the speed of the water determines how much resistance is created by friction. Further friction is created when water must go through or around other obstacles, such as through the filter, heater, solar panels, and plumbing fittings (yes, every plumbing elbow or bend creates head too).

Cavitation *Cavitation* refers to the vacuum created when the outflow capacity of a pump exceeds the suction intake. This happens, for example, when a pump is oversized for the suction line or when the distance from the body of water is too far. The result is bubbling and vibration.

Total dynamic head *Total dynamic head* (TDH) is the total of plumbing and equipment head for the entire system. Vacuum head (suction) plus pressure head (discharge) equals total dynamic head.

Shut-off head The amount of head at which the pump can no longer circulate water. It is 0 gpm.

Calculations

Here are a few general numbers to use in your calculations.

Pipe fittings To make it easier to calculate head in your plumbing system, it is measured for every 100 feet of pipe or the equivalent (see appendix A). Plumbing connections, fittings, and valves have different amounts of resistance than straight pipe, so these must first be converted to the equivalent length of straight pipe. Unions and straight connectors act like additional lengths of straight pipe, so no special calculations are needed. Going around corners is what creates head. Here are the values for the most common PVC fittings you will use:

- 1½-inch × 90-degree elbow = 7.5 feet of straight 1½-inch pipe
- 2-inch × 90-degree elbow = 8.6 feet of straight 2-inch pipe
- 1½-inch × 45-degree elbow = 2.2 feet of straight 1½-inch pipe
- 2-inch × 45-degree elbow = 2.8 feet of straight 2-inch pipe

It is interesting to see that three times as much resistance is created when a 90-degree fitting is used instead of a 45. That is particularly significant, because there are times when you have a choice between using two 45-degree fittings in a job rather than one 90. The combination of two 45-degree fittings creates less resistance than one 90.

Also note that in a T fitting, the turn around its 90-degree bend is sharper than the more gradual sweep through 90 degrees created by a typical 90 fitting. Thus, more resistance is created in a T fitting, even though the water is being bent 90 degrees in either case.

Filters Filters create 5 to 7 feet of head. The manufacturer will tell you in the literature that accompanies the product how many feet of head the unit creates. You can also measure the amount by placing a pressure gauge on the pipe leading into the filter and one on the pipe going out. The difference, measured in pounds per square inch (psi), tells you the feet of head.

As dirt builds up in a filter, however, head increases. A clean filter will have no more than 3 psi difference between the input pressure and the output pressure. Since 1 psi equals 2.31 feet of head, a new, clean filter should add no more than 6.9 feet of head.

As you will see in the chapter on filters, manufacturers recommend cleaning a filter when the operating pressure (as read on the pressure gauge, expressed in psi) builds up to more than 10 psi over clean operating pressure. Since 10 psi equals 23.1 feet of head, you can see that the resistance caused by dirt added to the normal amount of head for the filter itself can total over 30 feet of head for this component alone.

This is another factor that tends to make these calculations more art than science, and why providing a slightly larger horsepower pump than required on a system is always a good idea.

Heaters Heaters create 8 to 15 feet of head. Like filters, the manufacturer will tell you in the literature that comes with the unit what the head loss is for the unit at a given flow rate. Also like filters, as scale (lime and other minerals) builds up in the heat exchanger (explained in more detail in the chapter on heaters), more friction is created and therefore more head. This is another good reason to beef up your pump when doing the sizing calculations for an installation.

Poolside hardware Main drain covers, skimmers, and return outlets all add head. To know exactly how much, you must refer to each manufacturer's specifications. A general rule of thumb is to add 5 feet of head to allow for the total of such components in your system.

Pumps Pumps also create head, but the manufacturer's charts allow for this, so your calculations need not consider it. When you look at the TDH for the system on the pump curve, the pump head loss is already figured in the performance ability.

Turnover rate

The *turnover rate* of a body of water is how long it takes to run all the water through the system. It is desirable for the water to completely circulate through the filter one to two times per day, but local codes generally require a specific time period. In Los Angeles, for example, it is

- Pools must turn over in 6 hours,
- Spas must turn over in ½ hour, and
- Wading pools must turn over in 1 hour.

I have mentioned that various components offer more or less resistance at different speeds (expressed in gallons per minute). To calculate the TDH of a system, you must know that speed. To decide what speed is needed (and therefore what size pump is needed to deliver that speed in your system) you must establish a turnover rate.

Let's say you've calculated the volume of water in the pool (see chapter 1) as 18,000 gallons.

$$18,000 \text{ gallons} \div 6 \text{ hours} = 3000 \text{ gallons per hour}$$
$$3000 \text{ gph} \div 60 \text{ minutes} = 50 \text{ gpm}$$

Therefore, you need a pump capable of delivering a flow rate of 50 gpm under the TDH of the system. The manufacturer's pump curves described previously will tell us which pump can do this (Fig. 4-10).

Methods of calculating TDH

As I mentioned earlier, unless you are the pool or spa builder, you don't know exactly what plumbing is included in the system, so TDH is an educated guess at best. Here, however, are the three methods for calculating TDH.

Method 1: Exact values If you have the exact specifications of the pool as built or as proposed, measure all the pipe from the pool, through the equipment, and back to the pool. Add the equivalent feet of pipe for all the fittings. Add the feet of head at the desired flow rate for the filter, heater, and any other components to arrive at the TDH for the system.

Method 2: Estimated values

1. Suction-side head. Assume 2 feet of head for each 10 feet the equipment is away from the pool.
2. Discharge-side head. Estimate how many feet of pipe are in the system back to the pool. Double that estimate to allow for fittings.
3. Using the tables (Fig. A-2), calculate the feet of head for the total amount of pipe on the discharge side.
4. Equipment head. Consult manufacturer's tables and charts for the desired flow rate (in the example, 50 gpm).
5. Add these three parts together to get the TDH.

Let's try a simple example. Let's say our equipment is 30 feet from the pool (at 2 feet of head per 10 feet of distance, that makes 6 feet of head). There is about 60 feet of 1½-inch plumbing between the equipment and the run back to the pool (doubled is 120 feet). The table says 13.5 feet of head per 100 feet of pipe equals 1.2 ×

"A" performance curve

Capacity in gallons per minute
Curve based on 3450 rpm impeller speed
50 cycle units available

4-10 Actual pump performance curve. Aqua-Flo, Inc., Chino, Ca.

10.24 or 16.2 feet of head. The filter manufacturer says our sample filter has 7 feet of head; the heater manufacturer says 15 feet of head. Therefore

- Suction side estimate   6.0 feet
- Discharge side estimate   16.2 feet
- Main drain and skimmer estimate   5.0 feet
- Filter   7.0 feet
- Heater   15.0 feet

for a total dynamic head of 49.2 feet.

Now you can consult various manufacturer's pump charts to decide which pump will deliver the desired 50 gpm at 49.2 feet of TDH.

Method 3: Measured values An easier and more accurate way to estimate all of this, if the existing pump is operating, is to measure the vacuum on the suction side of the pump and the pressure on the discharge side. Plumb a vacuum gauge on the pipe entering the pump. It measures inches of mercury. Every 1 inch of mercury equals 1.13 feet of head. Plumb a pressure gauge on the pipe coming out of the pump. It measures pounds per square inch (psi). Every 1 psi of pressure equals 2.31 feet of head.

Multiply the gauges out accordingly and the sum of the two gives you the TDH of the system. This might sound like work, plumbing in two separate gauges, but it really isn't, and it gives you the most accurate TDH calculation because it takes into account the dirty filter, the limed-up heater, all the unseen plumbing . . . everything.

It also allows you to keep an eye on the TDH in the system at any time and more easily troubleshoot poor performance in the equipment.

Sizing

So now you've chosen a pump. Is bigger better? Well, yes, because as I mentioned, TDH estimating is more art than science and it changes every minute as the system gets dirty or clogged. So you do want a pump that offers at least a little more capacity than absolutely required.

The only caveat here is that running water not only encounters friction created by pipes and equipment, but the water itself is creating friction. This friction will strip copper from pipes and heater components causing all kinds of havoc (see the chemistry chapter), damages filter grids, and makes DE or sand inefficient (see the filter chapter).

Because of this, most building codes set maximum flow rates of 8 feet per second through copper pipe and 10 feet per second through PVC. Since heaters all use copper heat exchangers, use 8 feet per second even if the plumbing is PVC. Los Angeles County, for example, allows a maximum flow rate of 8 feet per second on suction pipes of any type. What is feet per second in terms of gallons per minute? Refer to Fig. A-2 to learn that (rounded to the nearest tenth):

- 50 gpm in 1½-inch pipe = 7.9 feet per second
- 50 gpm in 2-inch pipe = 4.8 feet per second
- 60 gpm in 1½-inch pipe = 9.5 feet per second
- 60 gpm in 2-inch pipe = 5.7 feet per second

By the way, there are a few exceptions to the rules. Los Angeles County requires pumps to deliver the desired gallons per minute at 60 feet of head. When sizing pumps, you must assume at least 60 feet of head regardless of the actual calculations. In filters, on the other hand, you must use the actual head as measured—go figure.

Altitude also affects these calculations. Over 3300 feet above sea level a motor runs hotter, so you will want to upgrade to the next horsepower.

MAINTENANCE AND REPAIRS

Since the pump and motor are the heart of the system, if they fail or don't perform efficiently, the abilities of the other components won't much matter. Keeping the motor in good working order is a matter of keeping it dry and cool. The best detection tool for motor problems is your ears. Laboring motors or those with bad bearings will let you know.

Keeping the pump in good order is also a matter of sight. Seeing leaks tips you that the pump needs attention. If the motor needs to stay dry, but problems with pumps most often result in leaks, the potential for pump and motor breakdown is high. Therefore, keeping an eye and ear on your pump and motor will pay dividends in a pool or spa that keeps running.

The basic repairs and maintenance of the pump and motor unit, starting from the front, the first place the water encounters, are discussed in the following paragraphs.

Strainer pots

Clean out the strainer basket often. Even small amounts of hair or debris can clog the fine mesh of the basket and substantially reduce flow. To be honest, this job is a pain in the valve seat. You have to shut down the system, struggle with tight cover bolts or clamps, clean out hair and filth from the basket, put the basket back, find a water source to fill the pot so the pump will reprime easily, check the O-ring, replace the cover, tighten the bolts or clamps, restart the system, and most often, reprime . . . whew!

But this, along with keeping a clean skimmer basket, are the two most simple and important elements to keep a pool clean and the other components working. If the water can't flow adequately, it can't filter or heat adequately either. It will turn cloudy, allow algae growth, and make vacuuming difficult.

The only up side to all this is that every pool person I know has found coins, gold chains, and jewelry in strainer baskets. So sometimes it pays to do your job right. Keep a clean strainer basket.

The only other problems you might encounter at the strainer pot are broken baskets or a crack in the pot itself. If the basket is cracked it will soon break, so replace it. If allowed to operate with a hole in it, the basket will permit large debris and hair to clog the impeller or the plumbing between the equipment components.

Cracks might develop in the pot itself, especially if you live where it gets cold enough to freeze the water in the pot. Again, the only remedy is replacement. Follow the directions described in the following paragraphs for changing a gasket, because it requires the same disassembly and assembly techniques.

Gaskets and O-rings

Most problems occur in strainer pots when the pump is operated dry. The air heats in the case as the impeller turns without water to cool it. The strainer basket will melt; the pot cover, if plastic, will warp; and the O-ring will melt or deform. Usually, replacement of the overheated parts solves the problem.

Gaskets

When gaskets leak, or in extreme cases, if the strainer pot itself must be replaced (Fig. 4-3, items 15 and 16 and Fig. 4-4, items 15 and 19), the replacement process is the same. Remove the strainer pot (take out the four bolts, usually using a ½- or ⁹⁄₁₆-inch box wrench. Clean out the old gasket thoroughly. Failure to do this will leave gaps in the new gasket that will eventually leak. Reassemble the new gasket and strainer pot the same way the old one came off. Tighten the bolts evenly (so the new gasket compresses evenly) by gently securing one bolt, then the one opposite, then the last two. Continue tightening in this crisscross pattern until each bolt is hand tight. When dealing with plastic pumps, do not overtighten because the bolt will crack the pump components or strip out the female side.

Sometimes the bolts are designed to go through the opening in the pot and volute and are tightened with a nut and lock washer on the other side. Still, do not overtighten, because you will crack the pump components. The key to this simple procedure, as with virtually all other mechanical repair, is to carefully observe how the item comes apart. It will go back together the same way.

O-rings

When removing and replacing the strainer pot cover, be sure the O-ring and the top of the strainer pot are clean, because debris can cause gaps in the seal. Sometimes these O-rings become too compressed or dried out and brittle and cannot seal the cover to the pot. In this case, replace the O-ring.

If no replacement is available, try turning the O-ring over. Sometimes the rubber is more flexible on the side facing the cover. Be careful to remove the O-ring gently. Too much stress will cause the rubber to stretch, making it too large to return to the groove in the cover. If the O-ring stretches, try soaking it in ice water for a few minutes to shrink it. Finally, coat the O-ring liberally with silicone lube. This can take up some slack and complete the seal if the O-ring is not too worn out.

Another emergency trick is to put Teflon tape around the O-ring to give it more bulk and make it seal. If you use this trick, be sure to wind the tape evenly and tightly around the O-ring, so loose or excess tape does not cause an even worse seal. If the O-ring has actually broken, it will almost always leak at that spot; however, I have used the Teflon tape trick successfully in these cases for a temporary repair when a new O-ring was not immediately available.

Changing a seal

As you have seen, all pumps have seals to prevent water from leaking out along the motor shaft. When these wear out (normally, or from overheating if the pump runs dry) they are easy to replace. The steps are shown in Fig. 4-11, based on the pump illustrated in Fig. 4-3. Before you start, turn off the electricity to the motor at the breaker.

4-11 Seal replacement. Step 1.

4-11 Step 2.

4-11 Step 3.

4-11 Step 4.

4-11 Step 5.

4-11 Step 5 continued.

4-11 Step 6.

4-11 Step 6 continued.

4-11 Step 6 continued.

4-11 Step 7.

4-11 Step 7 continued.

1. Figure 4-3 (item 8) shows a two-part seal. To access this seal for replacement, remove the four bolts that hold the pump halves together (item 5) (it is not necessary to remove the entire pump from the plumbing system).
2. Grasp the motor and pull it and the bracket away from the volute. Wiggle it slightly from side to side as you pull back to help break this joint. Do not

4-11 Step 7 continued.

wrestle with the equipment because you might bend the shaft—just be persistent.

3. Take your pliers or ⅞-inch box wrench and hold the shaft extender to prevent it from turning. Unscrew the impeller (it unscrews counterclockwise as you face it, just like a bottle cap) from the shaft extender using an impeller wrench (Fig. 4-11 step 4). You can also wrap a rag over the face of the impeller so you don't cut yourself and twist it off by hand.

As a last resort, hold a large screwdriver against the impeller and tap it gently with a hammer. Use care not to damage the impeller. Use even more care that the screwdriver doesn't slip and damage you.

4. Remove the four bolts (item 6) that hold the bracket on the motor. Use a hammer to gently tap the bracket away from the motor if needed.

5. Remove both halves of the old seal. Note how each half is installed so you get the new one back in the same way. One half is in the back of the impeller and is easily popped out with a flat-blade screwdriver. The other half is in the seal plate and motor bracket unit.

Lay the bracket on your workbench with the seal on the bottom. You will see the back of the seal through the hole in the seal plate. Use that trusty flat-blade screwdriver once again—put the tip on the back of the seal and tap it with a hammer. It will pop out easily.

6. Install the new seal. First, look up your pump in the manufacturer's literature or supply house catalog to determine what model seal you need. Failing that, you can take the old one to the supply house so they can identify it for you. There are only three commonly used seals and only another six or so less common types used in pool and spa work.

Clean out the seal plate and impeller where you have just removed the old seal. Use an emery cloth or a small wire brush and water. Dry each area and apply a small amount of silicone lubricant to help the new seal slide into place. Install each half of the seal the same way you removed the old one—white ceramic of one half facing the glazed carbon ridge of the other half.

7. Gaskets. (This information also applies to the gasket found on many pumps between the volute and strainer pot.) When you break apart a pump, the old gasket usually won't reseal (item 7). Clean all of the old gasket off of the seal plate and volute. Scrape it clean if needed with your trusty flat-blade screwdriver. Now reassemble the pump the same way you took it apart, placing a new gasket between the pump halves.

8. Check your work. Fire up the pump and watch for leaks. A new seal, improperly installed, might not leak for several minutes, so let the pump run awhile before deciding the job is done. A fresh paper gasket might leak for a few minutes until it becomes wet and swells to fill all the gaps, but it should stop leaking after a short time. If your job does leak, take it apart and go over each step again, making sure the seal halves are seated all the way and that there is no corrosion or debris left in the impeller or seal plate that might prevent the new seal from seating completely.

If the pump is old and corroded, you might need two gaskets to seal the uneven gaps between the volute and seal plate. Soaking the gasket in warm water before reassembly sometimes helps as well. The best gasket is a rubber one, but even with these you might need two to seal an old, warped pump. Finally, coat the pump halves with silicone where they contact the gasket. This helps to fill any tiny gaps that still might be present.

In Fig. 4-4, a Sta-Rite pump, you follow the same steps but the parts are slightly different. The clamp (item 12) is removed to disassemble the pump halves, and you must remove the diffuser (item 10) to get to the impeller.

To remove the impeller (Fig. 4-12) you can grip it with your hand and twist it off, but the trick with these units is to stop the shaft from spinning as you twist off the impeller. There are air vents in the motor on the end closest to the pump itself. Look in and you will see the motor shaft. Place a flat-blade screwdriver in one of the air vents and wedge it against the shaft to keep it from turning.

Alternatively, you can remove the end cap and look inside as you twist the impeller. You will see the back end of the shaft, with the start switch attached. Since this switch is fragile, you must remove it (one screw) to access the slotted screw in the back end of the shaft. Place that trusty screwdriver in this screw to keep the shaft from turning as you remove the impeller.

If all this fails and you still have a tightly screwed-on impeller, don't hold the shaft at all, but try this: lay the motor on a table, put the tip of the screwdriver on the edge of the impeller, and tap the screwdriver sharply with a hammer. Repeat this three or four times and sometimes the inertia of the shaft combined with the sharpness of the blow makes the impeller spin right off.

The drawing in Fig 4-12 shows an impact tool for removing a stubborn shaft extender (top) and removing a stubborn AH-8 impeller (bottom).

4-12 Impeller removal.

4-12 Continued.

Labels in top illustration: Shaft, Set screws, Hand grip, Sliding impact weight

Labels in bottom illustration: Shaft, Hand grip, Sliding impact weight, Adjustment bolt, Purex impeller puller adapter, Set screws

4-12 Continued. Pool Tool Co., Ventura, Ca.

Instead of a gasket, the Sta-Rite pump uses an O-ring. Clean this and lubricate it with silicone before reassembly. If it has stretched and it seems like there is too much O-ring for the channel in the volute, try soaking the gasket in ice water for a few minutes to make it shrink a bit. Other than these few differences, the Sta-Rite seal replacement is the same as any other pump.

One last problem. Some pumps use a plastic impeller with a housing that holds half the seal in place. If the pump has run dry and overheated the pot, this housing might be warped and the seal will not fit tightly. The only solution is to replace the impeller. This is a common problem with automatic cleaner pumps, which are not self-priming. These often run dry, for reasons discussed in a later chapter, warping the housing.

Remember to use only nonhardening silicone lube on all pool and spa work. Vaseline or other lubricants are made of petroleum, which eats away some plastics and papers. Get silicone lube at your supply house or any scuba diving shop—it is used for scuba equipment repairs for the same reasons.

If you master these few concepts, any pump seal will be easy for you.

Pump and/or motor removal and reinstallation

Sometimes it is necessary to remove an entire pump and motor unit to take it apart or complete a repair. If the pump is damaged beyond your ability to repair it, you might want to take the entire unit to a motor repair shop. They can rebuild it as needed, and you can reinstall it at the job site. Your local pool and spa supply house can recommend a rebuilder, or consult the phone book.

Generally, to remove the pump and motor as a unit you will need to cut the plumbing on the suction and return side of the pump. Cut the pipe (Fig. 4-13) with enough remaining on each side of the cut to replumb it later. Ideally, a few inches on each side allows you to use a slip coupling to reglue the unit in place later (see the chapter on basic plumbing).

When installing or reinstalling the plumbing between the pump and filter take a look at the equipment area. Keep bends and turns to a minimum. Remember, each turn creates head (resistance) in the system. Also, don't locate the pump close to the base of the filter. When you open the filter for cleaning, water is sure to flood the motor. Lastly, try to keep motors at least 6 inches off the ground. The bracket of the pump does this in part, but heavy rains or flooding from broken pipes and filter cleanings can flood the motor if it is too close to the ground.

4-13 Pump and motor removal and reinstallation.

4-13 Continued.

The other component of pump and motor removal is the electrical connection. You have already turned off the breaker (right?). Now remove the access cover (Fig. 4-14) to the switch plate area of the motor, near the hole where the conduit enters the motor. Remove the three wires inside the motor and unscrew the conduit connector from the motor housing. Now you can pull the conduit and wiring away from the motor and the entire pump and motor should be free.

There might be an additional bonding wire (an insulated or bare copper wire that bonds/grounds all of the equipment together to a grounding system). This is easily removed by loosening the screw or clamp that holds it in place.

Tape off the ends of the wires, even though the breaker is turned off, and put tape over the breaker switch itself. Leave a note on the breaker box to yourself, family members, or the customer to be sure no one accidentally turns the breaker back on while the pump and motor is away for service.

New installation

If you're lucky enough to install the pump and motor for the first time, do it right. I can't tell you how many installations look like they were done by someone who hated service technicians with everything plumbed together so tightly that later repairs were impossible knuckle-busters. You already know the plumbing and electrical techniques as discussed previously, so here are some tips on preparing for new installations.

Position the pump as close to the body of water and as near to water level as possible so it doesn't have to work so hard. Mount the unit on a solid, vibration-free base, not wood (which rots). Make sure there is adequate drainage in the area so that when it rains or if a pipe breaks the motor won't be drowned. Bolt or strap down the pump.

Plumb in both suction and return lines with as few twists and bends as possible, to minimize head. A gate valve on both sides is advisable to isolate the pump when cleaning other components. A check valve is essential if the unit is well above water level. Plumb the unit far enough away from the filter that it won't get soaked when you take the filter apart.

Replacing a pump or motor

Having learned how to remove and break down a pump and motor in the previous sections of this chapter, replacing any of the components is simply a matter of disassembling the pump down to the component that needs replacement, getting a replacement part, and reassembling the unit. Of course, if the entire pump and motor is to be replaced, you purchase the replacement as a unit and plumb it in as previously described.

Sometimes the motor will trip the circuit breaker when you try to start it. If this happens it is usually because there is something wrong with the motor; however, it could be a bad breaker or one that is simply undersized for the job and has finally worn out. The section on motor troubleshooting (later) and the chapter on basic electricity deal with checking the wiring, circuits, and breakers. However, to replace the motor depicted in Fig. 4-3, you follow the procedure of Fig. 4-14 as follows:

1. Break down the unit as described in the section on changing a seal. Remove the shaft extender by removing the allen-head setscrews and pulling the extender off the motor shaft. Sometimes this might need persuasion. Use your large flat-blade screwdriver to pry the extender away from the motor body. Sometimes corrosion will eat away at the setscrews and extender—if it is too tough to remove, replace it (they're only a few bucks).

2. Before sliding the shaft extender on the new motor, clean the motor shaft with a fine emery cloth such as you might have in your copper pipe solder kit. Apply a light coat of silicone lube to the shaft—no, the silicone won't make the extender slip loosely on the shaft. When you put the extender on the motor shaft, the setscrews go into a groove that runs along the shaft. This groove allows the screws to grip and not slide around the shaft.

3. Now slide the new extender in place, lining up the setscrews along the channel, but do not tighten the setscrews. When you have reassembled the bracket and seal plate, seal, and impeller, you can adjust the impeller to just barely clear the seal plate face, then tighten the setscrews. Be sure the impeller is screwed tightly onto the shaft extender before making this adjustment. If it is loose, when the motor starts it will tighten the impeller, by turning it tighter against the extender, thereby tightening it against the seal plate, seizing up the unit.

4. Secure the shaft extender with your pliers or ⅞-inch box wrench and lay a rag over the impeller. Firmly hand tighten it. Reassemble the remaining pump parts and/or replumb the entire unit back into place.

5. You can access the electrical connections through the switchplate cover in the front end bell.

4-14 Replacing a motor.

4-14 Continued.

Replacement of the motor for the Sta-Rite unit in Fig. 4-4 is the same process, but there is no shaft extender or adjustment to make when reassembling. All other pump and motor designs are variations on these themes and will be obvious once you have mastered these few steps.

Troubleshooting motors

The first and most common motor problem is water. Motors get soaked in heavy rain, when you take the lid off the filter for cleaning, when a pipe breaks, or when you look at it wrong. In all cases, dry the motor and give it 24 hours to air dry before starting it up—moisture on the windings will short them out and short out your warranty as well. The basic problems beyond this are as follows.

Motor won't start

Check the electrical supply and breaker panel, and look for any loose connection of the wires to the motor. Sometimes one of the electrical supply wires connected to the motor switch plate becomes dirty. Dirt creates resistance that creates heat which ultimately melts the wire, breaking the connection. Similarly, if the supply wire is undersized for the load, it will overheat and melt. Check the proper wire size in Fig. A-3, replacing the supply wiring if needed. Otherwise, clean dirty switch plate terminals and reconnect the wiring.

Motor hums but won't run

Either the capacitor is bad or the impeller is jammed. Spin the shaft. If it won't turn freely, open the pump and clear the obstruction. If it does spin, check the capacitor.

The best way to check a capacitor is to replace it with a new one. Check the capacitor (see Fig. 4-7) for white residue or liquid discharge. Either is a symptom of a bad capacitor. There is a screw or clamp bracket that holds it in place and two wires, connected to the capacitor with bayonet-type clips. When you see it, you'll realize that not much instruction is needed.

All of this assumes your internal motor switch connection is set for 120 volts, for incoming power of 120 volts, or set for 220 volts if the incoming is 220 volts. Check the wiring diagram and power supply.

A more rare condition that might cause the motor to hum but not run is that your line voltage is not what it should be. Your 120 volts, for example, might be coming in at only 100 volts because of a faulty breaker or a supply problem from your power company. Use your multimeter to test the actual voltage supply at the motor.

Loud noises or vibrations

This is most often caused by worn out bearings. Take the pump apart and remove the load (impeller and water). If the motor still runs loud or vibrates, it is the bearings. Take it to a motor shop, or better still, replace the motor (unless the motor is relatively new or is still under warranty). This problem can also be caused by a bent shaft, although that is not common.

The breaker trips

Disconnect the motor and reset the breaker. Turn the motor switch (or time clock on switch) back on and if it trips again, the problem is either a bad breaker or, more likely, bad wiring between the breaker and motor. Be very careful with this test. Switching the power back on with no appliance connected means you are now dealing with bare, live wires. Be sure no one is touching them and that they are not touching the water, each other, or anything else.

If the breaker does not trip when conducting this little experiment, the motor is bad. This usually means there is a dead short in the windings and the motor needs to be replaced. Water can cause this.

At Charles Bronson's house, I found a large lizard had crawled into the motor housing through the air vents and when the timer turned the system on, the poor reptile became the short across the winding wires, burning out the motor.

Priming the pump

Sometimes the most difficult step is getting water moving through the pump. *Priming* means getting water started, creating a vacuum so more will follow.

Basic priming

Let's go through the steps to prime most pools and spas.

1. Water level. Before starting a pump that you have had apart, always make sure there is enough water in the pool or spa to supply the pump. In taking equipment apart, water is usually lost in the process and there might not be enough to fill the skimmer. I have also encountered pools that seem to have enough water, but will not prime unless filled to the very top of the skimmer. Sometimes that extra inch or two is enough to change the hydraulics of the system and get it working. Factors such as distance of equipment from the pool and height above the pool also enter into the equation.

2. Check the water's path. Often, priming problems are not related to the pump, but to some obstruction. Check the main drain and the skimmer throat for leaves, debris, or other obstructions. Next, open the strainer pot lid, remove the basket, and make sure there are no obstructions or clogs in the impeller. Last, make sure that once the pump is primed it has somewhere to deliver the water. In other words, be sure all valves are open and that there are no other restrictions in the plumbing or equipment after the pump. If all this checks out, proceed.

3. Fill the pump. Always fill the strainer pot with water and replace the lid tightly so air cannot leak in. Keep adding water until the pot overflows so you fill the pipe as well as the pot. Sometimes the pump is installed above the pool water level so you will never fill the pipe (unless a check valve is in the line as well). Just fill what you can and close the lid.

4. Start the motor and open the air relief valve on top of the filter. Give the pump up to two minutes to catch. Carroll O'Connor's pool in Malibu takes 4 minutes and 30 seconds to catch prime, you can set your watch by it (this is because it is about 4 feet higher than the pool level). Most pumps will catch prime sooner and you don't want to overheat a dry-running pump.

Sometimes repeating this procedure two or three times will get the prime going. If there is a check valve in a long run of pipe, each successive filling of the pot pulls more and more water from the pool, which is held there each time by the check valve. Also, if it is warm outside, the air in the pipe might expand and create an airlock. The repeated procedures might finally dislodge the air.

The blow bag method

When basic priming fails, try a drain flush bag, also called a *blow bag* (Fig. 4-15). The drain flush is a canvas or rubber tube that screws onto the end of your garden hose. Slip this into the skimmer hole that feeds the pump and turn on the hose.

4-15 Using a drain flush bag.

The water pressure makes the bag expand and seal the skimmer hole so the water from the hose cannot escape and must feed the pump. After running the hose a minute or two, turn on the pump. When air and water are visible returning to the pool, pull the drain flush bag out quickly, while the pump is running, so pool water will promptly replace the hose water.

This method is not effective if the skimmer has only one hole in the bottom. Remember from chapter 2 that this hole is connected not only to the pump but also the main drain. The forced water from your drain flush bag will take the line of least resistance and flood through the main drain rather than up to the pump. In the two-hole skimmer, the hole furthest from the pool usually is plumbed directly to the pump. Your drain flush bag in this hole will give good results.

Filter filling method

Another method is what I call *filter filling*. Open the strainer pot, turn on the motor, and feed the pot with a garden hose. Open the filter air relief valve and keep this going until the filter can is full (water will spit out of the air relief valve).

Close the air relief valve, turn off the motor and garden hose, and quickly close the strainer pot. Open the air relief valve. The filter water will flood back into the pump and the pipe that feeds the pump from the pool. When you think these are full of water, turn the motor back on. The pump should now prime.

Drastic measures

Sometimes none of this works. Dick Clark's pool includes equipment that is about 10 feet above the pool water level and 40 feet away. I have had to put a drain flush in the skimmer and a rubber plug in the main drain to get it primed. Obviously this meant diving to the main drain to plug it off. I got so tired of this I found a better way.

I cut the pipe from the pool to the pump, a foot in front of the pump. I plumbed in a T to rejoin the pipe. In the third opening of the T, I plumbed in a garden hose bib (faucet). When I need to prime the pump, I attach a garden hose to this faucet, turn on the pump and the hose, open the faucet, and the hose water starts a suction that starts pool water up the line. When I see water and air bubbling out of the pool return lines, I close the faucet and turn off the garden hose. In really difficult systems, try this method.

Detecting air leaks

Okay, so none of this works. The problem might be that the pump is sucking air from somewhere, meaning it will not suck water (which is harder to suck).

Air leaks are usually in strainer pot lid O-rings, or the pot or lid itself has small cracks. The gasket between the pot and the volute might be dried out and leaking. Of course, plumbing leading into the pump might be cracked and leaking air.

If any of these components leak air in, they will also leak water out. When the area around the pump is dry, carefully fill the strainer pot with water and look for leaks out of the pot, volute, fittings, and pipes. Another way is to fill and close the pot, then listen for the sizzling sound of air being sucked in through a crack as the water drains back to the pool.

Sometimes there's just no easy answer—remove the pump and carefully inspect all the components, replace the gaskets and O-rings, and try again.

T-handles

Many pumps employ threaded T-shaped bolts that secure the lid to the strainer pot. Sometimes these corrode and snap off, with part of the bolt in the pot and the other part in your fist.

If part of the broken bolt extends above or below the female part on the pot, try using pliers, especially Vise-Grips, to grasp the broken section and twist it out. You can buy new T-bolts at the supply house. You can also take a flat-blade screwdriver and place it on the broken bolt end, tap it with a hammer to create a slot in the end of the broken piece, and twist out the broken piece like removing a screw.

If this doesn't work, take your tap and die set or electric drill and tap a small hole inside the broken piece, then use your Phillips-head screwdriver to grip inside the hole and twist out the broken piece. If all else fails, remove the pot from the pump and take it to a machine shop to be tapped out and rethreaded.

A hint: if one of these handles breaks off on Friday afternoon, you can clamp that side of the lid on the pot with your Vise Grips and run the system until Monday when you can get the parts to fix it.

Reverse flow problems

The following information from Charles Brown, of the motor manufacturer Century Electric, explains this rare problem better than I can. It is rare, but I have personally observed the phenomenon twice and have suspected it in two other cases. I'm not sure it was caused as outlined here or if there was some other electrophysical laws at work, but understanding the symptoms and being aware of the possibility is enough.

We have observed reverse flow in our test lab and found it can happen with several different manufacturers' motors but only under the following circumstances and only in the given sequence. We could not cause this reverse flow of action to occur any other way. It is important to note also, that we could not cause this action to occur in pumps with open-face impellers. The sequence that must take place is:

1. The pump is running, pushing a column of water out of the discharge.
2. A bubble of air is trapped in the filter. The air bubble becomes trapped in the top of a dirty sand filter for one of two reasons— the filter was not bled off properly in a new installation or there is an air leak in the system.
3. While the running pump is pushing water against the trapped air bubble in the filter, the power to the pump motor is interrupted. The interruption can occur by accidentally turning the switch off; by a power failure; or by the power company switching grids, which is done daily in some parts of the country.
4. The moment the power goes off the pump stops, instantly releasing the bubble of air compressed at the top of the filter. This instant release of air forces the column of water between the pump and filter to reverse and flow backwards toward the pump, entering the discharge side of the filter.
5. This reverse flow of water into the discharge side of the pump starts the impeller turning in the wrong direction. Within a few short seconds the pump impeller can be turning at high speeds driving the motor in reverse as well. The actual speed is determined by the amount of air that is compressed in the filter and the length of the column of water between the pump and the filter.
6. If power is restored to the motor at this precise point, the motor will start and may actually run in the wrong direction depending on the speed of the pump and motor at that instant. As we stated earlier, we witnessed this action in our test lab and it occurred with several different manufacturers' motors.

When power is restored and the motor does start while turning in the wrong direction, the torque of the motor can be great enough to loosen and spin the impeller from the shaft, stripping the threads.

As we stated earlier, reverse flow is a very rare occurrence and as far as we can determine it can be overcome simply by using a check valve between the filter and pump. This is the most effective way to prevent it.

Several large pump manufacturers have recognized the same occurrence and to prevent reverse flow have been securing the impeller to the motor shaft. This, of course, prevents the impeller from stripping its threads.

In the case of reverse flow, an "ounce of prevention" truly applies. Always bleed the air from the filter on new installations, check frequently for air leaks and remind your customers of the necessity to keep the sand filters clean.

As Brown noted, some manufacturers have secured the impeller to the shaft to prevent reverse spinning. Reverse spinning can occur when the impeller is jammed against the volute or diverter. The motor shaft turns but the impeller is stuck, effectively stripping the threads and damaging the impeller and other pump components.

Sta-Rite's Dura-Glas and Maxi-Glas pumps, with closed-face impeller and diverter shroud (Fig. 4-4), are sometimes retrofitted by pump and motor rebuilders or clever pool technicians. A small hole is drilled through the center of the impeller and into the tip of the motor shaft. The hole in the shaft is threaded to accept a bronze screw. By screwing the impeller in place, reverse flow is prevented.

Of course, if the impeller is really jammed by debris, the screw can't prevent the plastic center of the impeller from disintegrating under extreme torque. What usually happens, however, is that the motor can't overcome the jam and it blows a circuit breaker or the internal thermal overload switch.

A broken impeller is more likely to occur when the motor is running at full speed and an obstruction is suddenly sucked into the pump, jamming the impeller instantly. The torque of the motor is then enough to strip the impeller threads or break it outright. This should be another reminder of why it is important to keep an undamaged strainer basket in the pot and keep it clean.

MODELS AND MAKES

Throughout this chapter I have illustrated and dealt with two leading pump types, as shown in Figs. 4-3 and 4-4. A look at the manufacturer's literature or catalogs at your local supply house will reveal that all pumps are variations on these two basic concepts.

Motors to drive these pumps are made by several manufacturers as well, although I know of none who make both pumps and motors. Century, Franklin, A.O. Smith, General Electric, and others make all the various styles and horsepower of motors needed to run modern and older pump equipment. Some of the styles you will encounter (and need to specify when replacing) are as follows. Remember, the nameplate on the motor gives you this information, but it is often obscured or missing, so be familiar with these types.

C frame

If you look at the face of the motor, the end with the shaft, you will note a definite pattern in the casting—it looks like the letter C (see Fig. 4-3). The shaft is *threaded,* meaning it has male threads on the end of the shaft to receive the female threads of the impeller; or *keyed*, meaning the shaft has no threads, but rather a channel that runs the length of the shaft to accept the setscrews.

Square flange

Look at the end of the motor that has the shaft. This style of motor includes a flared bracket, square of course, that accepts the seal plate of the Sta-Rite pump (see Fig. 4-4, item 5) and the corresponding motor. These are only made with threaded shafts because they only work with the type of pump that uses a threaded impeller.

Uniseal

These are not in use much anymore. The uniseal motor includes a flange like the square flange for application with a uniseal-type pump.

48 frame

No, the face of this motor does not resemble the number 48. This is, however, the motor used on indoor spa booster pumps and fits that particular style. Jacuzzi makes a number of pumps that use the 48 frame. These also are all threaded shafts.

MOTOR CHARACTERISTICS

Now that I have reviewed the most common types of motors, let's examine some of their different characteristics.

End bell

The housing at each end of the motor is called the *end bell*. It is made of aluminum or cast iron, the latter being more expensive. Iron rusts and aluminum doesn't and aluminum more efficiently disperses heat away from the motor, thereby prolonging its life, but I can't figure out why the cast iron is more expensive. Some manufacturers give you a choice within each type of motor.

Full-rated or up-rated

The full-rated motor operates to its listed capacity (1 hp, 2 hp, etc.). The up-rated motor has a similar horsepower rating but functions to even higher standards if called upon (see the section on service factor). This might be useful if the pool or spa gets full of leaves, for example, and the motor is required to work against increasing loads until someone clears out the debris. By checking the service factors, you might be able to get a ½-hp up-rated motor that is able to perform as well as a 1-hp full-rated motor. Why would you do this? Well, the smaller horsepower motor might be cheaper or might be the only one available at the supply house that day.

Two-speed

As the name implies, this motor includes a switch box to wire it for two-speed operation. Spas, for example, often use a two-speed motor with the lower speed for circulating and heating, and the high speed for jet action. Wiring instructions are printed on the motor as described previously.

Energy efficient

As described previously in the section on motors, the so-called *energy efficient* motor is, in fact, a design that saves electricity over a comparable motor of standard design.

The special characteristics of pumps and motors for automatic cleaner boosters are discussed in the section on automatic cleaners.

MOTOR COVERS

Protective covers are made to fit over motors. Some are plastic, some metal, some foam rubber. In all cases, they are designed to keep direct sunlight and rain off the motor housing. In fact, the motor is designed to do this itself. You will notice that the motors in the illustrations all have air vents on the underside. The greatest danger to a motor is flooding of the equipment area in heavy rain or when opening a filter, or allowing water from the ground to get up into the motor. This will short out the windings and void any warranty. Believe me, a warranty repair station or motor rebuilder will know if the motor was flooded.

SUBMERSIBLE PUMPS AND MOTORS

There are essentially two types of submersible pump and motor combinations that you will encounter.

High-volume pump-out units

Sometimes you need to drain a pool or spa. Several manufacturers make pump and motor units with long, waterproof electrical cords, that can be completely submerged. The suction side is at the bottom of the pump, as if a regular pump and motor unit were stood on end with the motor on top and the pump on the bottom. The return line is sized to be attached to your vacuum hose. Smaller units are connected to a garden hose to feed water out of the pump.

Low-volume pump and motors

Fountains and small ponds use small submersible pump and motor units that contain all of the same components as their larger cousins, but range in size from no larger than a fist to about the size of a football. These have waterproof electrical cords so they can be submerged in the body of water.

Both high- and low-volume types contain the same components as the pumps and motors described earlier and are repaired in much the same ways. Submersibles have more crucial and tricky seals and gaskets, however, because leaks in these mean electricity in the body of water that can be fatal to both the motor and you.

I always leave repair of submersibles to a rebuilding shop, except for the wet-end parts, such as impellers and strainer housings, which are easily accessed and do not require breaking the seal that surrounds any electrical component.

COST OF OPERATION

As with any electrical appliance, you can easily calculate the cost of operation. I have heard many customers tell me their electric bills have gone up significantly and they can only attribute that to the pool or spa motors. Knowing how to calculate operating costs can help answer these questions—usually the customer is wrong.

Electricity is sold by the kilowatt-hour. This is, 1000 watts of energy each hour. You know that volts × amps = watts, so you can look at the motor nameplate and see that the motor runs, for example, at 15 amps when supplied with 110-volt service, and 7 amps when supplied by 220-volt service.

Let's say the pump in our example is running on 220-volt service—220 volts × 7 amps = 1540 watts. Looking at an electric bill, you learn that you pay 15 cents per kilowatt-hour. As noted, a kilowatt is 1000 watts, so if you divide 1540 watts by 1000, you get 1.54. That is multiplied by your kilowatt rate (15 cents), equaling 23 cents for every hour you run the appliance. If you run the motor eight hours per day, that means 23 cents × 8 hours = $1.84/day. Over a month, that equals 30 × $1.84 = $55.20/month.

BOOSTER PUMPS AND MOTORS FOR SPAS

Pump and motor units that provide only jet action for spas generally are not equipped with a strainer pot and basket, otherwise they are the same as other units discussed previously. Some units are designed to perform two functions and therefore run at two speeds. They run at high speed (3450 rpm) to provide jet action, but also circulate, filter, and heat the water (and then, of course, would have a strainer pot and basket) at a low speed (1750 rpm).

As a general rule of thumb, to operate efficiently, spa jets require 15 gpm running through each one. Therefore, if you have a system that delivers 60 gpm, you can install up to four jets. As another generality, each jet requires ¼-hp from its pump and motor, so again, that four-jet spa would need at least a 1-hp unit.

Remember, this assumes the pump is doing no other work. If it is pushing water through the filter and heater before getting back to the jets, or if the equipment is more than 20 feet from the spa, then some power will be lost and you will need to calculate more than ¼-hp per jet.

When planning a system or replacing equipment, many codes require that a spa to turn over completely two times per hour, so be sure the pump can handle that, especially when the filter gets dirty and head increases.

5
Filters

As the name implies, the filter is the piece of pool or spa equipment that strains impurities out of water that is pumped through it. There are few moving parts (in fact no parts should be moving when the filter is in operation) and simple components.

TYPES

Three basic types of filter are in common use and each is preferred for various reasons. Each type is more efficient for various adaptations and you might also discover regional preferences around the world. For example, in regions where sand is plentiful but diatomaceous earth is scarce, the sand or cartridge filter will be more commonly in use than the DE filter. Setting aside such regional prejudice however, as I review the types of filters, I will also discuss why each is preferred based on technical application.

Diatomaceous earth (DE) filters

In the diatomaceous earth type of filter, also called a DE filter (Fig. 5-1), the water passes into a metal or plastic tank, through a series of grids (also called filter elements) covered with fabric, and back out of the unit. The grids do not actually perform the filtration process, but instead are coated with a filter *media*, diatomaceous earth, that does the actual filtering work.

DE is a white powdery substance found in the ground in large deposits. It is actually the skeletons of billions of microscopic organisms that were present on earth millions of years ago. So, in essence, this powder is akin to dinosaur bones. If you look at DE under a microscope, you will see what appears to be tiny sponges, thus the filtration ability becomes more apparent. Just like a sponge, water can pass through, but the impurities in the water can't. Because the DE particles are so fine, they can strain very small, actually microscopic, particles from the water as it passes through.

So why not just dump a few pounds of DE in a tank and pass the water through? Good question. Because the DE will collapse together and cake, making it impossible for even the water to get through. Therefore, the tank is equipped with free-

Tank lid assembly
(includes)

Bushing	P03545
Gasket	P11355
Nut	P23820

Clamping ring

Knob

Tee bolt

Holding wheel

Wing nut

Filter elements

Washer

Retaining rod
assembly

Rubber O-ring

Manifold

Tank

5-1 Diatomaceous earth (DE) filter. Purex Pool Systems, Inc., City of Industry, Ca.

standing grids (also called elements) that are coated with DE to accomplish the filtration process. DE filters are usually located on the return side of the pump (a pressure filter), but some are designed for installation on the suction side. There are basically two types of DE filter: the vertical grid and the spin type.

Vertical grid DE filters

As can be seen in Fig. 5-1, the grids (elements) are mounted on a manifold and the resulting assembly fits into the tank. A retaining rod through the center screws into the base of the tank and a holding wheel keeps the grids firmly in place. The top of the tank is held in place by a clamping ring, and the two parts are sealed with a thick O-ring to prevent any leaking.

The water enters the tank at the bottom and flows up around the outside of the grid assembly. It must flow through the grids, down the stem of each grid, and into the hollow manifold, after which it is sent back out of the filter. This type of DE filter is called a *vertical grid*, for obvious reasons.

One way to clean this filter is called *backwashing*, a concept I will discuss in more detail later. As the term suggests, backwashing means the water is redirected through the filter in the opposite direction from normal filtration (accomplished with a backwash valve, also discussed later), thereby flushing old DE and dirt out of the filter.

Not all vertical DE filters are equipped with a backwash valve to allow backwash cleaning. These types must be disassembled each time for cleaning. Some are also equipped for *bumping* rather than backwashing. In this rather ridiculous process, the dirty DE is bumped off the grids, mixed inside the tank so the dirt is evenly distributed within the DE material, then recoated onto the grids.

The idea is that since most dirt sits on the outermost layer of DE as the DE rests on the grids, there is still relatively clean, unused DE available. By mixing DE and dirt together, this type of filter design reasons that somewhat cleaner DE can be brought to the surface to prevent the need to actually replace it. Because I find this type of filter to be too stupid for words, it will not be discussed further.

Why? Because dirty DE is dirty DE. DE is cheap (about 20 cents per pound and you use 5 to 10 pounds per filter) and organic (nonpolluting). Why, therefore, would you filter a pool or spa with dirty DE? As you will see later (with reasons why), when a filter is dirty, disassemble it and clean it completely.

In some areas, out of concern that DE will clog pipes, local codes require that DE not be dumped into the sewer system. In this case, a *separation tank* is added next to the filter (Fig. 5-2). When draining or backwashing the filter, the dirty water is passed through a canvas strainer bag inside this small tank before going into sewer or storm drains. The canvas bag strains most of the DE out of the water so it can be disposed of elsewhere.

5-2

Separation tank. Premier Spring Water, Inc., Pacoima, Ca.

Some rather tortuous vertical DE filter units were designed with the manifold on top and the holding wheel on the bottom. My experience is that these do not work any better but are **much** harder to get back together after disassembly for cleaning.

All filters are sized by the square footage of surface area of their filter media. In DE filters therefore, the total surface square footage of the grids would be the size of the filter. Typically there are eight grids in a filter totaling 24 to 72 square feet, designed into tanks that are 2 to 5 feet high by about 2 feet in diameter. Obviously the larger the filter, the greater capacity the filter will have to move water through it. Therefore, filters are also rated by how many gallons per minute can flow through them. More on this later.

Spin DE filters

There is also an obsolete, rather inefficient type of filter where the grids are wheel-shaped and lined up horizontally like a box of donuts. This design is called a spin DE filter.

The operating idea is the same as with vertical grid filters, but to clean this type you turn a crank on the tank that spins the grids, theoretically cleaning them. For obvious reasons, these are called *spin filters*. They don't work very well and you won't run across too many of them. If you do come across an old one in operation, you'll understand the mechanics if you first understand the basic vertical DE filter.

Sand filters

If the filter media is not DE, it must be sand. Sand and gravel are the natural methods of filtering water. There are two types of sand filters in general use.

Pressure sand and gravel filters

These were once called *rapid sand* filters (older-style metal tanks designed for flow rates less than 3 gpm per square foot), but are now called *high-rate sand* filters (fiberglass or other composit materials designed for flow rates of 5 to 20 gpm per square foot). The concept of the pressure sand and gravel filter is simple.

Water is passed through a layer of sand and gravel inside a tank, which strains impurities from the water before it leaves the tank. I designate these types of filters as *pressure* sand and gravel, because like their DE cousins, the water is under pressure inside the tank from the resistance created by trying to push it through the filter media. This differentiates it from another type of sand and gravel filter used especially in fish ponds where there is no such pressure (see free-flow filters).

In Fig. 5-3, the water enters the tank through the valve (item 7) on top and sprays over the sand inside. The water then runs through the sand, impurities being caught by the sharp edges of the grains, and is pushed through the manifold at the bottom where it is directed up through the pipe in the center and out of the filter through another port of the valve on top. The individual fingers of the drain manifold (item 20) are called *laterals* and the center pipe used to return clean water (item 19) is called a *stanchion pipe*. To drain the tank, a drain pipe is provided at the bottom as well.

Sand filters are also sized by square footage and gallons per minute. Does some engineer measure the surface area of the billions and billions of grains of sand in a given filter to arrive at the total square footage of filtration area? I don't know, but since I can't do it, I'll take their word for it.

1 Backwash valve handle
2 #8 retaining screw
3 Indicator plate
4 Handle bracket
5 Handle pin
6 Rotor shaft O-ring
7 Valve body with gasket
8 Plumbing adapter O-ring
9 Plumbing adapter
10 Plumbing adapter nut
11 ¼" assembly bolt with washer
12 0-60 psi pressure gauge
13 Backwash valve/tank O-ring
14 Rotor plate
15 Rotor plate O-ring
16 Diffuser
17 Distributor/strainer
18 Filter tank body
19 Center water flow pipe
20 Lateral underdrain (8)
21 Underdrain manifold hub
22 Filter tank stand

5-3 High-rate sand filter. Purex Pool Systems, Inc., City of Industry, Ca.

Knowing the volume of sand recommended for any given filter (expressed in cubic feet), the manufacturer arrives at a square footage value and a resulting gallons per minute rating. Sand filter tanks are usually large round balls, anywhere from 2 to 4 feet in diameter.

Free-flow sand and gravel filters

Used mostly in decorative ponds or fish ponds, the free-flow filter is created as a natural filtration device. The main drain of the pond is covered with a layer of gravel, then small stones, then sand. As the water is sucked toward the drain for circulation, it must pass through these layers of sand and gravel, filtering out impurities in the process.

Over several weeks, the fish waste, decaying plant material, and algae form a biological "soup" in the filter media. Enzymes in this decaying matter further break down impurities that the sand and gravel have trapped, making them a very effective way of filtering fish ponds and other heavy-debris bodies of water. It is similar to nature's own way of purification and this type of filter is often referred to as a *biological* filter.

Cartridge filters

A cartridge filter is similar to a DE filter except there is no DE filter media. In Fig. 5-4, water flows into a tank that houses one or more cylindrical cartridges of fine-mesh, pleated fabric (usually polyester). The extremely tight mesh of this fabric strains impurities out of the water.

1 0–60 psi pressure gauge and screen
2 Air relief valve T assembly
3 Retainer nut
4 Tension strap
5 Stainless steel tank body
6 Tank/base O-ring
7 Tank/base clamp assembly
8 Base with plumbing and drain ports
9 Drain plug
10 Clamp handle/nut
11 Cartridge element
12 Air bleed tube
13 Air bleed filter

5-4 Cartridge filter (single cartridge). Sta-Rite Industries, Delevan, Wisc.

Figure 5-5 also shows a cartridge filter, but instead of one large cartridge, this model uses 18 small ones. Why? To make life tough on the pool service technician, that's why. Well, okay, perhaps the manufacturer felt this design would filter better because water can circulate around 18 small cartridges more freely than around one large cartridge in the same size tank. If you've ever had to clean and reassemble one of these, however, you'll agree with me that it was a designer who hated pool service technicians.

Cartridge filters are classified by square footage of filter surface as are DE and sand filters. By pleating the cartridge material, a lot of square footage can fit into a very small package. This would not be possible if DE were added, because the tight pleating would prevent DE from coating evenly and effectively. Typically, cartridge filters for residential use range from 20 to 120 square feet in a tank not more than 4 feet high by 1½ feet in diameter.

MAKES AND MODELS

Now that you have had a brief introduction to typical pool and spa filter types, how do you make a selection and what size do you need for a given job? The intended use of the pool or spa and/or the local building and health codes will guide you in answering these questions.

Sizing and selection

In many jurisdictions, the turnover rate (if you've forgotten turnover rate, see the section on hydraulics) requires a pool to be turned over in 6 hours, a wading pool in 1 hour, and a spa in 30 minutes. Using this information and knowing the total gallons in your pool or spa, you will know how many gallons per minute the filter must be able to handle.

As an example, my pool is 40,000 gallons. To turn that over every 6 hours I need a filter that can handle

$$40,000 \text{ gallons in 6 hours} = 40,000 \text{ gallons in 360 minutes}$$
$$= 111.11 \text{ gallons per minute } (40,000 \div 360)$$

Now, regardless of the manufacturer's claims, many building or health codes also determine the maximum number of gallons per minute permitted for every square foot of a filter's surface area. Just for the record, many codes also specify the minimum rate of flow during backwash. These might differ for residential and commercial pools, but many jurisdictions require the following:

Filter style	Maximum flow, gpm/sq. ft.	Minimum backwash flow, gpm/sq. ft.
High-rate sand	15	15
DE	2	2
Cartridge	0.375*	No backwash

*Commercial installation.

Note that cartridge filters are rated at 1 gpm/sq. ft. maximum flow rate on most residential applications and the more stringent 0.375 gpm/sq. ft. on commercial installations. This is because cartridge filters are used primarily on spas where lots of bathers sit in a relatively small amount of water creating lots of bacteria, oil, and dirt for the filter to handle. The 0.375 rule simply means a commercial installation must have a larger filter than one at home where the bather load will probably be less.

5-5
Cartridge filter (multicartridge).
PAC Fab, Inc., El Monte, Ca.

So with this information, if you chose a sand filter, you divide the required gallons per minute for our sample pool, 111, by the maximum flow rate of the local code for sand filters, 15, giving you 7.4 square feet.

This means you need a sand filter of at least 7.4 square feet. Now the chances are that the manufacturer's claim of gallons per minute per square foot of filter area will exceed the local code requirement, but check to see that it does, at least, meet the code. In other words, the maximum flow rate permitted might be 15 gallons per minute per square foot of filter area, but if the maker says his 7.4 square foot filter can only handle 10 gallons per minute, then you will have to go to a larger size until you arrive at a filter that can do the job. Got it?

Using the same example, the 111-gpm rate with a DE filter is

$$111 \div 2 = 55.5 \text{ square feet}$$

So you need a DE filter of at least 55.5 square feet; again, checking the manufacturer's specifications.

Finally, the cartridge filter in the same application:

$$111 \div 0.375 = 296 \text{ square feet}$$

So you need a 296-square foot cartridge filter to do the same job as the 7.4-square foot sand filter or the 55.5-square-foot DE filter.

Another sizing and selection criteria is dirt. How dirty does the body of water get that this filter must service? You want to oversize the filter a bit so you don't have to clean it out every day, week, or hour (choose one). This time between filter cleanings is called the *filter run*. So you might as well get a filter 20, 50, or 100 percent larger than you actually need so it can hold more dirt and thereby leave more time between cleanings—right?

Well, not exactly. If the pump cannot deliver the gallons per minute for the square footage of this huge filter you have just installed, only part of the tank will fill with water, effectively giving you a smaller filter anyway. Also, the pump won't backwash the filter completely if it can't match the gallons per minute rating, which is why codes call for a minimum backwash flow rate. In the section on cleaning filters I tell you not to backwash filters anyway, so this might not be important to anyone except the building inspector (on new construction). More to follow. So the obvious question is, which do we use? You might also want to consider price, size, and efficiency.

Price is a factor of the current market and supplier availability in your area, so I'll let you check that one out for yourself. The size of the unit might be a deciding factor if the equipment area is small. To contain 7.4 square feet of filter area, a sand filter needs to hold about 8 cubic feet of sand. Wet or dry, that much sand weighs about 10 million pounds (well, okay, not quite that much, but you get the idea). In fact, for the filter to hold 8 cubic feet of sand it would need to be very strong and about 4 feet in diameter.

A cartridge filter with almost 300 square feet of filter area is also rather large and will be very expensive. But a DE filter of around 60 square feet is a fairly standard, serviceable tank in a modest price range.

If you're lucky enough to have a client like Barbra Streisand, and she just wants the filter that does the best job regardless of price, then how do you advise her which one to buy? The answer lies in the mighty micron. A *micron* is a unit of measurement equal to one millionth of a meter or 0.0000394 inches. Put another way, the human eye can detect objects as small as 35 microns; talcum powder granules are about 8 microns and table salt is about 100 microns.

So what? Well, each filter type has the ability to filter particles down to a particular size, as measured in (you guessed it) microns. Various references and manufacturers will give you different estimates and they might all be right as you will see in a minute. However, here's a rule of thumb:

- Sand filters strain particles down to about 60 microns,
- Cartridge filters strain particles down to about 20 microns, and
- DE filters strain particles down to about 7 microns.

I read an article recently (with data supplied by a manufacturer, of course) claiming sand filters at 25 microns, cartridges at 5 microns, and DE at 1 micron. Gentlemen, modesty, please!

Yet, as I said, both estimates are correct. Sand filter efficiency is based on the age of the sand. New sand is crystalline and many-faceted. These sharp-sided facets are what actually catches debris out of the water and filters it. As sand ages, the years of water passing over it cause erosion of those facets until each grain of sand becomes smooth and rounded, losing its ability to trap particles of the smallest size. So a new sand filter with fresh sand might strain particles of 25 microns, while a unit in service one year might be lucky to strain particles of 60 microns.

Efficiency of a DE filter is affected by the cleanliness of the DE. As mentioned previously, DE is composed of billions of tiny bone-like sponges. As these sponges get clogged with fine particles, they lose their ability to strain the smallest impurities.

Because cartridges do not rely on organic material like sand or DE, they are not affected the same way. As long as the cartridge surface area remains clean, it will filter the same size particles (although extremely old cartridges that have been acid washed many times will stretch out somewhat, creating a mesh that is not so fine as when new). However, dirt does affect the cartridge filter, as it affects all filters, but with the cartridge it actually has a positive affect.

Dirt makes the mesh of the polyester cartridge material even tighter, essentially acting like DE. I know some service technicians who add a small handful of DE to a cartridge filter after cleaning it to start this process. Thus, with each successive pass of water through the cartridge filter, it becomes more efficient as the mesh gets finer and finer from the retained particles. This is also true with sand filters, up to a point.

Regardless of these variations, the one consistent fact you will note in the estimates is that sand filters are the least efficient filters in terms of how fine the particles are that they can strain from the water; cartridge filters appear to be the second best and DE the best. I say appear to be because a slightly dirty cartridge filter will strain particles of 5 to 10 microns, while a dirty DE filter simply clogs up and doesn't allow as much water to pass through as it should. Although it too is straining particles of 5 to 10 microns in size, if the flow is reduced because the DE is clogged, what good is the rated straining capability?

This is why cartridge filters are generally regarded as the best overall filtration method. They are easy to clean and require no added organic media. As you saw in the example, however, the size required for a cartridge filter might make its use impractical or too expensive. Generally you will find cartridge filters used on spas and small fountains, DE filters on pools and larger fountains, and sand filters on fish ponds and decorative ponds.

Remember too that each filter adds resistance to your system. (Remember the old villain *head?* If not, refer to the section on hydraulics.) A smaller filter will restrict water flow more than a larger one, so consider the amount of head you are adding to the system when replacing an old filter and be sure the pump can handle that value if it is significantly different from the value of the old filter.

Manufacturers literature will tell you the amount of head their filter creates at various flow rates. You can read the pressure (in psi) from the filter pressure gauge and multiply by 2.31 to determine how many feet of head the filter is creating. Typically, filters run 12 to 20 psi, so lets use 15 as an average: $15 \times 2.31 = 34.65$ feet of head. A considerable amount. Remember too that as the filter gets dirty the pressure goes up. When it is only 10 psi higher, you are adding $10 \times 2.31 = 23.1$ more feet of head. A lot! Another reason to keep the filter clean.

Before I leave the subject of choosing a filter type and size, I noted that fish ponds use sand filters. Why not DE or cartridge filters if there are not a lot of gallons involved? Because the extreme amounts of dirt and waste generated by a fish pond or decorative pond with lilies would quickly clog a DE or cartridge filter, rendering it useless. You would be cleaning them daily. The loose, bulky quality of sand works well with heavy debris like that.

Backwash valves

As I noted previously, backwashing is a method of cleaning a DE or sand filter (cartridge filters do not backwash) by running water backwards through the filter, flushing the dirt out to a waste line or sewer line. There are basically two types of backwash valves—the piston and the rotary (the multiport being a variation of the rotary valve).

Piston valve

Figure 5-6 shows a piston-type backwash valve. In the normal operating position, the water enters the valve (marked Pump discharge) and, as you can see, the piston discs (item 7) only allow the water to go out to the filter tank inlet. The water is filtered through the sand or DE and returns at the top of the valve (marked Tank outlet). Again, you can see that the piston disc (item 7) only allows the water to flow to the outlet of the valve (marked Return to pool). Normal filtration has occurred.

To backwash the filter and flush out the dirt, the handle of the piston assembly (item 2) is raised. Figure 5-7 shows the same valve with the piston raised to the backwash position. Figure 5-6, item 3, shows two rolled pins that ensure exact placement of the piston inside the body. The outside pin acts as a stop when pushing the valve down, the inside pin acts as a stop when pulling it up.

Notice that the water still enters from the pump, but now the piston discs force the water to flow *in* to the filter tank through the *outlet* opening. The water flows backward through the filter, flushes the dirt out of the tank and *out* of the valve *inlet* opening. As you see, once inside the valve again, the dirty water is directed to the line marked Waste. O-rings on the piston discs ensure that water does not bypass the discs.

By the way, never change the piston position when the pump is running, because this puts excessive strain on the pump and motor and the O-rings of the valve,

1 ──
2 ──
4 ──
5 ──
3 ──
6 ──
◄ **Return to pool**

Tank outlet ◄ ── ── 3

7 ── ◄ **Pump discharge**

8 ──

Tank inlet ◄

9 ── ◄ **To waste**

1	Valve body cap assembly screw
2	Handle
3	Rolled pins
4	Valve body cap
5	Cap O-ring
6	Shaft O-ring
7,9	Piston disc with O-ring
8	Piston shaft assembly

5-6 Piston backwash valve (normal operating position). Val-Pak Products/Swimrite, Canyon Country, Ca.

causing it to leak. The piston-type backwash valve is usually plumbed onto the side of the filter tank.

Rotary valve

The rotary backwash valve, used only on vertical DE filters, does the same thing as the piston valve and in pretty much the same way. Figure 5-8 shows the typical rotary valve. Changing the direction of water flow is done by rotating the interior rotor (item 11 for the bronze, or item 11A for the plastic version). A rotor gasket seal (item 3 or 3A) or O-rings (item 4) keep water from leaking into the wrong chamber.

This type of valve is mounted underneath the filter tank. The tank has a hole in the bottom to accommodate the unit. The valve body (item 1) is held under the tank while a compression retaining ring (item 13) is placed inside the tank and the two halves are bolted together with cap screws (item 14). The handle underneath allows you to rotate the interior rotor to align it with the openings of the valve body as desired.

In normal filtration, water comes into the valve through the opening marked From Pump Discharge and flows up into the filter through the large opening on top

Tank inlet

Pump discharge

Tank outlet

To waste

5-7
Piston backwash valve
(backwash position). Val-Pak
Products/Swimrite, Canyon Country, Ca.,
modified by author

of the rotor marked Normal Water Flow. After passing through the DE-coated grids, the water flows back through the inside of the grids, into the manifold (see the DE filter section), down through the center of the rotor (marked Return Water), then out the effluent opening and on to the heater or back to the pool.

When the valve is rotated, water is sent up through the middle of the rotor, up inside the grids. The dirt and DE are pushed off the grids and the water flows from inside the grids to the outside. This is flushed back through the rotor, opposite the normal flow, and directed to the opening marked Backwash. This line is connected to a waste or sewer line. As with all backwash valves, do not operate it while the pump is running or you might damage the valve and the pump and motor.

Multiport valve

The multiport backwash valve is used on sand filters and looks just like a rotary valve when disassembled, except that there is one more choice for water flow. A rinse is added, so that after the water has backwashed through the filter, clean water from the pump can be directed to clean out the pipes before returning to normal filtration. This prevents dirt in the lines from going back to the pool after backwashing. If you understand piston and rotary valves, you will have no trouble with a multiport valve when you encounter it in the field.

Return water
to pool

Normal water
flow from pool

Backwash

Effluent

From pump
discharge

3A

11A

1 Valve body with 2" FIP plumbing ports
2 Filter tank/valve body O-ring
3 Port seal for brass rotor
3A Rotor seal gasket for Noryl rotor
4 Port seal O-ring for brass rotor
5 Rotor shaft O-ring
6 Rotor handle
7 Handle extension
8 Handle assembly cap screw
9 Lockwasher
10 Hex nut
11 Brass valve rotor
11A Noryl valve rotor
12 Manifold/rotor O-ring
13 Compression ring
14 Hex head compression ring assembly bolt

5-8 Rotary backwash valve. Purex Pool Systems, Inc., City of Industry, Ca.

Backwash hoses

To accomplish the noble purpose of backwashing, the dirty water has to go somewhere. Some waste or backwash openings are plumbed directly into a pipe that sends the water into a nearby deck drain and on toward the sewer. Many are not hard-plumbed and you must connect a hose to direct the waste water wherever you want it to go.

By the way, dirty filter water is an excellent fertilizer for lawns because it is usually rich in biological nutrients, algae, decaying matter, and DE (which, as you now know, is a natural material). You can run your backwash hose on the lawn or garden, providing the water chlorine residual level is not above 3 ppm (see the chapter on chemistry). Chlorine levels higher than that might burn the grass.

A backwash hose can be your pool vacuum hose, clamped onto the waste line of the backwash valve with a hose clamp. Normally, however, a cheap, blue, collapsible plastic hose is attached to the waste opening with a hose clamp. This 1½- or 2-inch-diameter hose is intentionally made flimsy because the water is not under much pressure when draining to waste and also to allow it to be easily rolled up and stored near the filter. This ease of rolling up is an advantage because backwash hoses normally come in lengths of 20 to 200 feet—you might have to route waste water into a street storm drain, so very long lengths are common.

Pressure gauges and air relief valves

Most filters are fitted with a pressure gauge, mounted on top of the filter (Fig. 5-4, item 1). Sometimes the gauge is mounted on the multiport valve (Fig. 5-3, item 12). These gauges read 0 to 60 psi and are useful in several ways.

First, when a filter is newly put into service or has just been cleaned, I make it a habit to note the normal operating pressure (in fact, I carry a waterproof felt marking pen and write that pressure on the top of the filter). Most manufacturers tell you that when the pressure goes more than 10 pounds over this normal operating pressure it is time to clean the filter.

The other value of the pressure gauge is to quickly spot operating problems in the system. If the pressure is much lower than normal, something is obstructing the water coming into the filter (if it can't get enough water, it can't build up normal pressure). If the gauge reads unusually high, either the filter is dirty or there is some obstruction in the flow of water after the filter.

When the pressure fluctuates while the pump is operating, the pool or spa might be low on water or have some obstruction at the skimmer—when the water flows in, the pressure builds, then as the pump sucks the skimmer dry, the pressure drops off again. This cycling will repeat or the pressure will simply drop altogether, indicating the pump has finally lost prime. In any event, a pressure gauge is a valuable instrument on the filter.

Some codes, especially for commercial pools or spas, require a pressure gauge be mounted on the incoming pipe and on the outgoing pipe to compare the differential. Obviously, as the water tries to push from the pump into the filter, the pressure is greater than as it leaves the filter. As you saw in the hydraulics section, even a clean filter creates this back pressure.

One way for a health department inspector to check the cleanliness of your filter system is to compare these two pressure readings and when that difference exceeds a certain amount, they know the filter is dirtier than whatever standard they have set. Normally the differential should be 2 to 4 psi; when it reaches 10 psi, the filter needs cleaning.

Mounted on a yoke or T fitting along with the pressure gauge, you will normally find an air relief valve (Fig. 5-4, item 2). It is simply a threaded plug that when loosened allows air to escape from the filter until water has fully filled the tank. When a system first starts up, particularly after cleaning or if the pump has lost prime, there is a lot of air present in the filter. If it is left like that, the filter might be operating at only half its capacity.

Figure 5-4 shows an inside view of a cartridge filter. The area above the cartridge is called the *freeboard*. This empty area is present above the filter media of all filters. For some of this to be air rather than water is okay, but if the air in the tank were down to, for example, the halfway mark of the cartridge and water were flowing around only the lower half of the cartridge, you would effectively cut the filter square footage in half.

So it is important to let the air out of the filter at every service call. Again, some air is normal, so don't be obsessive about it; just be sure that the filter media itself is covered with water, not air.

Sight glasses

A *sight glass* is simply a clear section of pipe (Fig. 5-9). The sight glass is actually composed of a metal frame with a glass interior, constructed to the same diameter as a standard pipe (1½ or 2 inches). A sight glass is installed anywhere in a line of pipe where you want to see the quality of the water.

A sight glass is used to see the effectiveness of your backwashing. Sight glasses are normally installed on the backwash line coming out of the backwash valve (going to the waste or sewer or going to a separation tank). When the dirty water starts to look clean, you know you have backwashed enough. A sight glass is handy if your waste pipe goes into a deck drain or if you use a long backwash hose where you can't see the clarity of the water being flushed out.

REPAIR AND MAINTENANCE

As noted previously, there are no moving parts on a filter when it operates and few when it is at rest. Therefore, there's not much to break down and, when they do, filters are easy to repair.

5-9 Sight glass.

Installation

Replacing a filter is one of the easiest repair jobs you can get. There is no electricity or gas to hook up and normally you are dealing with only three pipes.

1. Turn off the pump and switch off the circuit breaker. This way you can make sure it won't come back on (from the time clock, for example) until you're ready.

2. Drain the old filter tank by opening the drain plug or backwash valve. If you don't do this first, you'll take a bath when you cut the plumbing.

3. Cut the pipe between the pump and filter in a location that makes connecting the new plumbing easiest. There's no rule of thumb here, just common sense. If the original installation has more bends and turns in the pipe than needed, now is a good time to cut all that out and start over. Eliminating unnecessary elbows increases flow and reduces system pressure. Cut the pipe between the old filter and the heater using the same guidelines. Last, cut the waste pipe if there is one plumbed into a drain.

4. Remove the old filter. Even without water, the old filter will be heavy, so you might want to disassemble it for removal. Sand filters are the worst, and you will have to scoop out the heavy, wet sand before you will budge the tank. Save any useful parts. Old but still working grids, valves, gauges, air relief valves, lids, lid O-rings, cartridges, and other components make great emergency spares to carry in your truck. There are some companies that recondition old filters or parts, so when you make the deal with the homeowner to replace the filter, make it like a car battery or tire sale—you get the trade-in. The customer doesn't want it anyway and their trash service probably will refuse to pick it up for disposal, so agree to do the hauling away in exchange for any usable parts.

5. Set up the new filter. After removing it from the box, make sure all the pieces are there—grids, pressure gauge, etc. Most new filters come with instructions and it really pays to read these. In fact, if the homeowner doesn't want the booklet, add it to your library for reference. While the unit is out in the open and easy to work on, screw into place the appropriately sized MIP fittings (1½ or 2 inch), after applying a liberal coating of Teflon tape or pipe dope (see the plumbing section). Some manufacturers include their preference of tape or pipe dope right in the box.

   On a filter with threaded openings in a plastic base (like most cartridge filters) or a plastic multiport valve (like most sand filters), be careful not to overtighten—you'll crack the plastic. On a vertical grid DE filter with the rotary valve on the bottom, lay the filter down and screw the MIPs into place. Some manufacturers make this a knuckle-busting experience. Be sure you have the fittings tight—if they leak, you'll have to cut out all your work to fix it.

   Some manufacturers still provide filters with bronze plumbing openings on the backwash valve for sweating copper pipe. If you encounter this type of plumbing, remove any parts that might suffer from heat when you sweat the fittings in place. Plastic grids and manifolds will melt if they are not removed, and with the lid on the filter, the heat inside will build up quickly. Just as the backwash valve conducts water efficiently, so too will it conduct heat to those plastic parts.

6. Place the new filter in the location by the pump and figure out the plumbing between the pump and filter, between the filter and heater, and between the filter and waste line (if appropriate). Some creativity and planning here will save many service headaches later. Basically you want to avoid elbows and you want to leave enough room between the pump, filter, heater, and pipes for service access. Remember, you'll have to clean this filter someday. Can you easily access the lid? Will water flowing out of the tank as you clean it flood the pump? Can you access the backwash valve and outlet as needed?

7. Plumb it in. Use the plumbing instructions in the chapter on basic plumbing and make careful connections. Who was it that said, "There never seems to be enough time to do it right, but there's always enough time to do it over when it leaks"?

   If a check valve was not a part of the plumbing between the filter and the heater, it is not a bad idea to add one now. If there is any chance of hot water flowing back into the filter, you run the risk of melting the grid or cartridge materials. Also, you don't want the heater to sit without water in the heat exchanger (see the heater chapter).

Starting up the newly installed filter is just like restarting after cleaning, so read further for start-up procedures and hints.

Filter cleaning and media replacement

The most important thing you can do for a pool is to keep the filter clean. This is also the simplest way to ensure the other components work up to their specifications and you end up with satisfied clients. Let's review the process for cleaning each type of filter.

DE filters

I do not believe backwashing a DE filter is of any value as a regular practice and, in fact, I know it can be harmful to the filter and pool cleanliness. Obviously the makers of backwash valves and those who have bought into their technology over the years will disagree with me, but here's what I believe based on years of experience. When you backwash, some dirt and some DE are flushed from the filter. The remainder drops off the grids and falls to the bottom of the filter in clumps. The manufacturers say that after backwashing 70 percent of the DE has been removed, so you need to replace that amount. I have opened up filters that were just backwashed and have seen as little as 10 percent of the DE washed away, while others had virtually 100 percent washed away.

If you don't know how much went out, how do you know how much to put back in? If you add too little, the filter grids will quickly clog with dirt and the pressure will build right back up, even stopping the flow of water completely. If you add too much, you will get the same effect by jamming the tank with DE.

Moreover, backwashing cannot remove oils from the grids, which get there from body oil, oil in leaves, and suntan lotions. You can backwash for hours and when you open up the filter, you will find the grids clogged with oils and a layer of DE and dirt that sticks to these oils.

Finally, backwashing wastes water. If you break down the filter and clean it completely, you will use some water to wash the grids and tank, but you will not have to clean it again for weeks or even months.

When you backwash you really are not cleaning the filter thoroughly. You'll be backwashing again in a few days or weeks and when you get tired of that you will break down and clean the filter anyway.

Okay, I've had my say! Now I'll tell you the one time when backwashing a DE filter is useful. You have a pool that has been trashed by winds, mudslides, algae, or other heavy debris. You start to vacuum it and quickly the filter can't hold any more dirt. To save a lot of time you backwash, add a little fresh DE, and get on with the job. You repeat this process until the big mess is cleaned up, then you break down the filter and clean it properly.

The other time you might backwash is when you're vacuuming a normally dirty pool, but the filter hasn't been cleaned in awhile and is just about full of dirt. You're getting no suction because the filter is clogged. It's Friday night at 4:45 PM and you have a hot date. Okay, okay. Backwash, add some fresh DE, finish cleaning the pool, and make a note to do a breakdown (clean the filter) on Monday.

Another important fact about backwashing. Since the water is going inside the grid and flowing outward, any debris in the water from the pool will clog the inside of the grids (or laterals on sand filters), rendering them useless. On a new pool start-up where a lot of plaster dust or gunite debris might be in the water, don't backwash. If you must, open the strainer pot and turn on the pump. Flood the pot with water from a hose and backwash as needed that way. Obviously, never vacuum a pool with the filter on backwash because the dirt and debris you vacuum will flow directly inside the grids (or laterals).

So how do you properly break down (or tear down) and clean a DE filter? I'm going to describe the Purex/Hydrotech style of vertical grid tank DE filter (Fig. 5-1). They are common in the field and if you can do these, you can do them all.

1. Turn off the pump and switch off the circuit breaker.
2. Remove the lid of the filter. That might sound easy, but depending on the design, it can be real work. On some filters, it is as easy as removing the clamping ring and applying light pressure under the lid with a screwdriver (be careful not to gouge the lid or O-ring—if you do, leaks will develop in these spots).

   Some filters make such a tight seal with the O-ring and lid that a nuclear bomb will not remove them. Draining the tank first won't help—as the water drains out, it sucks the lid on even tighter. What I do is, after removing the clamping ring, turn on the pump for a few seconds. The pressure from the incoming water will pop the lid off. When it does, grab it quickly, otherwise when the pump is shut off and the water recedes, it will suck the lid right back onto the tank.

   Some manufacturers do not approve of this procedure. I have "popped" literally thousands of filter lids and never had one pop more than a few inches off the tank. I have also never observed damage to the equipment by this technique. The pressure is applied evenly as the lid pops off, so the tank or lid doesn't warp or bend. Be sure not to grasp the lid by the gauge assembly—they snap off easily.

The only caveat to this practice is to be sure that you don't stand in the water that will inevitably flow out of the tank as the lid pops while you are holding onto the pump/motor switch. Water and electricity can be deadly. Also, if the motor is installed directly adjacent to the filter tank, the water flowing out might flood the motor. In that case, get a screwdriver and crowbar to remove the lid, or wrap the motor in a plastic bag.

3. Open the tank drain and let the water run out. Remove the retainer's wing nut and remove the retainer (also called the holding wheel). Now gently remove the grids (elements). One design flaw of many grids lies in the fact that they are made like small aircraft wings—large, curving units—but they are set into the manifold on stubby little nipples. Applying a reasonable amount of force on the rather large wing part of the grid won't hurt it, but the resulting torque on the flimsy nipple will snap it right off. Therefore, to remove the grids, wiggle them gently from side to side as you pull them straight up and out.

   Tennis star John McEnroe fired his pool man of many years and called my company. I went over and found the pool very dirty and the filter pressure almost off the scale. I popped the lid and found so much dirt and DE in the tank that there was literally no room for water. This condition is called *bridging* because the DE and dirt bridge the normal gaps between the grids, clogging and effectively reducing the amount of filter area. It was so packed that it took me three hours of archaeologist-like excavation with a small spade, stick, and lots of water to finally get the grids free. This, by the way, was the result of a pool guy who backwashed and added fresh DE about once a month—for 6 years. He never once opened the filter. The cumulative result was that he was adding more DE than was being flushed out, the pool stayed dirty because it couldn't filter or be vacuumed, and, best of all, he lost the customer.

   The point of telling you this is that to some lesser degree you might find the same condition when you open a filter. Be prepared to hose out the tank while the grids are still in place (if the drain hole isn't also clogged with DE) or patiently excavate the dirt and DE until you can free the grids.

4. Remove the retaining rod. It threads into the base of the rotary valve like a screw, so just unscrew it. Sometimes it is corroded in place, so have pliers (Vise-Grips work best) handy to grip the rod and unscrew it. A word of caution—the rod might be corroded enough that if you force it with your pliers, it snaps off at the bottom, leaving the threaded end in the rotary valve.

   If you find too much resistance to your effort to remove the rod, leave it in place and clean the tank as best as you can. If it has broken off and you don't have time to disassemble the entire tank and rotary valve to get out the stub, lay a brick on top of the holding wheel when you put the unit back together to keep it and the grids in place. Then come back and fix it when you have time.

5. Reach in the tank and remove the manifold. It rests just inside the rim of the rotary valve; it is not threaded in place.

6. Hose out the inside of the tank, the manifold, and the holding wheel. Hose off the grids. You might need to scrub them lightly with a soft bristle brush to loosen the grime. If the grids are still dirty, soak them in a garbage can of

water, trisodium phosphate (1 cup per 5 gallons of water), and muriatic acid (1 cup per 5 gallons) (acid alone will not clean grids because it does not affect the oil). After 30 minutes, try scrubbing them clean again. Don't use soap—you won't get it all out, no matter how well you rinse the grids, and when you start up the circulation again you'll have soap suds in the pool.

7. Inspect the manifold for chips or cracks. DE and dirt will go through such openings and back into your pool. Cracks can be glued. If chunks of plastic are missing, buy a new manifold—they're only about $30. Particularly inspect the joint between the top and bottom halves of the manifold. Where these two parts are glued together, they often start to separate. Replace the manifold as you took it out. Reinstall the center rod.

8. Carefully inspect the grids before putting them back inside. Look for worn or torn fabric, cracked necks on the nipples, or grids where the plastic frame has collapsed inside the fabric. Replace any severely damaged grids. When you reinstall the grids, notice that inside each hole in the manifold is a small nipple and on the outside of each grid nipple is a small notch. By lining up the nipple and notch as you reinsert each grid, the grids will go back as intended.

   Now lay the retainer over the tops of the grids and spin it around until it finds its place holding down and separating the grids. Screw on the wing nut and washer that holds down the retainer holding wheel.

9. Getting the lid back on can be as tough as getting it off. Make sure the O-ring on the tank is free of gouges and has not stretched. If it is loose, soak it for 15 minutes in ice water and it might shrink back to a good fit. If not, replace it. Apply tile soap as a lubricant to make it slide on easier (or silicone lube if you can afford it) to the inside of the lid around the edge that will meet the O-ring. Don't use Vaseline or petroleum-based lubricants because these will corrode the O-ring material. Don't use that green slime called Aqua Lube—it sticks to everything and comes off of nothing.

10. Now close the tank drain, turn the backwash valve to normal filtration, and turn on the pump. Let the tank fill with water. Turn off the pump and turn the valve to backwash. The water will drain out, sucking the lid down. Don't be afraid to help it along by getting on top of the lid. Your weight will finish the job. Be careful not to hit the pressure gauge assembly—they snap off very easily.

11. Replace the clamping ring, return the valve to normal filtration, and start the pump/motor. Open the air relief valve and purge the air until water spurts out the valve.

12. Never run a DE filter without DE, even for a short time. Dirt will clog the bare grids. Remember, it's not the grids, but the DE that does the actual filtering. The label on the filter will tell you how much DE to add, or refer to the table on the bag of DE. It tells you how many pounds of DE to add per square foot of filter area. As a convenient scoop, use a 1-pound coffee can; but remember that because DE is so light and powdery, a 1-pound coffee can holds only ½ pound of DE. One pound of DE covers 10 square feet of grids.

   DE is added to the system through the skimmer. Do not dump it in all at once. It will form in clumps at the first restricted area, like a plumbing elbow or the inlet on the filter tank. Sprinkle in one can of DE at a time, mixing it in the skimmer water with your hand. It should appear to be dissolving in the water. In fact, it is not dissolving, just freely suspending itself in the water, but this will keep it from clumping.

Add one can and wait about a minute. If you have any gaps in the manifold or holes in the grids or you didn't assemble the unit correctly, DE will get through these areas and flow back into the pool. If that happens, it is better to have one can of DE flowing back into the pool rather than all 10 or 15. On most start-ups you will see a little milky residual entering the pool from any DE or dirt that settled in the pipes during cleaning. This is normal, but if you see great clouds of DE returning to the pool, shut the system down and take the filter apart. You missed something.

I must emphasize this problem of adding DE too fast (or using too much—follow the DE package directions). Comedian Rich Little's system was sluggish and the filter pressure unusually low. I cleaned the pump strainer, skimmers, blew water through the suction lines with a drain-flush bag, and cleaned the filter twice . . . and no luck. I took apart the pump to make sure the impeller was clean and operating properly. I checked the power supply, thinking a bad breaker was perhaps delivering low voltage and making the motor work too slowly. No luck.

Finally, after hours of hunting, I emptied the filter tank and fed hose water directly into the pump and watched it trickle into the tank. I knew the obstruction was in the plumbing between the pump and the rotary valve. I cut open the plumbing and sure enough, at one of the 90-degree elbows I found clumps of DE. The pressure had crystallized it and made it rock hard. I had to replace that section of plumbing and later in the shop I took a hammer and chisel to the DE to see just how hard it had become.

A good way to add DE to prevent this problem is in a *slurry*. On large commercial installations that require large amounts of DE, there is actually a slurry pit, but you can use a bucket. The concept is that you thoroughly mix the DE in water (achieving that suspension I spoke of) before pouring the solution into the skimmer.

One last tip. Most pools have skimmers where you can add DE. But what if there isn't one or if a hot tub or other body of water that has no skimmer uses a DE filter? Again, make a slurry in a bucket. After cleaning the filter, take the lid off the pump strainer pot and turn on the pump. Add the slurry to the strainer pot, followed by clear water (have a hose handy) to make sure all the DE gets to the filter and completely and evenly coats the grids. Turn off the pump, fill the pot with water, replace the lid, and reprime the system. Again, remember to let the air out of the tank.

Sand filters

Sand filters are designed to use #20 silica sand, a specific size and quality of sand. Larger sand will not filter fine particles from the water and finer sand is pushed through the slits in the laterals, clogging them.

Sand filters do need regular backwashing and, unlike DE backwashing, it is effective. Although wet sand is heavy and lays on the bottom, when the tank is full of circulating water the sand is suspended in the tank. In fact, you can reach into a tank of sand and water and get your hand all the way to the bottom of the tank. Try that without water in the tank. Your hand will push about 6 inches into the thick sand.

The sand is light enough to stay floating in the tank, but heavy enough that it does not flow out with the backwash water. This is why the multiport valve is located on top of sand filters, so as the backwash flows from the bottom toward the top the dirt flows up and out while the sand stays put. So backwashing is an effective way of cleaning a sand filter. Most rotary valves have the steps printed right on them, and they are very simple.

1. Turn off the pump. Rotate the valve to Backwash. Roll out your backwash hose or make sure the waste drain is open.
2. Turn on the pump and watch the outgoing water through the sight glass. It will appear clean, then dirty, then very dirty, then it will slowly clear. When it is reasonably clear, turn off the pump and rotate the valve to Rinse.
3. Turn the pump back on and run the rinse cycle for about 30 seconds to clear any dirt from the plumbing. Turn off the pump, rotate the valve back to Filter, and restart the pump for normal filtration.

Backwash as often as necessary. When the filter gauge reads 10 psi more than when the filter is clean, it is usually time to backwash. A better clue is when dirt is returning to the pool or when vacuuming suction is poor.

When backwashing, be sure there is enough water in the pool to supply the volume that will end up down the drain. It is usually a good idea to add water to the pool or spa each time you backwash.

At every third backwash, I recommend opening the filter. Sand under pressure and with the constant use of pool chemicals or dissolving pool plaster will calcify, clump, and become rock-like over time. Passages are created through or around these clumps, but less and less water is actually filtering through the sand and more is passing around it. This is called *channeling*. To correct or avoid this problem, regular teardown is the answer.

1. Turn off the pump. Disconnect the multiport valve plumbing by backing off the threaded union collars. Some valves are threaded into the body of the tank, others are bolted on. Remove the valve.
2. Some sand filters have a large basket just inside the tank. Remove this and clean it out. The sand is now exposed. Push a garden hose into the tank and flush the sand. As noted previously, it will float and suspend in the water. Use a broom handle to bust up clumps. As the water fills the tank, it will overflow, flushing out dirt and debris. Be careful not to hit the laterals on the bottom of the tank because they are fragile and break easily.
3. When the sand is completely free and suspended in the water, not clumped, turn off the water and replace the basket (if any), multiport valve, and plumbing. Backwash briefly to remove any dirt that was dislodged by this process but not yet flushed out.

This teardown process also allows you to check to see if the regular backwashing has flushed out too much sand. You might need to add some fresh sand. Most sand filters need to be filled about two-thirds with sand and have one-third freeboard. Backwash after adding any new sand to remove dust and impurities from the new sand.

If channeling is a problem because of hard water or pool chemistry (which speeds up calcification of the sand), introduce aluminum sulfate (alum) through the skimmer just like you would add DE to help prevent this problem. Use the amounts recommended on the bag.

Every few years you need to replace the sand completely because erosion from years of water passing over each grain makes them round instead of faceted and rough. Smooth sand does not catch and trap dirt as efficiently and it slowly erodes to a smaller size than the original #20 silica, allowing it to clog laterals and pass into the pool. Some manufacturers suggest adding a few inches of gravel over the laterals first. This keeps the sand separated from the laterals so the sand cannot clog them. To replace sand, or add sand to a new installation:

1. Open the filter as described previously. Remove the old sand by scooping it out with your hands or with a sand-vac (Fig. 5-10).
2. Fill the bottom third of the tank with water to cushion the impact of the sand on the laterals.
3. Slowly pour the sand into the filter, being careful of the laterals. Fill sand to about two-thirds of the tank. Reassemble the filter parts and backwash to remove dust and impurities from the new sand, then filter as normal.

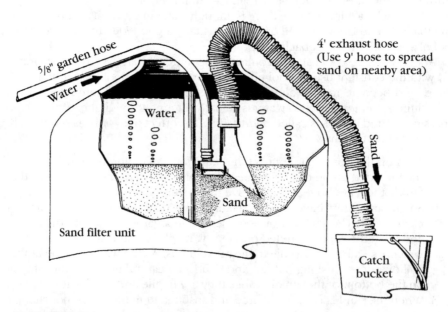

5-10 Sand filter vacuum. Lass Enterprises, Altamonte Springs, Fla.

Cartridge filters
Cleaning a cartridge filter is perhaps easiest of all.

1. Turn off the pump. Remove the retaining band and lift the filter tank or lid from the base. Remove the cartridge.
2. Light debris can simply be hosed off, but examine inside the pleats of the cartridge. Dirt and oil have a way of accumulating between these pleats.

Never acid wash a cartridge. Acid alone can cause organic material to harden in the web of the fabric, effectively making it impervious to water. Soak the cartridge in a garbage can of water with trisodium phosphate (1 cup per 5 gallons) and muriatic acid (1 cup per 5 gallons). About an hour should do it. Remove the cartridge and scrub it clean in fresh water. Don't use soap. No matter how well you rinse, some residue will remain and you will end up with suds in your water.

3. Reassemble the filter and resume normal circulation.

A good idea is to sell your customer a spare cartridge so you can install it and take the dirty one back to the shop for soaking instead of wasting time on the job site while it soaks there. Replace cartridges when they won't come clean, when the webbing of the fabric appears shiny and closed, or when the fabric has begun to deteriorate or tear.

Leaks

Filters, being the rather simple creatures they are, don't have many repair or maintenance problems beyond cleaning as discussed. The biggest general complaint, however, relates to leaks of various kinds.

Backwash valves

Backwash valves leak in two ways: internally and externally. Internally, O-rings deteriorate and allow water and DE or dirt to pass into areas not intended.

Piston backwash valves The valve in Fig. 5-6 has piston discs equipped with O-rings (item 7). As these wear out, water or dirt bypasses the intended direction. Similarly, the O-rings on the shaft (just under the handle) wear out from regular repeated use. If you suspect the disc O-rings or see water leaking from the top of the shaft, tear down the valve as follows:

1. Turn off the pump. Remove the screws on top of the valve cap. Pull the handle up as if you were going to backwash, but keep pulling straight up to remove the entire piston assembly.

   Replace the O-rings on each disc. They pull off like rubber bands and the new ones go on the same way. Apply silicone lube to the O-rings.

2. Remove the handle from the piston stem. It is held in place by setscrews or allen-head screws. This also allows you to slide the cap off the stem. Look inside the cap. You will find two small O-rings. Pull these out with the tip of a screwdriver and replace them. Apply silicone lube.

3. Clean (as needed) the stem and disc assembly and flush out the inside of the valve body. Grit or sand can create leaks or cause your new O-rings to wear out sooner than necessary. Reassemble the unit the same way you took it apart.

Rotary or multiport backwash valves Rotary and multiport valves are similar in construction, so if you understand one you will not have trouble with the numerous other designs. Figure 5-8 shows a rotary valve normally mounted under a vertical grid DE filter. As with piston-type units, these leak either externally or within the chambers of the unit itself. If water appears under the filter, use a flashlight to inspect underneath as carefully as possible. If you can see or feel a leak where the plumbing enters the valve openings, you can repair that without disassembling the entire filter. If the leak appears to be at the joint of the valve and filter tank, or if the problem is DE and dirt bypassing the normal flow and getting back into the pool, you will need to tear down the filter and valve.

Another typical symptom of an internal leak is drips coming from the backwash outlet even though the valve is turned completely to the normal filtration position. Figure 5-8 employs a rotor seal (item 3A) that can compress or wear out. When the body gasket wears out (and it will wear out prematurely by rotating the valve with the pump on) and water bypasses the normal flow, some leakage gets to the backwash side and appears as a leak under the filter. If the backwash outlet is plumbed directly into a waste or sewer drain, this leak might not be visible.

Remember this when you're looking for a pool leak—sometimes the problem is not in the pool or spa itself, but in some hidden area within the system plumbing. If the gasket completely wears out, the leaking can be substantial and, as noted previously, if the plumbing prevents you from seeing it you will never know. Such a hidden problem can also cause the system to lose prime overnight when the pump is off. The leak drains the water from the filter tank, then siphons the water out of the pump.

On start-up the next day, the pump has no prime. If the pump runs dry for several hours, overheats, loosens or melts the plumbing fittings, you will attribute the loss of prime to the damaged plumbing. You repair the plumbing and the same problem occurs the next day.

The moral of the story is that it pays to have a sight glass on the backwash outflow line so you can see any leaks and/or have a shutoff gate valve on that line that stays closed when the valve is in the normal filtration position.

To tear down this type of valve (Fig. 5-8), use the following procedure.

1. Cut the plumbing to isolate the filter, and take the unit apart as described previously.
2. Reach inside the bottom of the filter and remove the bolts (item 14, usually ¼-inch hex-head bolts) that hold the compression ring (item 13) with a nut driver. This ring holds the valve in place as well, so the valve will now fall away from the filter tank.
3. You now have the valve body (item 1) with the rotor inside. Remove the handle (items 6 and 7) on the underside of the valve by removing the bolt assembly (items 8, 9, and 10) that holds it on the rotor shaft and slide it off the shaft. Pull the rotor out of the body. Bronze rotors are very hard to remove and you might have to take the valve to a pump rebuilding shop. Most shops have special rotor pulling tools or they can carefully heat the valve body so it expands and releases the rotor. I wouldn't try to do this yourself. Without a lot of experience, you will probably warp or destroy the components.
4. Pull the old rotor seal gasket (item 3A) from the rotor with needle-nose pliers. Clean the rotor and inside the valve body. Put a new gasket on the rotor, being careful not to overstretch the new gasket.
5. Lube the gasket with silicone lube and replace it in the valve body. On bronze rotors, each port has an O-ring instead of one body gasket seal as you will find on the plastic versions. Before reassembling the filter, replace the O-ring (item 2) that sits between the tank and valve and the O-ring

(item 5) that seals the shaft as it passes through the valve body to the handle. Also replace the O-ring (item 12) on the neck of the rotor. The grid manifold sits on this neck and the O-ring seals that joint, so to prevent dirt from bypassing the correct direction of flow, you need a good seal here. Lube all O-rings with silicone lube.

6. Reassemble the valve and tank the way you took it apart. Be sure the tank itself is clean (dirt or sand will prevent the O-ring from sealing tightly) and that the opening in the bottom shows no rust or cracks. If it does, you should clean it thoroughly and have the cracks welded. Such weak spots will come back to haunt you, so it might be time to suggest to your customer a new tank (or new filter). Replumb and restart the filter as described previously.

Lids and gauge assemblies

Lids on filters leak in two places: the O-ring that seals them to the tank and/or the pressure gauge air relief valve assembly. The lid O-ring can sometimes be removed, cleaned, turned over (or inside out), and reused. I have not recommended that for other O-rings on the filter, such as in the backwash valve, because if they are too worn or compressed to reseal, you must tear down the entire filter again to replace them. Hardly worth the price of an O-ring. But the lid O-ring is thick and expensive and easy to remove and replace. So try the cleanup/turnover method and if you still have leaks, then replace it.

Some filters will crack on the rim of either the lid or the tank where the O-ring is seated. Obviously, the problem in this case is not a bad O-ring, but a bad lid or tank. Inspect these stress areas carefully for hairline cracks that might be the source of the leak.

Air relief valves sometimes leak if they become dirty or they simply wear out. Some are fitted with an external spring that applies tension to create the seal, and when the spring goes, so does the watertight seal. Others have a small O-ring on the tip of the part that actually screws in to create the seal. Unscrew this type of valve all the way. The screw part will come out to reveal the O-ring on the tip that makes the seal, and you can easily replace that. Air relief valves themselves simply screw out of the T assembly. Apply Teflon tape or pipe dope to the new one and screw it back in place.

The pressure gauge also threads into the T assembly. If you have a leak there, unscrew the gauge, apply Teflon tape or pipe dope to the threads and screw it back into place. If the gauge doesn't register or seems to register low, take it out and clean out the hole in the bottom of the gauge. Dirt or DE can clog this small hole, preventing water from getting into the gauge.

Remember, when removing an air relief valve or pressure gauge, you must secure the T with pliers or a wrench while removing the component. The T assembly can easily snap off the filter lid or come loose if you fail to hold it securely when removing or replacing a valve or gauge.

The T assembly itself can come loose and create a leak where the close nipple passes through the hole in the lid. In this case you must remove the lid and tighten the nut from the underside of the lid. Some makes of filters have a nipple welded to the lid, so you won't have this problem unless you crack the weld.

Dirt passing back into pool

I have reviewed most of the ways dirt or DE gets through the filter and back into the pool and the methods to make repairs. Just as a summary, however, if you see this condition when vacuuming, check the following:

- Damaged grids, laterals, or cartridges;
- Backwash valves with bad gaskets or O-rings; and
- Broken manifolds or retainers.

Prevention is always the best cure, so when you feel a backwash valve getting hard to turn, do a teardown and lubrication before the leaks occur. Examine grids, laterals, cartridges, and manifolds carefully each time you break down a filter for cleaning, and always take your time when reassembling. Sloppy reassembly after cleaning is the cause of more leaks than anything else in filters.

6

Heaters

Let's begin the discussion of pool and spa heaters with a point that seems obvious to the technician but which serves to create confusion with many homeowners. The pool and spa heater doesn't work like the water heater in your home. Customers have asked me countless times why the water coming out of the return line isn't hot (like tap water in the home). The assumption in that question is that the pool heater holds a large reservoir of preheated water. They don't realize that the pool and spa heater heats the water as it passes through copper coils, creating a mixture of the heated water with cold water so that the resulting flow out of the heater is not more than 10° to 25°F warmer than the water that originally went in.

The basic principle of the pool and spa heater (Fig. 6-1) is simple. A gas burner tray creates heat. Heat rises through the cabinet of the heater, raising the temperature of the water that is passing through the serpentine coils above.

6-1 The concept of the pool and spa heater.

SAFETY

Heaters are unquestionably the most potentially dangerous component of the pool or spa equipment group. They combine water under pressure and heat, gas or other combustible fuel, and electricity. The point is simply that whatever care you exercise normally must be doubled when working with heaters. Therefore, I have a simple safety checklist for working around heaters.

- Never bypass a safety control and walk away. Jumping controls is a good way to troubleshoot, but do not operate the unit this way. Always remove your jumpers after troubleshooting.
- Never repair a safety control or combination gas valve. Replace it. You will notice that your supply house doesn't even sell parts for gas valves. They should never be repaired, because future failure could be catastrophic.
- Never hit a gas valve—it might come on, but it might stay on.
- Keep wiring away from hot areas and the sharp metal edges of the heater.
- Communicate. Tell your customer about heater part failures and repairs. Disable the heater and tape a shutdown notice on the unit until repairs are made. **You can be held liable** if you are the last person to work on a heater and it causes damage or injury by firing incorrectly or before repairs have been made.
- When jumping a safety control or otherwise trying to fire a heater that will not come on, keep your face and body away from the burner tray, where flashback might occur. It might be awkward to squat alongside a heater and jump a control through the opening in the front, but awkward is better than burnt. Double that warning with LP-fueled units.

GAS-FUELED HEATERS

Figure 6-2 shows a typical gas-fueled heater (those that use natural or propane gas as the heating fuel). The water passes in one port of the front water header (item 23), then through the nine heat exchanger tubes (item 45). The water reaches the rear header (item 24) and is returned through other exchanger tubes to the front header and out the other port.

Most modern exchangers are four-pass units, meaning the water goes through at least four of the tubes, picking up 6° to 9°F on each pass, before exiting the heater. Generally, these are self-cleaning unless extreme calcium (scale) is present in the water.

The heat exchanger tubes are made of copper which conduct heat very efficiently. The tubes have fins (about eight per inch) to absorb heat even more efficiently and are topped with sheet metal baffles (item 53) to retain the heat. The heat rising from the burner tray (item 55) is effectively transferred to the water in the exchanger because of the excellent conductivity of copper; however, improper water chemistry can easily attack this soft metal and dissolve it into the water. More on that later.

Notice that there is a flow control assembly on the front header (items 38–42). This spring-loaded valve is pressure sensitive, designed to mix cool incoming water with hot outgoing water to keep the temperature in the exchanger from becoming excessive. This design keeps the outgoing water no more than 10° to 25°F (depending on manufacturer) above the temperature of the incoming water to prevent condensation and other problems that greater differentials would create. Temperature control is achieved by flow regulation rather than direct temperature regulation. Maintaining a constant flow through the heat exchanger results in a constant water temperature.

6-2 Detail of a typical millivolt, standing pilot heater. Teledyne Laars, Moorpark, Ca.

Key No.

1 Pilot generator assembly	31 Drain grommet
2 Visoflame lighter tube	32 Drain grommets
2A Pilot tube	33 Drain valve
3 Automatic gas valve	34 Drain plug
4 Burner orifice	35 Drain valve
5 Burner w/pilot bracket	36 Bushing, drain valve
6 Burners	37 Brass plug
7A Plate assembly	38 Flow control cap
7B Thermostat dial	38A, 43 Bolt & washer
8 High-limit switch (135°F.)	39 Flow control gasket
9 High-limit switch (150°F.)	40 Flow control shaft
10 Redundant limit	41, 41A Front control spring
11 Temperature control	42, 42A Flow control disc
12 Protective sleeve, bulb	43 Bolt, front & rear header
13 Wire harness	44 Header nut, $3/8"$ hex
14 Thermostat knob	45 Heat exchanger
15 Temp-lok	45A Syphon loop
16 Pressure switch	46 Fiberglass blanket
17 High-limit switch retainer clip	47 Insulation block, side
18 High-limit switch cover	49 Insulation block cover, front & back
19 O-ring	49A Insulation block cover, end
20 On-off switch	50 Door
21 Fusible link	51 Jacket assembly
22 Fusible link bracket	52 Gap closure
23, 23A, 23B Front water header	53 Heat exchanger baffle
24 Rear water header	54 Burner tray shelf
25 Header gasket	55 Burner tray assembly
26, 29 Flange packing collar w/copper sleeve	57 Rear deflector
27 Water header flange	58 Lower deflector
28 Water header flange bolt	59 Noncombustible floor base (optional)
30 Heat exchanger baffle retainer	

6-2 Continued.

When reaching temperatures over 115°F, water breaks down, allowing minerals suspended in it to deposit in the heat exchanger. Also, water is designed to flow through the unit at no more than 100 gpm with 1½-inch plumbing or 125 gpm with 2-inch plumbing. Above that, a manual bypass valve is installed.

The other major component of the gas-fueled heater is the burner tray. This entire assembly can be disconnected from the cabinet and pulled out for maintenance or inspection. Depending on the size of the heater, there will be 6 to 16 burners (item 6), the last one on the right having a pilot (item 5) mounted on it. Individual burners can be removed for replacement. The combination gas valve (item 3) regulates the flow of gas to the burner tray and pilot and is itself regulated by the control circuit.

Gas-fueled heaters are divided into two categories based on the method of ignition.

The millivolt or standing pilot heater

As the name implies, the standing pilot system of ignition uses a pilot light (burner) that is always burning. The heat of the pilot is converted into a small amount of electricity (0.75 volt or 750 millivolts) by a thermocouple which in turn powers the control circuit. The positive and negative wires of the thermocouple (also called the pilot generator) are connected to a circuit board on the main gas valve.

When lighting the pilot, it is necessary to hold down the gas control knob to maintain a flow of gas to the pilot. When the heat has generated enough electricity (usually a minimum of 200 millivolts) the pilot will remain lit without holding the gas control knob down. The positive side of the thermocouple also begins the electrical flow for the control circuit. When electricity has passed through the entire control circuit, the main gas valve opens and floods the burner tray with gas which is ignited by the pilot.

The control circuit

The control circuit is a series of safety switches—devices that test for various conditions in the heater to be correct before allowing the electrical current to pass on to the main gas valve and fire up the unit. Figure 6-3 shows a millivolt control circuit; Fig. 6-4, and electronic control circuit. Following the flow of electricity (not all manufacturers follow the same routing of their control devices, but they all include the same devices), a control circuit includes the following items.

Fusible link

The fusible, or fuse, link is a simple heat-sensitive device located on a ceramic holder near the front of the gas burner tray.

6-3 The control circuit.

6-4 Components of the control circuit. Top row (left to right): transformer, automatic combination gas valve; middle row: intermittent ignition device (IID), pressure switch; bottom row: fusible link, high-limit switch, on/off switch, mechanical thermostat.

If the heat becomes too intense, the link melts and the circuit is broken. This would most commonly occur when debris (such as a rodent's nest or leaves) is burning on the tray or if part of a burner has rusted out causing high flames. Other causes are improper venting (allows excessive heat build up in the tray area), extremely windy conditions, or low gas pressure causing the burner tray flame to roll out toward the link.

Figure 6-5 diagrams a fusible link. A wax pellet, designed to melt at certain high temperatures, melts and allows the spring to break its normal contact, thus cutting power in the circuit. My early experience was that these links often need to be replaced even when the heater was new or operating normally; however, current engineering has overcome the oversensitivity of these devices. When the fusible link burns out, it pays to examine the other components of the heater or the installation to find the cause.

6-5
Detail of the fusible link.

Improper venting is not the only cause of overheating. Rusted components, improper installation, low gas pressure, and insect or rodent nests more frequently cause the problem, so the fusible link is one more safety device in a component of your pool and spa equipment that needs a lot of safety.

Not all manufacturers included a fusible link in their control circuit when they were first introduced by Teledyne Laars. Today, virtually all heaters use one.

On/off switch

As the name implies, the on/off switch (Fig. 6-2, item 20) is usually a simple, small toggle-type switch on the face of the heater next to the thermostat control. Sometimes, particularly on older Teledyne Laars models, the switch is located on the side of the heater in a separate metal box that also contains the thermostat. In older Raypak models, the switch and thermostat might be located on the side, mounted directly through the sheet metal side of the heater cabinet.

Often the switch is remotely located so the user can switch the unit on and off from a more convenient location than where the equipment itself is located. Manufacturers recommend that a remote on/off switch for a millivolt heater be located no more than 20 to 25 feet from the heater. This is because with less than 0.75 volt passing through the control circuit, any loss of power caused by heat loss from running along extended wiring means that there might not be enough electricity left to power the gas valve when the circuit is completed. This is particularly true on cold days when heat loss is greater.

Also, as the thermocouple wears out and the initial electricity generated decreases, the chance that there won't be enough power becomes very real. Therefore, I suggest from experience that remote switches be located no more than 10 feet from the heater and that they be run through heavily insulated wiring to avoid heat loss. Better yet, run a remote switch off a relay so the control circuit wiring does not have to be extended at all (see the chapter on basic electricity).

If the heater has two thermostats, the switch has three positions: high, off, and low. Why would you have more than one thermostat? Let's say you have a pool and spa, both operated by the same equipment. You can set one thermostat for your desired pool temperature and the other for your desired spa temperature. Then, instead of having to reset the thermostat when you run the spa, you simply flick the switch to read the second thermostat. Many manufacturers now sell only dual thermostat heaters, because it costs supply houses too much to stock heaters of both kinds when the difference is only a few bucks for one additional thermostat.

It is also worth noting that if the switch has been remotely located or duplicated in a remote on/off system, the factory-installed unit might be left in place but not be operative. Factory technicians will place a sticker above such a nonfunctioning switch to alert future users or technicians. However, they are often lost or not used by pool builders or other technicians and as a result, you might be confused when troubleshooting the heater.

Thermostats

Thermostats (Fig. 6-2, item 11), also called temperature controls, fall into two categories: mechanical and electronic. The mechanical thermostat is a rheostat dial connected to a metal tube that ends in a slender metal bulb. The tube is filled with oil and the bulb is inserted in either a wet or dry location where it can sense the temperature of the water coming out of the heater. These thermostats are precisely calibrated, but many installation factors affect the temperature results. In other words, setting the dial at a certain point might result in 80°F water in one pool while the exact same setting might result in 85°F water in another pool. Therefore, pool heater thermostats generally are color-coded around the face of the dial, showing blue at one end for cool and red at the other end for hot. Settings in between are by trial and error to achieve desired results.

Usually, as shipped from the factory, thermostats will not allow water in the pool or spa to exceed 103° to 105°F, although they can be set higher. Also, they do not generally register water cooler than 60°F, so if the water is cooler than that you might turn the thermostat all the way down and the heater will continue to burn. Therefore, the only way to be sure a heater is off is to use the on/off switch (or turn off the gas).

The electronic thermostat uses an electronic temperature sensor that feeds information to a solid-state control board. These are more precise than mechanical types; however, because of the same factors noted previously they are also not given specific temperatures, but rather the cool to hot, blue to red graduated dials for settings. Some manufacturers of spa controls make specifically calibrated digital thermostats, but my experience is that no matter what the readout says, the actual temperature will vary greatly.

High-limit switches
High-limit switches (Fig. 6-2, items 8, 9, and 10) are small, bimetal switches designed to maintain a connection in the circuit as long as their temperature does not exceed a predesigned limit, usually 120° to 150°F. The protection value is similar to the fusible link and often two are installed in the circuit, one after the other, for safety and to keep the heater performing as designed.

The first high-limit switch is usually a 135°F switch, and the other is a 150°F switch. Where the fusible link detects excessive air temperatures, the high-limit switch detects excessive water temperatures. They are mounted in dry wells in the heat exchanger header. Sometimes a third switch, called the *redundant high-limit*, is mounted on the opposite side of the heat exchanger for added safety.

Pressure switch
The pressure switch (Fig. 6-2, items 16 and 45A) is a simple switching device at the end of a hollow metal tube (siphon loop). The tube is connected to the header so that water flows to the switch. If there is inadequate water flow in the header there will not be enough resulting pressure to close the switch. Thus, the circuit will be broken and the heater will shut down.

Although preset by the factory (usually for 2 psi), most pressure switches can be adjusted to compensate for abnormal pressures caused by the heater being located unusually high above or below the water level of the pool or spa.

Automatic gas valve
The automatic gas valve (Fig. 6-2, item 3) is often called the combination gas valve because it combines a separately activated pilot gas valve with a main burner tray gas valve (and sometimes a separate pilot-lighting gas line combined with the pilot gas valve). After the circuit is complete, the electricity activates the main gas valve

which opens, flooding the burner tray. The gas is ignited by the pilot and the heater burns until the control circuit is broken at any point, such as when the desired temperature is reached and the thermostat switch opens, if the on/off switch is turned to off, if the pressure drops (such as when the time clock turns off the pump/motor) and the pressure switch opens and breaks the circuit.

There are several different designs of automatic gas valve, but all have aspects in common. Figure 6-6 shows various valve designs. For millivolt-powered units, the terminal board will have three terminals (or four, where terminals 2 and 3 are connected by a common connection or fusible link). Terminals 1 and 2 are the neutral and electric hot (positive) lines from the pilot generator to start the control circuit. When the circuit is complete, power arrives at terminal 3 (or 4) which opens the main gas valve.

On 25-volt units, there is a pair of terminals to power open the pilot valve and to return the current to the common (or neutral) line of the intermittent ignition device (IID) (detailed later). Another pair of terminals power open the main gas valve and return the current to the common (or neutral) line of the IID.

If you are unsure how a valve should be wired, look for markings near each terminal. They are often marked *PP* meaning powerpile (the pilot generator) or *TP* meaning thermopile (which are the same thing and, just to confuse you, they're also called thermocouple, but I have never seen *TC* on a gas valve terminal); *TH* meaning thermostat or other connection to the control circuit; and *TR* meaning transformer.

The gas plumbing of the automatic gas valve is self-explanatory. The large opening (½ or ¾ inch) on one end, with an arrow pointing inward, is the gas supply from the meter. Note that it has a small screen to filter out impurities in the gas, like rust flakes from the pipe. The hole on the opposite end feeds gas to the main burner. The small threaded opening is for the pilot tube and a similar hole is for testing gas pressure. These are clearly marked. Teledyne Laars heaters employ an additional small tube to assist in lighting the pilot (see visoflame tube).

Automatic gas valves are clearly marked with their electrical specifications, model numbers, and most important, Natural Gas or Propane. Black components or markings usually indicate Propane.

All combination gas valves have on/off knobs. On 25-volt units, the knob is only on or off. With standing pilot units, there is an added position for *pilot* when lighting the pilot. As a positive safety measure in most, you are required to push the knob down while turning.

Honeywell, General Controls, and Robertshaw make most of the combination gas valves, pilot assemblies, and IIDs in use today.

Electronic ignition heaters

When a heater (Fig. 6-7) with electronic ignition is turned on, an electronic spark ignites the pilot which in turn ignites the gas burner tray in the same manner as described previously. In all other respects, these heaters operate the same way as those already discussed. Where the control circuit on the standing pilot heater is powered by millivolts, the electronic ignition heater is controlled by the same kind of circuit but is powered by 25 volts ac (Fig. 6-8). Regular line current (at 120 or 240 volts) is brought into the heater and connected to a transformer that reduces the current to 25 volts.

Square flange for
easy wrench grip

Gas
flow

Coil
connection

TH-TP   TH
       TP

Gas pressure adjustment screw

6-6
The automatic (combination)
gas valve. Teledyne Laars, Moorpark, Ca,
Raypak, Inc., Westlake Village, Ca.

To pilot generator
To limit switches
To thermostat

General Controls gas valve

To pilot
generator

To limit switches

To thermostat

Honeywell gas valve

Stack top

Low profile

Top filler plate
(low profile outdoor)
Drafthood/vent cap.
adapter plate
Flue transition ring
indoor/outdoor
Grate top assembly

Vent cap/stack
(outdoor)

Heat exchanger
baffle retainer

Flue collector

62

42

41

Heat exchanger
support clip

Insulation & retainer

20
13

45
46
46A
47
48
49
50
9

29

34, 37

30
51
38
52
61
56
33
52A
40
53
55
66
10
8
28A
28
16
15
28B
21
28C
26
24
22
4
2
4
Burner manifold
3
28
1
58
64
25

35

36
60
18
19

12

11

14

Header nut, ³/₈" hex
65
43
44
39
57
17

Blind grommet
54
27
23
59
5
6
63

Noncombustible
floor base (optional)

6-7 Detail of a typical electronic ignition heater. Teledyne Laars, Moorpark, Ca.

Key No.
1 Pilot burner electric assembly
2 Pilot tube
3 Ceramic insulator assembly
4 Automatic gas valve
5 Burner w/pilot bracket
6 Burners
7 Burner orifice
8 Gasket, temperature control
9 Temperature control bulb
10 Temperature control (type EPC)
11 High-limit switch (135°F.)
12 High-limit switch (150°F.)
13 Protective sleeve, bulb
14 Wire harness temp. control
15 Thermostat knob
16 Temp-lok
17 Pressure switch
18 High-limit switch retainer clip
19 High-limit switch cover
20 O-ring
21 Transformer
22 Ignition control
23 High voltage lead
24 Fuse pack
25 Fusible link assembly
26 Electrical fuse assembly
27 Wire harness
28 Fusible link bracket
28A On-off switch
28B Plate assembly
28C Rain guard
29 Front water header
30 Rear water header
33 Header gasket

34, 37 Flange packing collar w/copper sleeve
35 Water header flange
36 Flange bolt
38 Clip for tube baffles
40 Drain grommets
41 Drain valve
42 Drain plug
43 Drain valve
44 Bushing, drain valve
45 Brass plug
46 Flow control cap
46A Bolt
47 Flow control gasket
48 Flow control shaft
49 Flow control spring
50 Flow control disc
51 Bolt, front & rear header
52 Heat exchanger
52A Syphon loop
53 Fiberglass blanket
54 Insulation block, side
55 Insulation block front & back
56 Insulation block cover, front & back
57 Insulation block cover side
58 Door
59 Jacket
60 Gap closure
61 Heat exchange baffles
62 Gap closure
63 Burner tray shell
64 Burner tray assembly
65 Rear deflector
66 Lower deflector

6-7 Continued.

This voltage is first routed into an electronic switching device called the IID (intermittent ignition device), which acts as a pathway to and from the control circuit. From here the current follows the same path through the same control circuit switches as described previously.

When the circuit is completed the current returns to the IID, which sends a charge along a special wire to the pilot ignition electrode creating a spark that ignites the pilot flame. The IID simultaneously sends current to the gas valve to open the pilot gas line.

When the pilot is lit, the heat generates a current that is sensed by the IID through the pilot ignition wire. This information allows the IID to open the gas line to the burner tray, which is flooded with gas ignited by the pilot.

Natural versus propane gas

The differences between heaters using natural gas and those using propane gas are nominal. Most manufacturers make propane heaters in standing pilot/millivolt models only. Because of different operating pressures, the gas valve is slightly different (although it looks the same as a natural gas model), as are the pilot light and the

6-8 Electronic ignition heater control circuit (with manual thermostat).
Raypak, Inc., Westlake Village, Ca.

burner tray orifices. The gas valve is clearly labeled *Propane*. The heater case, control circuit, and heat exchanger are all the same as for a natural gas model.

Natural gas is lighter than air and will dissipate somewhat if the burner tray is flooded with gas but not ignited for some reason. Similarly, the odor added to natural gas will be detected if you are working nearby as the gas floats out and upward. Make no mistake, this is still a serious situation and explosions can occur.

With propane, however, the gas is heavier than air and if it floods the burner tray without being ignited it tends to sit on the bottom of the heater. Because it remains undissipated and because you are less likely to smell it because it is not floating out and upward, if it does suddenly ignite it will do so with violent, explosive force. Rarely is the heater itself damaged—the explosion takes the line of least resistance, which is out through the open front panel. Never position your face in front of the opening to try to learn why the heater hasn't fired. Remember to follow your safety checklist, and treat propane with great respect.

ELECTRIC-FUELED HEATERS

I have thus far only discussed gas as a fuel for heating. Electricity is also used in some small heaters, usually for spa applications. Because of the cost of operation, the slower recovery and heating time, and the high amps required with the corresponding heavy wiring and electrical supply, electric heaters are useful only where gas is unavailable. They are also used in small portable spas where gas hookups would be impractical.

Figure 6-9 shows a typical electric spa heater. The components are similar to gas heaters except the heat is derived from an electric coil that is immersed in the water flowing through the unit. This is also true of the small in-line electric heaters used in small spas. Often these in-line units have no control circuits or they might have only a thermostat control because the other controls are built into the spa control panel itself.

Several sizes of electric heater are manufactured, rated by the kilowatts consumed and, therefore, the Btus produced. Here is a comparison of the energy use and output of the most common models:

- 1.5-kilowatt (1500-watt) heater = 5119 Btu
- 5.5-kilowatt (5500-watt) heater = 18,750 Btu
- 11.5-kilowatt (11,500-watt) heater = 37,500 Btu

Each of these generally consume about one-third more power to start up than to run at the designated wattage.

SOLAR-FUELED HEATERS

I could write an entire book dedicated solely to solar heating systems. The pool and spa technician is normally not required to service these systems and believe me, be glad of it. Unfortunately, when the federal government was paying people large tax credits to install solar heating systems, innumerable fly-by-night companies designed, built, and installed them. The problems with these systems are legion—leaking panels or plumbing on the roof, control systems that no longer control, unavailable replacement parts, wiring and sensors installed by someone apparently paid per foot of wire used, etc.

Needless to say, there are many fine systems installed by legitimate companies who are still around to service them. My opinion, as explained below, is "great, let them."

Refer back to Fig. 3-7, which shows a typical solar installation. The concept to understand here is that the water should go through the solar panels before it passes through the heater. In this way, whatever heat can be gained from the sun is obtained first, then the gas heater adds additional heat if desired. Sensors detect if the panels are warm enough to heat the water, and if so, open motorized valves to divert the water to the panels before it gets to the heater. When the panels are cold, the normal flow is to bypass the solar panels and go directly to the heater.

Solar heating systems are controlled by time clocks and/or thermostats because, in summer, the panels might add too much warmth to the water and some means of regulation is needed. Also, they typically have simple on/off toggle switches to completely disable the system.

Solar heating systems are becoming more user friendly. Panels are made of lighter materials than just a few years ago, including some made from doormat-like rubber that simply tacks onto a roof or hillside in large, flexible sheets. Most manufacturers of this technology or traditional panels sell complete kits—systems composed of the necessary panels, controls, and installation instructions. Because there

1 Element
2 Element gasket (not shown)
3 Element tube assembly
4 Sensor well
5 Well retaining clip
6 Pressure switch
7 High limit
8 Contactor
9A Thermostat control (Honeywell)
9B Thermostat control (Sunne)
10 Toggle switch
11 Indicator light
12 Knob
13 Knobstop
14 Dial plate
15 Wire kit (complete)
16 Jacket top
17 Upper front panel
18 Lower front panel

6-9 Detail of a typical electric heater. Raypak, Inc., Westlake Village, Ca.

are so many manufacturers, some of whom will be out of business before this page is printed, I will not discuss solar heating in more detail here. That is also why I leave installation and repair of solar heating systems to companies that specialize in that work. Not that the technology is difficult to comprehend, but you might spend more time trying to find replacement parts than actually making a repair.

Having said that, I can tell you this, however—if you learn the water and related technologies described in this book, any solar heating system you buy will present no great challenge. They use the same concepts outlined for all other pool equipment, including plumbing, water circulation, electronic controls, and safety devices. There's no reason not to make basic repairs to the plumbing or related equipment, but be careful not to invest so much time trying to repair an obsolete control system that you can't recover enough on the job to pay for your time. But if solar is in your customer's plan, check out what is available in your area (very important because you'll want to provide after-sale service and support) and go for it.

HEAT PUMPS

An old technology, used in refrigeration and air conditioning, is employed in the latest craze for pool and spa heating. It is the heat pump. The heat pump does not pump any more heat than any other design of pool and spa heater. Like the others, pool and spa water circulates through the unit (Fig. 6-10) and heat is transferred to that water. Instead of using gas, electricity, or solar heat as a fuel, the heat pump takes warmth out of the air that is created by compressing a gas.

6-10 Detail of a typical heat pump. Heat Siphon® Swimming Pool Heat Pumps, Latrobe, Pa.

As you all remember from high school physics (you don't?), when you compress a gas, it increases in temperature. A compressor in the unit exerts pressure on a gas (usually Freon) and heat is generated (Fig. 6-10, #1). The water is circulated through a heat exchanger (#2) that is warmed by contact with the hot gas. The gas cools (#3) from contact with the water and is recompressed and heated to start the cycle all over again.

By the way, Freon belongs to a family of chemicals you might have heard about—fluorocarbons, a combination of fluorine and carbon. It is a nonflammable, noncorrosive gas, which makes it suited to this application. Some fluorocarbons contain chlorine as well and are called chlorofluorocarbons, which are in part responsible for depletion of ozone in our atmosphere. The Freon used in heat pumps does not contain the chlorine component that makes it environmentally hazardous.

Heat pumps are energy efficient and last a long time. They can repay their high initial cost after years of use, especially where the heater is used regularly. They are not effective spa heaters because they take a long time to do the job. Because they rely in part on taking warmth from the air (#4), the hotter the surrounding (ambient) temperature, the better and quicker they work.

Unlike gas-fueled heaters, heat pumps are rated like air conditioners, expressed in tons. In this rating, a *ton* is the amount of energy required to keep one ton of ice at 32°F for 24 hours. As a rough rule of thumb, one ton equals 15,000 Btu (described in the section on sizing).

OIL-FUELED HEATERS

Less common in residential applications are oil-fueled heaters. These heaters are designed identically to gas-fueled units, but they burn #2 diesel fuel instead of natural or propane gas. Because these are not very common, I will not describe much about their operation, but the plumbing, electrical, control circuits, and many of the actual components of the heater are identical to those of gas-fueled units. So if you encounter one you should have no problem troubleshooting or repairing it if you understand the other lessons in this chapter.

MAKES AND MODELS

Three firms dominate the pool and spa heater market—Raypak, Teledyne Laars, and Hydrotech (formerly Purex)—with products that are remarkably similar. Several smaller companies market spa heaters. A.O. Smith formerly made pool and spa heaters and still provides replacement parts.

You'll learn more about the sizes of heaters and their functions in the next section. For now you need to know that heater models are based on their size as expressed in output of heat (measured in Btus). Each manufacturer produces models of similar size; for example, 50,000, 125,000, 175,000, 250,000, 325,000, and 400,000 Btu. After this range, you enter the realm of commercial heaters.

Because the products, prices, parts availability, and service and warranties are about the same, let me give you some highly personal observations to distinguish the manufacturers. These observations are based on models that are most prevalent at the time of this writing; newer designs might change some of the comments.

Teledyne Laars makes a good reliable unit with almost everything in a convenient place inside the heater for easy service work. I say almost because they design the pilot assembly to be removed without pulling out the entire burner tray, but in practice this is a knuckle-busting near-impossibility (but possible—I have done it many times).

The wiring for the transformer on the Teledyne Laars electronic ignition models is awkwardly located in tight quarters behind the intermittent ignition device. This is only inconvenient when installing the original power hookup and will not be accessed again unless you are replacing the transformer (which does happen, especially since a dead short of a wire to the cabinet will burn out the transformer very quickly).

Raypak makes good heaters also. On their electronic ignition models, the IID is hidden in a small compartment that is accessed by opening the main panel first, then unscrewing two or three upside-down-mounted screws to get into the second compartment. This makes quick diagnosis a longer process when troubleshooting repairs. Recently they have made this more accessible, but there are a lot of the older models still in service.

Raypak burner tray design requires removing the gas burner tray completely to work on the pilot assembly, and I have had to pull it out and turn it upside down to get at tough screws. Replacing the pilot is even tougher because the screws holding it in place are very short.

Hydrotech/Purex heaters have not been as popular as Teledyne Laars and Raypak, but their assembly and quality is comparable. Numerous other companies make heaters in addition to other equipment, but I find something comforting in working on heaters made by companies that focus their attention only on heaters.

In my opinion, A.O. Smith made the best pool and spa heaters ever built. Many are still in service and parts seem to be readily available. They were overbuilt— everything was twice as beefy and high-quality as it needed to be. As a result, they are no longer made. Legend in the pool business has it that they couldn't compete with the prices of the less expensive units on the market today and they refused to compromise their quality, making it no longer financially viable to build their units. Too bad, they were great heaters.

I have one very wealthy customer who routinely spends hundreds of thousands of dollars adding new cars to his million dollar collection, but wouldn't part with his 20-year-old A.O. Smith heater when the frequency of repairs suggested it was indeed time to do so. Instead, he wanted us to keep soldering the old heat exchanger wherever it leaked and keep that baby running his pool at 85°F year-round. He still has it.

A few years ago it was true to say that each manufacturer employed different designs but basically the same components and concepts. In the past five years or so, they have gone off in substantially different engineering directions so that they can no longer be viewed generically. The basic concepts are, however, the same, and Raypak, for example, has recently redesigned its cabinets so that plumbing and gas connections are in the same place as on Teledyne Laars (obviously to be more competitive when replacing an old heater).

Other manufacturers, particularly the many making small spa heaters or in-line portable electric spa heaters are not mentioned here specifically because I have no particular likes or dislikes among them, and many of them will be in or out of business before the ink is dry on this page. As noted elsewhere however, if you understand the concepts of one, you should have no trouble with the others.

Because of frequent changes in engineering and design, what I say here about a particular manufacturer might be true about some models and not others. I have tried to make comments based on the heaters I find most commonly in use today. While it is important to understand the latest technology, it is more important to understand the technology found in the most installations at the present time.

SELECTION

Beyond manufacturer preferences or price, there are two basic parameters to consider in selecting a heater: sizing and cost of operation.

Sizing

Sizing of heaters is fairly easy, particularly if you follow a simple guideline. Your customer will buy a heater once every decade or two, so for the relatively few extra dollars, why not buy the next size larger than you think you need? This is not some conspiracy to give the pool technician a shot at making a few extra bucks profit, but rather common-sense advice from someone who's heard too many customers complain that their pool or spa doesn't heat up fast enough.

Starting out with a heater that is not large enough for the job is the first mistake, which is quickly made worse by other factors. Is the pool located in an extremely windy area which causes rapid cooling of the water? Is the surface area of the pool very large (the greater the surface exposure, the faster the heat loss)? How much water are you heating? What average temperature will you start at and how much temperature rise is needed? All of these factors, and changes over the years in them, will determine how large a heater you need.

Although each manufacturer supplies heating capabilities in their literature, here's some guidelines to help make that determination. One Btu is the amount of heat needed to raise the temperature of 1 pound of water 1°F. There are roughly 8 pounds of water in every gallon.

Let's say your pool is 15 feet by 30 feet, with an average depth of 4 feet. Remember, there are 7.5 gallons in each cubic foot of water (see chapter 1). Let's also say the average temperature of the pool is 70°F and you want to get it to 80°F—a 10°F increase.

$$15 \text{ ft.} \times 30 \text{ ft.} \times 4 \text{ ft.} = 1800 \text{ cubic feet}$$
$$1800 \text{ cu. ft.} \times 7.5 \text{ gal/cu. ft.} = 13{,}500 \text{ gallons}$$
$$13{,}500 \text{ gal.} \times 8 \text{ lb./gal.} = 108{,}000 \text{ pounds of water}$$
$$108{,}000 \text{ lb.} \times 10 = 1{,}080{,}000 \text{ Btu needed}$$

If you are planning to circulate the pool 8 hours per day, and you have just calculated that you need 1,080,000 Btu in the entire day, then your heater output needed each hour is 1,080,000 ÷ 8, or 135,000 Btu per hour. Therefore, you need a heater rated at least 135,000 Btu.

Rated? Here's a simple trap in heater sizing—there is a difference between the input and output ratings of a heater. Obviously some heat is lost up the venting, so you need to know how much is actually going into the water. Look at the heater label or manufacturer's literature to determine each. Typically with gas- or oil-fueled heaters, the output (the amount that actually heats into the water) is 70 to 80 percent of the input.

Just for the record, the American Gas Association (AGA), which sets standards for gas pressure, pipe size, etc., classifies a *heater* as a unit with 70 percent efficiency and a *boiler* as a unit of 80 percent efficiency or more. With electric-fueled heaters almost all of the energy consumed actually heats the water, there is virtually no waste. Heat pumps are also energy efficient, because the energy they use is only the electricity used to run the compressor that heats the Freon.

So the point is that when calculating Btu heating needs, you will arrive at an output amount. When buying the heater, you must add 20 to 30 percent to arrive at the input amount, which is the nominal (or advertised) rating of the heater. In the example, where you wanted 135,000 Btu *output* to your water, you will need to buy at least a 175,000 *input* Btu rated heater.

Before going on I must point out that the example of a 10°F increase might be true in California in summer (or Florida at any time), where the water temperature might average 70°F without help. But when sizing a heater, consider the coldest average temperature you might start with and the hottest average temperature you might want to reach. For example, let's say you want to swim in spring or fall when the average water temperature is 50°F. You might want to heat the pool to 85°F. That's a 35°F increase, far above our 10°F example. Keep overall heater use in mind when selecting a heater.

As I said, estimating heater size is more art than science because of all of the factors that can affect the job. So ballpark estimates are generally good enough because you will want to up-rate your choice anyway. My examination of manufacturers heater sizing charts reveals that there is a constant rate used in the calculations. This method of calculation tells you how many Btus per hour, rather than the method just described, which gives you total Btus in 24 hours. Here it is.

You need 100 Btu every hour for every 10 square feet of pool surface for every 1°F temperature rise desired. So, using the previous example:

$$15 \text{ ft.} \times 30 \text{ ft.} = 450 \text{ square feet}$$
$$450 \text{ sq. ft.} \div 10 \text{ sq. ft.} = 45$$
$$45 \times 100 \text{ Btu}/1°F = 4500 \text{ Btu for each degree rise desired}$$
$$4500 \text{ Btu} \times 10 \text{ degree rise} = 45{,}000 \text{ Btu needed per hour to hold temperature}$$
$$45{,}000 \text{ Btu} \times 24 \text{ hours/day} = 1{,}080{,}000 \text{ Btu total}$$
$$1{,}080{,}000 \text{ Btu} \div 8 \text{ hour circulation per day} = 135{,}000\text{-Btu heater}$$

So either way you get the same answer: that a minimum 135,000 output Btu heater is required, regardless of the heating fuel used. It was noted that about 15,000 Btu equals 1 ton of heat pump rating, so 135,000 ÷ 15,000 = 9-ton heat pump.

Now remember, this calculation shows the heater needed under ideal conditions. One engineer told me that a 10-mph wind would require you to double the heater size to get the same results. Suddenly your 135,000-Btu heater is a 270,000-Btu unit if you are in a consistently windy area. That is what I mean about always going up in size when recommending a heater.

Let's say you already have a heater and you want to know how quickly it will heat your pool or spa. Here's a formula that works, not including any wind or other cooling conditions:

Heater Btu output/(gallons × 8.33) = degrees temperature rise/hour

So going back to the sample pool of 13,500 gallons, let's say you have a 175,000 input Btu heater already installed and want to know its expected abilities. First, you must determine the true Btu output. Using a 75 percent efficiency, that would be

$$175,000 \text{ Btu} \times 0.75 = 131,250 \text{ actual Btu output}$$
$$131,250 \div (13,500 \times 8.33) = 1.16°\text{F/hour}$$

Therefore, in eight hours of circulation you might expect a 9°F rise in temperature (8 hours × 1.16).

To calculate heater size for spas you use the same calculations. But with much less water in a typical spa than in a typical pool, a much smaller heater will do the trick. Or will it?

There might be less water, but if you have a hot date (pun intended) and a cold spa, do you want to wait eight hours to raise the temperature 10°F? Oh, that's right, there's less water, but you want it much hotter and much faster. My spa is 6 feet round and 4 feet deep. What heater do I need to get it from it's usual 70°F to 100°F in 1 hour? (Refer to chapter 1 if you've forgotten how to measure the surface area of a circle.)

$$28 \text{ sq. ft.} \times 4 \text{ ft.} = 112 \text{ cubic feet}$$
$$112 \text{ cu. ft.} \times 7.5 \text{ gal./cu. ft.} \times 8 \text{ lb./gal.} = 6720 \text{ pounds of water}$$
$$6720 \text{ lb.} \times 30°\text{F rise} = 201,600 \text{ Btu total}$$

Remember, heaters are rated in their Btu capacity per one hour, and since I want that spa hot in one hour, then I need at least a 201,600 output Btu heater for this little spa. Many customers don't want to wait an hour either, more like half that, meaning they need twice that size when selecting a heater.

For the record, the American Red Cross suggests pool temperatures of 78° to 82°F and spas not more than 104°F. Most building and safety codes follow these guidelines.

Cost of operation

The cost of operating a heater is simple to figure out if you know what your customer pays for a therm of gas. A *therm*, the unit of measurement you read on the gas bill, is 100,000 Btu/hour of heat. My last gas bill showed I pay about 50 cents per therm. The heater model tells you how many Btus per hour your heater uses. Divide that by 100,000 to tell you how many therms per hour it is. Next, determine how many hours of operation are needed to bring the temperature up to the desired level.

Let's use the example from before. You need at least a 135,000 output Btu heater, but to allow for the other factors noted, you buy a 250,000 input Btu model. You decided it would run eight hours a day, remember? Well, using the facts you have, here is your operating cost:

$$250,000 \text{ Btu} \div 100,000 \text{ Btu/hour} = 2.5 \text{ therms/hour}$$
$$2.5 \text{ therms/hour} \times 8 \text{ hours/day} = 20 \text{ therms/day used to run the heater}$$
$$20 \text{ therms/day} \times 50 \text{ cents/therm} = \$10/\text{day}$$

In the spa example, let's say I use the same 250,000-Btu heater to heat my spa and I figured it would take an hour to do so. Using the same calculations, it costs $1.25 per day to heat the spa. Of course, the heater will run while the spa is in use and the heat loss will be considerable as jets and blowers stir up the water.

Once you have reached your goal temperature, if you want to know what it will cost to keep the spa at that goal you must know how much heat the water is losing. In an hour without the heater, if the spa lost 10°F, you would use the same calculations using a 10°F heat rise to arrive at the cost. Obviously it will not be as much per hour as the original 30-degree rise in one hour.

Someone will ask you one day how much it costs to keep a standing pilot burning. Well, it uses between 1200 and 1800 Btu per hour. Figure it out from there. By the way, the temperature of that little flame is over 1100°F, so when removing a pilot assembly for repair, don't grab one that has recently been lit.

To calculate the true cost of operating your pool or spa however, don't forget the cost of electricity for pumps and motors, blowers, and other appliances. Of course you could also get into the cost of depreciation of the equipment, depletion of chemicals, water use, etc., but then no one would have a pool or spa.

Fuel oil for an oil-fueled heater is sold by the gallon. For comparison, one gallon equals 140,000 Btu. Electricity is sold by the kilowatt-hour (1000 kilowatts consumed in one hour) which equals 3412 Btu.

To measure the exact volume of gas used, you can read the gas utility meter before and after a certain time period when you know the heater is the only gas appliance in operation. (See appendix D. "Meter Reading.")

INSTALLATION, REPAIRS, AND MAINTENANCE

A checkerboard has eight rows of eight squares or 64 squares total and, I am told, over five billion possible combinations of moves. That's in checkers, imagine how many there are in chess. So what?, you say. The point is that heaters have only about 10 main components and 10 basic things that go wrong, but about 5 billion combinations too. That's because there are so many possible combinations of cause and breakdown and so many different types of heaters and manufacturing designs, that to cover all the possibilities is virtually impossible—that's another book.

What I'm saying is that heater troubleshooting is both art and science. After years in the field, most good service technicians can sense what is wrong by the customer's description or by a look at not just the heater, but the overall pool and spa and the equipment area. Since nothing replaces experience in this field, I will limit my comments to the most common heater failures and how to repair them. Beyond that, refer to the reference section to get and absorb as much information as you can from manufacturer's seminars, literature, pool and spa seminars, trade publications, and other service technicians.

I have learned a lot over the years by watching the technicians that come out on factory service calls. If I couldn't fix a heater, I would call the factory on behalf of the customer and make the appointment at a time when I could be there to watch them work. Most of these people really know what they're doing and they have a wealth of knowledge for you to tap.

Having said all that, let's get on with some basic guidelines for any heater repair. My Golden Rule of heater repair is that you always turn the heater off when making repairs. Preferably turn the pump off as well and disconnect or turn off any source of electricity. If you don't, you might complete the repair or touch some wiring together causing the heater to restart when you don't really want it to. By shutting everything down, you control the entire process—you check your work and control the test-firing when you think you're done. Otherwise the heater controls you.

Next, it is generally better to replace components rather than repair them and, if it's a well-worn heater with other parts that will soon give out, replace the entire heater instead of adding new parts every month ad infinitum. Associated with this point is the fact that each part in a heater fails for a reason, rarely old age. When you replace or repair something, find out why it failed in the first place or it will happen again. I have been told that I was a genius when I got a heater running again and told what an idiot I was when I had to come back two weeks later for the same repair.

Last, most heater repairs are not the heater at all. The Teledyne Laars service manager once told me something I too have observed—the majority of heater failures are the result of dirty filters (and add to that low or obstructed water flow from the pool). In short, nothing to do with the heater at all. Moral of the story? Look around at the entire installation before starting on what has been called in as a heater problem.

One other small point that should be obvious but perhaps isn't. If the heater has been running prior to any repair you are making, watch out for hot components. As noted previously, pilots generate over 1100°F, so they stay hot for a long time after they've gone off. Cabinets and other metal parts get hot too, so watch what you grab.

Installation

Before worrying about how to fix a heater, you will want to know how to install one. A lot of heater failures result from improper installation, so you need to understand what's proper in the first place. Any heater installation is composed of four basic steps: location (including ventilation), plumbing, gas, and electrical.

Location

Hot air rises. Very hot air rises very quickly in large volumes, requiring replacement by adjacent cooler air. Burning fossil fuel, such as gas or oil, results in by-products such as carbon monoxide, which is deadly. These simple concepts are at the heart of your decision about where to locate a heater.

Most residential heaters are designed to be installed along with the pump, motor, and filter at an outdoor or indoor location. When purchasing the heater, you ask for a *stackless* heater (Fig. 6-2 shows both the stack and stackless tops) for an outdoor installation, meaning there will be a draft hood on top of the heater, but no additional vent pipe or stack to remove excess heat and the products of combustion (carbon monoxide).

If your installation is to be indoors, ask for a *stack* heater, which comes with a vent hood for attachment to the stack pipe. Heaters used to come with no top and you would be given one designed for your setup when you bought the heater. Now, most manufacturer's tops are adjustable to either indoor or outdoor installation. Never use a stackless top on an indoor installation. Burning gas produces carbon monoxide, and even if the heater is small and the indoor location large and well ventilated, this is a deadly gas.

Take the heater out of the box and look for instructions. **Read them**. I have installed over 500 heaters, but constantly changing designs mean I still have to read those instructions before working with any new unit (okay, maybe I only skim the booklet looking for new information, but I can skim now because I have read so many of these booklets cover to cover).

Open the front panel on the heater. Usually packed in here is a plastic bag with the hardware you will need. Set the heater near the filter, but remember you will need to do future service work on both the filter and heater, so leave enough space to work. Other spacing guidelines are provided by the American National Standards Institute (ANSI) and are designed to keep carbon monoxide from entering the closed living spaces of your home and to allow enough draft area around the heater. Remember, it needs a large supply of air, which it will heat, and enough distance from flammable walls or fences so that the hot air and hot metal cabinet don't burn the house down.

Refer to Fig. 6-11 for clearances, but here are the highlights.

- The rear and nonplumbed sides must have at least 6 inches of clearance (12 inches is better); the plumbed side 18 inches.
- The front must have 24 inches of clearance. If that is all you allow, you might be safe with ANSI, but the next service technician who comes to work on the unit will curse you. To effectively work on a heater, which requires squatting and laying in front of it and sometimes pulling out the burner tray, I would allow at least 4 feet. Obviously on a small heater you might allow less, but I would say never less than the depth of the heater cabinet (measured front to back).
- Leave at least 3 feet from the top of the heater to the underside of any overhang and at least 5 feet below any window (or 4 feet away from such a window, door, or vent). Again, I think ANSI is shy about this. For example, trees that overhang by only 3 feet above a large residential heater that operates for long periods will dry and burn. I think this clearance should be at least 6 feet.

The rest of the clearances, including vent stack requirements, are shown in the diagram. The stack pipe should have no 90-degree angles, and the stack itself should not be altered. The vent stack is an integral of the design that each manufacturer has engineered for compensation between indoor and outdoor atmospheric pressure differentials.

Set the heater on a solid, level, noncombustible base. Although heat rises, the metal cabinet will get quite hot underneath as well, so a concrete slab over a wood floor is insufficient protection. Heaters have their own feet or runners to hold them off the surface, but even these get warm and would not be compatible with a wooden floor or wall to wall carpet. The best flooring is a concrete pad, hopefully the pad all of your equipment is resting on.

Alternatively, you might use a precast concrete slab available at your supply house or an air-conditioning supply house. These come in slabs of 3 or 4 feet square by 1½ to 2 inches thick. All you need do is level out the ground, lay down the slab, and add the heater.

The last choice would be brick or cinder block, with the holes lined up for ventilation, with a piece of sheet metal over the top to keep the heat from cracking them. In any event, refer to local building codes for what is required in your area.

6-11 Indoor heater installation. Teledyne Laars, Moorpark, Ca. Raypak, Inc., Westlake Village, Ca.

Reversible water connection

Most heaters are right-handed; that is, as you face the heater, the plumbing is on the right side of the unit. What if your installation would be more convenient if the plumbing were on the left? Good question.

Answer—just reverse the heat exchanger on the heater so the plumbing (and gas and electrical) connections are on the left side of the cabinet. Each make of heater is slightly different, but if you disassemble one, the others will be easy to figure out. Each heater comes with a booklet that describes the exact manufacturer's procedure for reversal. It will take you no more than 30 minutes the first time you try it and about 15 minutes each time thereafter.

Flipping the heat exchanger from right to left is simple and it will help you understand many important components of the heater at the same time. Flipping is actually a misnomer; you don't actually turn the heat exchanger over, you turn it around.

1. As with any disassembly, make careful note of how you remove components so you can put them back together the same way. In this procedure, it is especially true of the wiring. Note how it is routed to keep it away from heat or sharp metal edges. Refer to Fig. 6-2 to identify the components discussed in this procedure.

2. Remove the front faceplate (access door) of the heater. Remove the vent top (unscrew the retaining screws) and flue collectors underneath (simply lift straight up and off). Remove the faceplates around the plumbing header (unscrew the retaining screws) and the corresponding plate on the left side of the heater.

3. Remove the thermostat tube or sensor from the drywell in the header (some lift out, some are held in place by a single screw and small bracket). Disconnect the wires to the high-limit switches. Unscrew and remove the tube feeding water to the pressure switch.

4. Remove the heat exchanger drain plugs or air relief valves. Most designs have these passing through the heater cabinet, so they must be removed to free the exchanger unit. Lift out the heat exchanger unit and rotate it so the plumbing connections are on the left side of the cabinet. Again, remember you are not flipping the unit over like a pancake. You simply rotate the unit 180 degrees.

5. While the exchanger unit is out, you have a great opportunity to examine the interior of the heater. Replace any broken firebrick, and examine the inside of the cabinet and top of the burner tray for soot (if this is not a new heater). You will see now how simple the heater really is.

6. Replace the components in the reverse order of their removal. The pressure switch will need to be repositioned on the left side of the heater so the water tube feeding it will reach. Reroute the wiring as needed for the pressure switch, high-limit switches, thermostat sensor (or bulb and tube). Make sure that wiring is not routed over hot or sharp metal surfaces or edges. The only other difference in assembly from right to left is that the faceplates that were on the right side of the heater (around the plumbing connections) now move to the left, while the plates from the left move to cover the right side. Hole patterns are drilled by the factory to accommodate this right/left switch.

I mention reversible connections at this point because before you plan your installation, it pays to know that you can hook up from either side and, of course, you would do it now before setting up the heater, venting, etc.

Venting

As noted, outdoor installations require no additional venting, they are vented sufficiently within their stackless tops. In windy areas, however, you might want to consult the manufacturer's recommendations and add a short stack and cap (about 3 feet total) to cut down on excess drafting. Refer to each manufacturer's specifications on this.

For indoor installations, follow the guidelines in the heater booklet. Working with stack pipe (made of flexible sheet aluminum), which you get at your supply house, is no tougher than working with PVC, except there is no glue, they just snap together. Vent pipe comes in lengths of 2 to 6 feet and diameters of 3 to 12 inches. You also buy preformed 45- and 90-degree elbows and connector sleeves just like PVC plumbing.

Going through walls or a roof to get outside, however, is trickier because you need to use doublewall pipe, flashing, and waterproofing to make the passage safe and rainproof. Some local codes require a permit and a licensed building contractor to handle this part. If your indoor installation is a replacement of a heater that otherwise worked well (and your replacement is the same Btu output), you really only need to connect your new heater venting to the old vent that goes up and out of the building. But if it is a new vent stack that means cutting walls and a roof, I'd hire a contractor to handle that. It is well worth the few extra bucks.

In either case, it is your responsibility to make sure the clearances meet the ANSI code so the heater works properly. Familiarize yourself with the guidelines in Fig. 6-11 and supervise the contractor's work. One simple way to perform a rough test of the venting is to light a match and hold it under the draft hood—the smoke should be drawn up into the vent system and out of the building.

Some howling, whistling, or other ventilation harmonic noise is normal and acceptable. As mentioned, a very large volume of air is rushing up through the heater and as it passes over vent fins it might howl. Notice what is normal for the heater and then compare that to any future noises. Changes in the sound might denote problems.

A few minutes after the heater fires, a knocking noise that might actually rock the heater is not normal. This denotes overheating for some reason and is caused by the superheated water expanding and trying to escape. Some high-pitched whining is caused by debris in the gas line. It should be disassembled, cleaned, and reassembled.

Plumbing connections

Modern heaters offer basically two kinds of plumbing connections: compression fittings and threaded. No other aspect of installation could be easier, as there is one pipe in and one out. The manifold (the end where the plumbing enters the heat exchanger) is set up for 1½- or 2-inch pipe, sometimes allowing either.

Compression fittings The heater comes with two flange gaskets (Fig. 6-2, items 26 and 29; the flange is item 27). Very simply, the pipe slides through the flange and gasket and into the header as far as it will travel (Fig. 6-12). The flange is then bolted to the header, in the process applying pressure to the gasket which tightens around the pipe, sealing it.

As noted, some heaters supply gaskets for 1½- and 2-inch pipe, your choice, while larger heaters might require 2-inch pipe only.

Threaded fittings To screw a threaded pipe directly into the flange, you first assemble the flange gaskets with the plastic or metal sleeves provided. The sleeves act like the pipe in the compression fitting (otherwise when you bolt down the flange, the gasket will collapse and leak or restrict water flow). With the gaskets in place, bolt the flange to the header (Fig. 6-12). The flange is female threaded, usually for 2-inch pipe, so you can directly thread fittings or threaded pipe into the flange.

Use of plastic pipe
(where permitted by code)

6"

Check valve

24"

Manual bypass valve

Plastic pipe

Note: Metal pipe to act as heat sink. When joining metal and PVC, use metal male/ PVC female connection.

Plastic sleeve

Method for installing threaded pipe with plastic sleeve

For nonthreaded pipe or tube, discard plastic sleeve

6-12 Heater plumbing installation. Teledyne Laars, Moorpark, Ca.

With either compression or threaded fittings you must provide the heater with a means of dissipating the heat as the water comes out of the header. Your supply house will carry *heat sink* pipes as one method of accomplishing this.

Heat sink pipes (also called risers) are stainless steel pipes of 1½ or 2 inches in diameter designed to be attached to the heater with compression flange gaskets as described. They are available in lengths of 4 to 8 feet. Obviously, an 8-foot pipe coming straight out of the heater would make plumbing difficult in small spaces, so these pipes are curved or looped so that the entire assembly will fit into 1 foot of space adjacent to the heater. The other end is fitted with a threaded union so you can glue PVC pipe directly to the riser and continue plumbing.

Another method of heat dissipation is to screw threaded galvanized pipe directly into the flange. On heaters up to 175,000 Btu, I have found this to be adequate. For easier plumbing, I usually use a 12-inch galvanized nipple (a 12-inch galvanized pipe that is male threaded at both ends) on the outflow side, which is hotter, and a 9-inch nipple on the inflow side of the flange. I use pipe dope on these installations because I have found that the heat shrinks the Teflon tape over time and allow leaks. Again, after the nipples, you can continue with PVC fittings and plumbing.

The last type of plumbing for direct attachment to the heater is CPVC. This is a type of PVC pipe designed to withstand high temperatures and it can be attached directly to the heater with either the compression or threaded methods just described. Not all local codes permit use of CPVC for this purpose and not all heaters are designed for it either, so check with your local building code and manufacturer's requirements before using it. CPVC will glue together with PVC, so once you are away from the heater you can continue plumbing with PVC.

Of course, for die-hard plumbers, you can also plumb copper pipe directly to the heater, using either compression or threaded fittings, and continue plumbing in copper back to the pool. I am not a fan of copper plumbing because of the expense, difficulty of working with it, and, as you will see in the chemistry chapter, if the water becomes corrosive, you have more materials to replace and more copper on the walls of your pool.

Remember, a check valve should be installed between the filter and heater so that when the pump shuts down, the extremely hot water does not backflow into the filter and damage the plastic grids or cartridge fabric. If your system includes a chlorinator (which is always installed in the plumbing after the heater), a check valve between the heater and chlorinator is also a good idea so that when the pump shuts down there cannot be backflow of the highly chlorinated water into the heater where it might corrode the soft copper of the heat exchanger.

Suction lines for automatic pool cleaner pumps should also be installed in the plumbing after the heater. To put them before the heater could result in too much water being pulled away from the plumbing before it enters the heater, resulting in overheating or low-pressure shutdown of the heater.

Heaters require a minimum water flow to operate efficiently. Sometimes, however, the system provides a flow that is too fast. If the flow is more than 125 gpm, a manual bypass should be installed (as shown in Fig. 6-12). This is a simple gate valve that allows some of the flow to bypass the heater.

Some codes require that a pressure safety valve be installed on the heater, just like on your household water heater. On the front water header where the plumbing is attached, you will find a ¾-inch threaded plug that can be replaced with a ¾-inch pressure relief valve available at your supply house.

Gas connections

When replacing a heater, you will generally hook the new heater up to the old gas line and that is that. It is important to understand the factors that influence original installation, however, so you can be sure that the old supply is adequate. Also, you might be making a new installation or moving the heater to a new location and need this information.

Gas pressure is measured in inches of water column pressure (WCP). Simply speaking, 28 inches WCP is equal to 1 psi. Natural gas-fueled heaters require about 8 to 10 inches WCP and propane requires 14 inches or more. Figure 6-13 describes testing for gas pressures of either type at the supply pipe and inside the heater itself.

Testing mainline gas pressure	Testing manifold-regulated gas pressure

1. Attach slack tube manometer to heater jacket.
2. Open both valves on manometer.
3. Shut off gas to heater by using shut-off cock ahead of heater controls.
4. Remove $\frac{1}{8}$" NPT test plug in upstream shut-off valve as illustrated.
5. Screw in $\frac{1}{8}$" NPT fitting from manometer kit. Attach manometer hose to fitting and to one of the manometer valves.
6. Open all gas valves, light pilot and bring on main gas burners.
7. Mainline gas pressure will register on the manometer. With burners on, readings should be as follows:
   5" to 10" WC – Natural gas
  11" to 14" WC – LP gas
	1. Attach slack tube manometer to heater jacket.
	2. Open both valves on manometer.
	3. Shut off manual main gas valve.
	4. Remove $\frac{1}{8}$" NPT plug on valve outlet face and screw in $\frac{1}{8}$" fitting from manometer kit. Connect manometer hose to fitting and to one of the manometer valves.
	5. Wait five minutes. Relight pilot as instructed on rating plate and bring on main burners.
	6. Manometer should register as follows:
	  4" WC – Natural gas
	  9" WC – LP gas
	7. To adjust gas pressure: Remove regulator cap screw on top of valve marked "Reg. Adj.". Turn screw adjustment clockwise to increase or counterclockwise to decrease gas pressure.

6-13 Testing gas pressure. Teledyne Laars, Moorpark, Ca.

Heaters are engineered to operate at altitudes of 2000 feet or less, and modifications are required to compensate for the changing density of gas and oxygen at higher altitudes. Manufacturers provide specifications and instructions for such modifications.

Gas plumbing must follow rigid guidelines because a gas leak is a greater hazard than a water leak. Having said that, the procedures are the same as described in the chapter on basic plumbing and require no special skills or tools, just lots of care.

Each manufacturer specifies the required size of pipe to ensure an adequate supply of gas, depending on the size of the heater and the distance from the source of the gas (natural gas meter or propane tank). Generally, these will be ¾- to 1½-inch diameter pipes.

Gas plumbing can be steel or special PVC pipe. Some building codes permit special PVC gas pipe for underground runs (a heavy-duty PVC pipe material that is green-tinted to denote gas). It can be argued that especially with underground pipe, PVC might be better because it doesn't rust. The disadvantage is that anyone digging in the yard can more easily rupture a PVC pipe than a metal one. If you do run PVC gas pipe, it must be accompanied in the ground by a 16-gauge tracer wire so the gas line can later be found by a metal detector if needed. Most codes also require that PVC gas lines be buried at least 18 inches below ground (12 inches under concrete). Finally, any risers (lines coming up out of the ground) or lines above ground must be metal, so you must transfer from PVC to metal before leaving the trench.

Metal gas pipe must be 12 inches below the ground or 6 inches below concrete. Metal gas pipe is painted green (to denote gas) and any underground portions must be wrapped with waterproof 10-mil thick gas tape to at least 6 inches above the ground. Lines should be run as close as possible to the meter at one end and the heater at the other. Flexible gas lines are not permitted in any building code I know of.

Gas hookups must include a shutoff valve (Fig. 6-14) just before the heater, followed by a sediment trap (just a drop of pipe off of a T that can collect impurities in the gas before it gets to the heater), followed by a threaded union. When you need to service the heater, you can close the valve, break the union, and remove any parts of the system without shutting off the gas to the entire house. Similarly, if a leak or other emergency develops in the heater you can quickly shut off the supply of gas in a convenient location. Because of this, I recommend an easy shutoff valve with a handle rather than a style that uses a nipple on the valve that requires a wrench to close.

6-14
Gas connection plumbing. Raypak, Inc., Westlake Village, Ca.

Be sure to check the ability of the gas meter to supply the required amount of gas. Many are rated at no more than 175,000 Btu, which the gas company will tell you actually delivers twice that amount. Even so, twice 175,000 is only 350,000 Btu, not enough for a 400,000-Btu heater or a smaller one when your house is simultaneously using a gas clothes dryer, gas water heater, gas stove, etc.

Gas lines must be pressure tested. Building codes determine how much pressure the line must hold and special gauges are available at the supply house, but I suggest hiring a licensed plumber to conduct this test for you. They have the tools and knowledge to conduct this important safety check. When putting a heater into service or when connecting a replacement heater to existing lines, you can conduct a simple test of your own.

Turn on the gas supply to the heater and wipe liquid soap over any joints or unions (ammonia in a spray bottle also works well). Escaping gas makes the soap or ammonia bubble so you can detect any leaks quickly. Never leak test with a match—you might end up in the next county.

Electrical connections

If the heater has an electronic ignition, you must supply the electricity. Wire should be 14-gauge copper, run in it's own waterproof conduit (not shared with wiring for other purposes). For either 120- or 240-volt supply, run three wires. One wire, the ground wire, is green. With a 120-volt supply, the other two wires should be black or red (hot line) and white (neutral). With a 240-volt supply, both wires might be black or red.

Electronic ignition heaters are designed to operate on either 120 or 240 volts. The heater control circuit works on 25 volts regardless of the voltage supplied. Carefully study the wiring diagram provided by the manufacturer to make sure you are connecting 120 volts to 120-volt connections or 240 volts to 240-volt connections. If you set up the heater's transformer wires to 120 volts and supply 240 volts, you will be eating fried transformer.

Where should the electrical supply come from? I prefer to run the wiring from the system time clock so that the heater can't operate unless power is also being supplied to the pump and motor. This doesn't guarantee that the heater will only come on when water is flowing, but it is one more safety measure. The ground should be attached to the ground lug or bar in the clock and to the ground lug inside the heater cabinet (you'll see the lug near the transformer wiring, usually with a big sticker that says Ground).

As you will see in the chapter on time clocks, there is a line side and a load side. Always connect the heater (and pump and motor) to the load side, otherwise the wiring to the heater will always be hot.

Some manufacturers suggest including a *fireman's switch*. This switch is mounted in the time clock and is designed to turn the heater off about 20 minutes before the time clock turns off the pump. The idea is that by running cool water through the heater for 20 minutes, you cool the heater components and extend their life. Most time clocks have predrilled holes for standard fireman's switch installation (and directions are provided in the switch package).

Teledyne Laars recommends a fireman's switch. Raypak advises that their *unatherm governor* keeps the internal temperatures between 105° and 110°F so that a fireman's switch is not needed. Also, they say that their firebrick dissipates heat sufficiently so that there is not a dangerous heat buildup in the cabinet. In short, follow each manufacturer's suggestions.

Electric-fueled heaters need heavier gauge wire because they consume large amounts of electricity as the heating fuel. Follow the manufacturer's guidelines for these installations or bring in an electrician to make the hookup.

Basic start-up guide

When you first fire up your heater, a few pointers are worth noting.

1. If the gas supply line is new, bleed the air out of the pipe by opening the line at the union near the heater and opening the shutoff valve. When you smell gas instead of air coming out of the pipe, shut off the valve and reconnect the union. There is still air in the remaining foot or two of pipe and in the combination gas valve inside the heater, but this will bleed quickly if gas is present in the line up to the shutoff valve. Open the gas shutoff valve.
2. Make sure the heater on/off switch is off. Turn on the pump and motor and make sure the air is out of the water system. Check for leaks.
3. Light the pilot (follow the instructions that follow or those provided with the heater) if the unit is a standing pilot type. It might take 60 to 90 seconds to get a flow of gas in the pilot tube as the air bleeds out and is replaced by gas. Never try to light an electronic ignition pilot with a match or other fire source. Believe it or not, I have gone on heater service calls where other service technicians have left the evidence of their stupidity in the heater— burnt matches or burnt paper. After lighting the pilot, turn on the on/off switch and turn up the thermostat as needed to fire the heater.

If it is an electronic ignition unit, turn the valve on the combination gas valve inside the heater to on. Turn on the on/off switch and turn up the thermostat as needed to fire up the heater. It might take a minute or two to bleed the air out of the system and replace it with gas, so don't worry if it seems to take awhile for the pilot to ignite, and the unit to fire the first time.

4. When the heater first fires, the heat will burn off the oil that is applied by the factory to the heat exchanger as a rust preventive. Light smoke for a few minutes is normal. Also normal is some moisture condensation as very cold water runs into the very hot heat exchanger. The condensation will drip down onto the burner tray and sizzle—a little of this is normal too.

5. Observe the heater for the first 10 minutes. Make sure the smoke and condensation stop and that there are no leaks. Use your eyes and nose (gas leaks?). Turn the heater off and on a few times to be sure it operates properly. Do this from the on/off switch and the thermostat a few times. Remember, if the water is colder than about 65°F, the thermostat won't turn the unit off because it doesn't register that low. Now turn the pump off. The heater should shut off within 5 seconds. If it doesn't, the pressure switch needs adjustment.

Use common sense. As the heater is working, look around and make sure you have allowed sufficient ventilation and air supply and that any overhanging building areas or trees are far enough away to stay cool.

6. With the pump running, touch the inlet pipe to the heater and then the outlet. The temperature differential should not be more than 10°F. If it is much more than that, refer to the discussion of bypass valves that follows.

Terry's troubleshooting tips

This is the only section of the book I name after myself, because after years of troubleshooting heaters I have developed a method that seems to work for me. I think it will work for you too.

To initiate repairs, you first need to understand how to identify the problem. I suggest you read over the entire repair section to familiarize yourself with component operation and repair, then the troubleshooting tips will make more sense.

While troubleshooting heaters is fairly simple, remember that symptoms are similar for several different problems, so you need to be sure you have isolated the correct malfunction and also be sure there's not more than one malfunction. As mentioned previously, there are so many potential combinations of causes of heater malfunction that you must look at the entire system and installation and be flexible in your thinking.

What follows is by no means comprehensive, but is the way I approach heater problems. This outline will get you through most of the main failures and their causes so that at the least, even if you can't fix the problem yourself, you will be able to communicate intelligently with the factory service technician.

If you do end up calling the factory service office, be prepared to give them a concise description of the problem, the model and serial number of the heater, and the date it went into service if you know. Warranties differ between manufacturers and even within one heater some components are guaranteed for one year, others for five. In the reference sources section at the back of the book you will find the phone numbers of major heater makers for technical help over the phone or factory service (and training seminars) in your area.

I think the simplest way to approach this section is to provide you with my checklist, then to explain any tests or repairs that are not self-explanatory in more detail at the end. This list deals mainly with gas-fueled heaters, because these are the most complex and common in the field. Some of the technology and diagnostic tips apply to any heater, however. So, whatever the problem reported, the steps I follow are in about the same order. Think like a doctor—whatever the patient reports to be wrong, the doctor gives the entire body an exam to see if there are interrelated disorders.

With heaters, there are three basic complaints: the heater won't come on, the heater won't stay on long enough, or the heater won't shut off. Regardless, it pays to check out the entire flow of water, gas, and electricity through the heater because, as mentioned before, one problem might be a symptom of a larger problem. In each area, use your eyes, ears, and nose—look for leaks or soot, listen for odd sounds, and smell for burning components or leaking gas.

Here's how I approach the checkup.

Terry's troubleshooting checklist

1. Check the water system.
   - A. Is the pump primed and running without interruption?
     ~Is there enough water in the pool or spa?
     ~Has the air been purged from the system?
   - B. Is the water flow too low?
     ~Are the skimmer and main drain clear?
     ~Are the pump strainer pot and impeller clear?
     ~Is the filter clean?
     ~Are all valves open?
   - C. Is the water flow too great?
     ~Is the internal flow control valve operating?
     ~Are the in and out water temperatures within 10°F?
     ~Is the external bypass valve set correctly?
   - D. Is the water chemistry correct?
     ~High pH could mean scale in the heater.
     ~Low pH could be causing leaks.
   - E. Are there any visible leaks?
     ~At the exterior plumbing connections?
     ~At the interior heat exchanger components?
2. Check the gas system.
   - A. Is gas getting to the heater?
     ~Look for the pilot flame.
     ~Conduct the open gas line test at the heater.
     ~Is the propane tank full?
     ~Are the gas supply lines adequate and unobstructed?
     ~Is the supply dedicated (only used for the heater)?
     ~Is the heater large enough for demand and expectation?
     ~Is the combination gas valve turned to on?

- B. If the pilot and/or burner are working. . .
  ~Is the flame 2 to 4 inches and steady?
  ~Is the flame a healthy blue?
  ~Are all the burners fully lit?
  ~Is pilot ignition within 5 seconds?
  ~Is burner ignition within 10 seconds after pilot?
  ~Is the tray ignition without flash or a loud boom?
- C. Is the ventilation adequate?
  ~Is there adequate air supply to the heater?
  ~Is there adequate venting away from the heater?
- D. Smell for leaks.
  ~Sniff around the outside connections and unions.
  ~Sniff around the combination gas valve and joints.

3. Check the electrical system.
- A. Is the on/off switch on? Is the remote switch (if any) on?
- B. Is the thermostat turned up high enough?
- C. Is 25 volts coming out of the transformer?
  ~Is 120/240 volts coming into the transformer?
  ~Is the heater grounded properly?
  ~Are all the connections tight and clean?
  ~Is the reset breaker button tripped (electric-fueled heaters)?
- D. Check the pilot.
  ~Light the standing pilot and look for a healthy flame; or
  ~Electronic ignition has fired the pilot and there is a healthy flame.
  ~Is the pilot clear of rust, dampness, or insects?
- E. Check the control circuit. Follow the path of the electricity.
  ~Is there power from the pilot generator or transformer?
  ~Is there power to each switch. . .
    -On/off?
    -Thermostat?
    -Fusible link?
    -Pressure switch?
    -High-limit (2)?
    -In-line fuse (some models)?
    -Fireman's switch?
- F. Is there power to the gas valve?

Repairs

Now that I have reviewed the most common problems that can occur, let's turn to more detailed troubleshooting and repair. For simplicity, I will go down the list in the same order as the troubleshooting checklist. Items that are self-explanatory or which are detailed in another chapter are not repeated here.

Is the water flow adequate?

If you can answer yes to all of the water flow questions posed in the checklist, yet the flow of water is still too low, the problem might be inside the heat exchanger.

Over time and with improper water chemistry, scale (calcium, liming) can build up inside the tubes of the heat exchanger slowly closing the passages, somewhat like fat closing the arteries in your body. To correct this, proceed as follows (Fig. 6-15):

400-500 rpm drill motor

Be sure tubes are dry to make reaming easy.

City water pressure

Note: When pressure relief valve is used it must be removed and hole plugged as shown. Important—remove the plug before reinstalling heat exchanger.

6-15 Heat exchanger reaming (cleaning). Teledyne Laars, Moorpark, Ca.

1. Make a visual inspection to verify the problem. On one or both sides of the heater, depending on the make, you will find heat exchanger drain plugs. Some are fitted with air relief valves similar to those on a filter. Remove the drain plugs and make a visual inspection inside. Scale formation will be obvious.
2. Remove the heat exchanger assembly from the heater as described in the installation section. Remove the headers from the exchanger by removing the bolts on the top and bottom at each end.
3. The tubes of the exchanger can be cleaned out with a special reaming tool available at your supply house. The reamer is attached to your power drill. Withdraw the reamer frequently to clear lime deposits from the tool and prevent binding. For best results, work with a dry heat exchanger. When finished, flush each tube with water to be sure it is thoroughly cleaned.
4. Install new gaskets and replace the headers on the heat exchanger. Tighten the bolts from the center first, working toward the outside, alternately tightening bolts on top and bottom to ensure even compression of the gaskets and a solid seal. Working the reverse of the disassembly procedure, reinstall the exchanger and the components removed.

Is the internal flow control valve operating?

As noted, if the temperature differential between the incoming and outgoing pipes exceeds 10°F, the internal flow control is not operating correctly. I have also seen water flow blocked when the plunger of one of these units has broken apart and lodged in the header. Age and heat can destroy the stem or plunger on these valves.

Figure 6-16 shows the simple disassembly of these units for inspection. The Raypak Unatherm Governor acts like your car thermostat, opening and closing in response to a specific temperature. The spring-loaded type responds to water pressure. In either case, unless the heater is several years old, these units should not appear to be disintegrating or failing. Failure indicates another problem—poor water chemistry, unit overheating from some other cause, or scale buildup. Check further.

6-16 Flow control valves. Raypak, Inc., Westlake Village, Ca. Teledyne Laars, Moorpark, Ca.

Is the water chemistry correct?

Improper water chemistry can cause heater problems. Refer to the chemistry chapter for more details; however, heater-friendly water should be

- pH 7.4 to 7.8.
- Alkalinity 120 to 150 ppm.
- Total dissolved solids of no more than 2500 ppm (pools) or 1500 ppm (spas).

Is gas getting to the heater?

If this is not apparent (from a working pilot, for example), turn off the gas valve and open the union near the heater. Make sure there are no open flames or sparks nearby that might ignite the gas, then turn the gas valve back on and listen for gas flow (you will hear the hissing before you will smell the gas). If there is no flow, go to the source—the main meter or propane tank. Notice if the gas supply pipe to the heater is dedicated to the heater alone, not also servicing a nearby gas barbecue or household water heater.

Lighting (or relighting) the pilot

This procedure applies to standing pilot units only, because the electronic ignition is automatic unless something is broken. Instructions are almost always printed on the heater itself. Look for these and follow them. If the directions for your particular heater are obscured or missing, the following procedure is most common:

1. Turn the gas valve control to Off and wait five minutes for the gas in the burner tray or around the pilot to dissipate (safety consideration). Turn the on/off switch to Off.
2. Turn the gas valve control to Pilot and depress. If the area is quiet, you should hear a strong hissing sound as the gas escapes from the end of the pilot. If not, it might be clogged by rust or insects.
3. Light the pilot and continue to depress the control. If the heater is equipped with a pilot lighting tube (Raypak), you should be able to hold your match in front of the tube and ignite the gas from the pilot (Fig. 6-17). Check to see that it is clear of obstructions.

Depress gas control knob

Visoflame tube

Pilot lighting tube

6-17 Pilot lighting systems. Teledyne Laars, Moorpark, Ca. Raypak, Inc., Westlake Village, Ca.

If the heater is equipped with a visoflame tube (Teledyne), you should be able to hold your match over the mini-burner of this tube which ignites its own gas supply and in turn ignites the gas supplied to the pilot (Fig. 6-17). Visoflame tubes sometimes flare up because they have their own gas supply, so keep your hands and face as far back as practicable and light the tube promptly after depressing the control.

4. Hold the control down for at least 60 seconds. This allows heat from the pilot to generate electricity in the thermocouple to power the gas valve (which will then electronically hold the pilot gas valve open and power the control circuit which ultimately opens the main burner gas valve). Release the control and the pilot should remain lit.

5. Verify—this step is not usually included in instructions printed on the heater—that the pilot flame is healthy; that is, look to see that a strong-burning blue flame of 2 to 3 inches extends toward the burner tray and a secondary flame of equal value is heating the thermocouple. Make sure the pilot is securely in place; it might be burning properly, but rusted fasteners or poor installation from a previous repair might have left it dangling near the burner tray, but not in close contact with it.

As mentioned previously, burner tray design leaves a lot to be desired when servicing the pilot assembly of most heaters. A common problem is that a lazy service technician previously left the pilot assembly hanging in the general area by wire, tape, a screw not fully in place, etc. The result is that the pilot burns just fine, but the burner tray needs to be flooded with excessive amounts of gas before any reaches the pilot flame to be ignited. The resulting ignition is explosive. Therefore, always make this visual inspection.

6. Turn the gas valve control to On. Stand back from the heater (and to one side—not in front of the open front of the heater) in case of *flashback* (the explosive ignition described previously). Make sure the pump is operating and water is flowing freely, turn the on/off switch to On, and turn the thermostat up. The heater should fire normally.

Repair or replace burner tray components

Rusted out burners or tray components will be obvious to the naked eye. Following Fig. 6-2, remove the burner tray assembly and replace any defective burners (Fig. 6-18). The burner tray assembly includes the combination gas valve,

6-18
Replacing burners in the burner tray. Teledyne Laars, Moorpark, Ca.

burner tray, and pilot assembly. Figure 6-2 shows a standing pilot type heater, but the electronic ignition unit looks the same except you must also disconnect the high-tension wire that serves the pilot. You also need to disconnect the three wires connected to the IID and the ground wire in order to completely remove the tray.

If the pilot fails to light it might be clogged. If the heater "booms" when it lights, it means the pilot is not igniting the gas in the tray soon enough. Is the flame too small because of an obstruction in the pilot? Is the pilot bent, rusted, or not positioned close enough to the burner tray? Take it apart.

There are several types of pilots used in various heaters, but once the tray is removed, the disassembly should be obvious. Let's review typical installations and problems with standing pilot units and electronic ignition units.

Standing pilot units Disconnect the wires of the pilot generator from the combination gas valve. Remove the pilot gas tube from the combination gas valve and remove the entire pilot assembly from the tray. The pilot generator either clips in place or is held by a threaded ring (Fig. 6-19). Remove the pilot generator. If it is rusted or swollen, replace it.

6-19 Types of standing pilot assemblies. Raypak, Inc., Westlake Village, Ca. Teledyne Laars, Moorpark, Ca.

The pilot itself will further disassemble into two or three sections, depending on the make. Examine each part for obstruction or rust. Blow through one end of the pilot gas tube and be sure it is unobstructed. The pilot can be cleaned by soaking it in muriatic acid for 30 to 60 seconds, then cleaning thoroughly with fresh water and blowing dry (do not leave acid or water on metal parts—they'll rust).

Natural gas is actually odorless, and to make leaks more detectable, the gas supplier adds a perfume. This perfume has been found to attract insects who build nests in pilot assemblies. Insects and rust are the two main enemies of the pilot.

Electronic ignition pilot units The ignition electricity feeding the pilot is between 10,000 and 20,000 volts, so be sure the power is off before servicing. Remove the high-tension wire that supplies electricity to the pilot. Remove the pilot gas tube from the combination gas valve and remove the pilot assembly from the burner tray.

On Teledyne Laars units, the orange insulated wire terminates at the burner tray and is connected to a bare wire which itself is held in place by a ceramic in a bracket. This wire then attaches to the pilot electrode with a threaded end and two nuts. Sometimes, this bare wire can get too close to the metal burner tray and the electricity jumps or shorts out to the tray instead of getting to the electrode. This is solved by bending the bare wire (with the power off) so it remains clear of any other metal. Be careful when you replace the burner tray, because this wire can easily be bent back into contact with the burner tray.

The end of this wire on the electrode is also subject to failure. Working with installations on the Malibu beach, this connection rusts easily and is often coated with salt from the moisture in the air of the ocean. This is not enough to stop the flow of the high voltage coming to the pilot for spark ignition, but once the pilot lights it is corroded enough to prevent the IID from sensing that the pilot has indeed been lit.

The IID senses the pilot by the electricity generated from the heat of the flame—a reverse current of 0.00002 amps (2 milliamps) is generated along the line. It is so sensitive that any electrical obstruction such as corrosion prevents this signal from reaching the IID. I have cleaned and reassembled these units and successfully put them back into service, but they usually soon fail again. In this case, replace the pilot assembly.

Raypak pilots are not subject to this problem because they do not use the bare section of wire. Instead they use insulated wire all the way to the pilot electrode. Teledyne Laars uses the bare wire because they say heat in the burner area would melt insulation. Raypak tells me that their ventilation system keeps temperatures in this area cool enough to use the insulated wire all the way. It seems to be true and I have no corrosion problems with wires to the pilot on Raypak heaters.

As described previously with standing pilot units, disassemble the pilot assembly and supply tube and clear any obstructions or rust. Reassemble the opposite of the way it came apart.

Automatic combination gas valves

There are no in-the-field repairs that can be made to the automatic combination gas valve. If you determine that it is failing, replace it. They are sometimes rebuilt at the factory, but the cost of doing so is greater than the replacement cost ($70 to $120).

The plumbing of the valve is female threaded and screws directly onto the gas pipe of the burner tray. When replacing, apply Teflon tape to the male gas pipe of the burner tray. I do not like pipe dope in this case because it can easily squeeze off the threads and into the opening of the gas valve, obstructing gas flow.

Note that on the end of the valve, a flat ridge has been provided (Fig. 6-6) around the female threaded opening so a standard box wrench can be used to tighten or loosen the valve from the gas pipe of the burner tray. This is easier than trying to get a huge pipe wrench around the entire combination gas valve unit. When tightening, be careful to support the pipe so it does not bend or snap off of the burner tray.

The gas tube that supplies the pilot is prebent by the factory. Trying to bend these in other ways usually leads to crimping and leaks. Your supply house now carries a universal supply pipe in a thinner, more flexible pipe style, which is coiled in sufficient length to serve any make of burner tray. Buy these for replacements.

Having said you cannot service a gas valve in the field, there is one adjustment that can be made—gas pressure to the pilot or main burner. Figure 6-6 shows the typical location for the small screw that adjusts these pressures, sometimes beneath a threaded cap which must first be removed (black caps denote propane). Clockwise turns increase pressure; counterclockwise decrease pressure.

This adjustment should only be done when you have a manometer (see Fig. 6-13) and can test the results of your efforts. Otherwise call a factory service technician. Personally I have never adjusted the factory preset pressure adjustments and I suggest you don't either. If your installation is over 2000 feet above sea level, you will want to consult with a factory service technician anyway about overall adjustments and operational expectations. If your installation is at a lower altitude, there should be no reason to adjust the valve. More likely, the supply pressure is at fault or the gas valve is shot. Personal opinion—don't fool with this adjustment.

Is ventilation adequate?

You would be surprised how many problems are created by improper ventilation. You need only place your hand about 2 feet over the top of a gas-fueled heater to feel the power and volume of hot air moving through the ventilation system to understand why any restrictions under such heat result in damage of some kind.

Sooting, which is black carbon buildup on the heat exchanger, is a symptom of improper ventilation. Carbon starts as a dirty black coating and builds up to the point where chunks of coal burn and break off, falling on the burner tray below. The burners become clogged, shutting down the heater's full capacity. The heater smokes when it operates.

Here's a perfect example of a problem that is easy to repair, but the repair will be of no value if you do not also address the cause. Refer to the installation section and to the manufacturer's guidelines about proper ventilation. Correct any problems in this area first.

The remainder is easy. Remove the burner tray as described previously, cleaning all the components with soap and water, and drying out everything before reinstallation. While the burner tray is out, remove the draft hood and any vent stack to reveal the heat exchanger. Using a stiff bristle brush and soapy water, clean the exchanger. Be sure to get inside the cabinet to clean the underside of the exchanger as well (it might be easier on small heaters to remove the heat exchanger for cleaning).

Be careful not to soak electrical components or the firebrick. After reassembling the heater, small amounts of carbon that you might have missed will burn away over time, providing that you have corrected the ventilation problem and no new sooting is occurring.

Do not use a wire brush for soot removal. It can cause sparks that might ignite the carbon. Use a stiff natural or plastic brush.

Check the control circuit

Replacing the various control circuit switches as needed is very easy—the on/off switch and thermostat (mechanical or electronic) come off the cabinet with the removal of two screws; the high-limits are easily removed from the header; the pressure switch simply unscrews from the end of the water tube feeding it (Fig. 6-20); the fusible link is held by a ceramic holder that is held in place by a screw or two; and the wires to each switch simply unclip or come off by loosening a screw. No, the problem is not in replacing a faulty control circuit switch, the problem is determining which switch is faulty . . . and why.

6-20
Pressure switch adjustment. Teledyne Laars, Moorpark, Ca.

Whether the circuit is powered by 750 millivolts or 25 volts, the simple troubleshooting procedure is to follow the path of the electricity. Using your multimeter (discussed in more detail in the chapter on basic electricity), you check to see if electricity is getting to the switch (see Fig. 6-8) and, if so, is it getting out? If not, then that is the faulty switch.

On electronic ignition heaters, check the transformer to see that 25 volts is leaving the transformer (Fig. 6-22). Set your meter to the appropriate scale (on mine, 50-volts ac). Touch the negative (–) lead of your meter to the negative side of the transformer and the positive (+) lead to the positive side.

To save time, check if there is 25 volts at the end of the circuit (Fig. 6-22 item 2). Leave the negative lead touching the negative of the transformer and move only the positive lead to test each part of the circuit. If there is current, then obviously all of the switches in the circuit are closed and you need to look elsewhere for failures. If there is no power at item 2, then one of the switches is open for some reason.

On standing pilot heaters, use the same technique. Check the beginning of the circuit (Fig. 6-21) to determine if the pilot generator is delivering 400 to 700 millivolts. As discussed previously, set the meter to the correct setting (on mine, 1-volt dc) and touch the negative lead of your meter to the negative terminal on the gas valve and the positive lead to the positive terminal (item 1).

The heater will work on as little as 200 millivolts, but the electricity required to power the circuit and the combination gas valve will use almost all of that, so the heater might fire, but it will soon shut down. A healthy pilot generator delivering over 400 millivolts will withstand the 200-millivolt operational drop and still retain enough power to continue the job.

6-21 Millivolt wiring schematic. Teledyne Laars, Moorpark, Ca.

6-22 Electronic ignition heater control circuit (with electronic thermostat). Raypak, Inc., Westlake Village, Ca.

Now check the end of the circuit (item 2). As noted before, leave the negative lead touching the negative terminal on the gas valve and move the positive lead around the circuit. Again, if there is 200 millivolts or more there, the circuit is complete and the heater should fire. If not, start checking the control circuit switches.

Be aware that the switches themselves might be fine but there could be a short in the wiring. Sometimes rodents will nest in a heater and for some reason gnaw at the wiring. They rarely cut all the way through the wire, but by stripping the insulation, the bare wire sometimes comes in contact with the metal cabinet and creates a dead short—the electricity just flows through the cabinet and is dissipated or is sent through the ground wire.

On a millivolt heater, the symptom of this is usually that even though the pilot generator is producing 400 to 700 millivolts, the pilot goes out when you let the gas valve control knob up after lighting. To test if this is the problem, remove both wires of the control circuit from the gas valve. If the pilot now stays lit, you have a short in the wiring. If it still goes out, the gas valve is bad and should be replaced. On a 24-volt system, there is no simple way to test for bad wires, except visual inspection.

For both millivolt or 24-volt systems, the control circuit switches are similar. As noted previously, follow the path of the electricity and test for voltage into and out of each switch until you find the one that is open. Also as noted, to test a terminal or switch connection, keep the negative lead of the meter on the negative of the transformer and the positive lead on the wire or connection being tested.

Another way to test a switch is to *jump* it. Take a short wire and connect the two terminals of a switch, in essence, completing the circuit by bypassing the switch itself. If the heater fires, then obviously the switch is open.

As a time saver, you might want to test the last one in the circuit first and work back to the first—when you find a switch with power to it, you know everything prior to that one is working and it might save you from checking several switches unnecessarily. Another time-saver is to start with the most accessible switches first, because you might find the problem without digging into the cabinet (some high-limit switches and thermostats require faceplate removal to test). You will develop your own preferences, but I start with the switches that are most accessible (and most often the problem), as follows.

Pressure switch failure Most pressure switch failure is caused by obstructions in the circulation—low water in the pool or spa, a clogged skimmer basket or main drain, a clogged pump strainer basket, or a dirty filter. Check all of these before touching the pressure switch.

Sometimes the obstruction is in the water supply to the pressure switch. Remove the wires to the switch and unscrew the switch from the tube (Fig. 6-20). Turn on the pump. Water should flow vigorously from the tube. If it does, look in the hole at the end of the pressure switch to be sure it is clear of obstruction.

If no water is coming out of the tube, remove the heater vent top, front header faceplate, and flue collectors to expose the end of the tube in the header. Unscrew the tube and remove it. Blow through the tube to clear any obstructions. Turn on the pump. Water should shoot out of the hole in the header if there is no obstruction. I recall clearing lime from this location on Dick Clark's spa heater more than once (the spa heater always seemed to fail when he had house guests and I was always the villain). Finally I corrected the water chemistry and reamed out the heat exchanger and had no such problems thereafter.

If the problem is not in the water supply, adjusting the pressure switch to be more or less sensitive is done with the screw or knob on the switch (Fig. 6-20). On the Hobbs-type unit, loosen the nut that holds the screw in place (some units have a spring on the screw to provide tension and keep the adjustment). Tighten the screw to make the switch more sensitive (the switch closes and the heater comes on with less pressure), or loosen it to make it less sensitive (needs more pressure to come on). The heater burns with no water circulating.

When adjusting either way, turn the screw only a one-quarter turn at a time because these switches are very sensitive. Check the operation after each quarter turn until the heater operates correctly. The heater should fire and stay lit when the circulation is running; it should shut down within three seconds when the circulation stops.

Remember, the pump might be off but the circulation might still run if the heater is substantially above or below the pool or spa water level (or if solar panels are part of the system). Therefore, a lag in the heater shutoff might be several seconds after the pump is off. Retighten the nut to keep the screw at the adjustment made. The other type of pressure switch is adjusted with a knob in the same manner as the setscrew on the Hobbs-type unit.

If the switch can no longer be adjusted or is rusted out, replace it as follows:

1. Turn off the pump.
2. Remove the two wires and unscrew the unit from the line.
3. Reverse these steps to install the new unit.
4. Test the new unit by turning on the pump and verifying that the heater will fire. Leave the heater switched on and the thermostat on High and turn off the pump. The heater should go out promptly. If it continues to run, refer to the previous section and adjust the new switch accordingly.

Fuses and fusible links As described previously, some 24-volt heater designs include an in-line fuse, like a car fuse, in a bracket on the positive wire coming out of the transformer. Make a visual inspection or use your multimeter to test for current at a point after the fuse. The fusible link is easy to test with your multimeter, following the procedures outlined previously. If current is not passing through the link, it must be replaced, but you want to know why.

Is the venting adequate or is there an overheating condition? Is there soot or debris on the burner tray causing overheating somewhere? Is the gas pressure low, causing *lazy flame* (yellow flames that seem to lick out of the burner instead of strong, blue flames burning straight up)? Replace the link, but also address the cause of the failure.

High-limit switches High-limit switches are slightly different in each make of heater, but the principles are the same. In Teledyne heaters, two high-limits are located side by side beneath a small protected cover (as seen in Fig. 6-2). Remove the cover (usually one screw) and pull the switches with their retainer bracket from the header area. This operation can be done with the pump on or off. Pull the switches from the bracket, noting how this setup is assembled (pay attention or it can be a jigsaw puzzle to put back together). Pull the switch from its socket and replace it.

I usually replace both. If one has worn out, the other one will usually fail shortly and the customer will not be understanding about two labor charges for what appears to be the same repair within a few days or weeks of the first job. Start the pump or turn the heater back on and verify that the heater operates.

In Raypak heaters, the switches are most often located on each side of the heat exchanger. Turn the pump off. Follow the wiring to locate the switch, clip the wires, and unscrew the switch from the header. Put a little Teflon tape or pipe dope on the new switch to prevent leaks and screw it into the header. Connect the wires and use wire nuts or tape to keep the live unprotected wires from touching metal.

With Raypaks, I don't replace both switches if just one has gone out. Because they're on opposite sides of the heat exchanger, they are heated and worn down at different rates, so if one is still working it might continue to do so for a long time. Start the pump and verify that the heater operates.

On/off switch and thermostat   The mechanical thermostat can be tested for current just like the others. The electronic version looks more complex, but is actually easier to diagnose because it is connected through a terminal block where all the connections are in one place for easy testing.

To avoid heat-related injuries, manufacturers limit their thermostats to 104°F (maximum temperature recommended for spas by most local codes and health laws). Because the mechanical thermostat has an accuracy of +/-3°F, these units are calibrated for 101°F maximum to allow for the extra 3°F. Of course that means it might be 3°F the other way and you might only get your spa to 98°F. Now add bubbling water and a cool evening breeze and the water only goes to the low 90s. This is a common complaint with mechanical thermostats, so be prepared to educate your customer.

Electronic thermostats are calibrated to go no higher than 104°F, but their accuracy rate is within 0.25°F. Still, on a cool, windy night with the jets and blower going full bore, the water might not get higher than the high 90s and your customer might complain of insufficient heat. Be aware that the problem might be in the design and not in the components or the heater itself.

Figure 6-22 shows the terminal block of a typical electronic thermostat. Twenty-five volts travel from the transformer to terminals 6 and 7. Check here to verify current supply.

From here the current goes to terminal 4 and into the temperature sensor (called a *thermistor*), then back to terminal 3. If no current is present at terminal 3, the sensor is bad.

From terminal 3, the current goes to terminal 2 and into the on/off switch. The switch sends the current into one of two thermostats (also called potentiometers or pots), and the thermostat sends current to terminal 1. If no current is present at terminal 1, the switch or thermostat is bad.

From terminal 1, the current runs to terminal 8 which feeds the high-limits, fireman's switch, and pressure switch. As noted earlier, a time-saver is to look for current at terminal 8 first. If 24 volts is there, everything in the line before it is good.

You can also test the terminal block by jumping connections. If you jump terminals 1 and 2 and the heater fires, you know the problem is in the switch or thermostats. If you jump terminal 8 to the IID's terminal 2 and the heater fires, the problem is in the high-limits, fireman's switch, or pressure switch. Examine the diagrams and see if you can find several other jumps that will quickly tell you without a meter the condition of various switches.

With new models or different manufacturers, the terminal block might not be set up exactly like the one in Fig. 6-22, but if you understand this concept you can read the wiring diagram for the heater you are working on and follow the same procedures. The wiring diagram can be found inside the cabinet door or printed on a panel inside the heater. If it is missing or obscured, refer to the booklet that comes with the heater.

Replacing a mechanical thermostat requires removing the draft hood top of the heater to expose the oil-filled bulb and tube that are attached to the thermostat dial. Remove this along with the thermostat and be sure the replacement is the same model, not one with a shorter tube that might not reach the dry well.

Replacing the electronic type means replacing the unit on the panel in the front of the heater, unless you also suspect the thermistor. If so, the thermistor is found in the same location as the tube or bulb in the header.

Fireman's switch A fireman's switch as previously explained is simply an on/off switch attached to the time clock that shuts off the heater 20 minutes before the pump. The result is a cool-down of the heater before the water circulation stops, prolonging component life. Teledyne Laars recommends the use of the fireman's switch, Raypak does not, saying their ventilation is adequate to complete cooling of the heater upon system shutdown. You can troubleshoot this switch by jumping across the electrical wires leading to it at any convenient place. If it is defective, replacement is self-explanatory.

Intermittent ignition device The IID is actually a sophisticated switching device, not an electronic brain as some call it. There are no hidden computer chips in this device

All IIDs have the same terminals, though perhaps not in the same location on the box. Figure 6-22 shows that power is supplied to the device by the transformer after passing through the control circuit. The IID sends power to the pilot to spark ignition and simultaneously opens the pilot gas valve by sending power (from the pv terminal) to the appropriate terminal on the combination gas valve. When the pilot lights, the heat creates a voltage that is detected by the IID and acts as a signal to send power to the main gas valve on the combination gas valve, opening it to send gas to the burner tray.

The IID must be grounded to operate properly, so that is a good place to start your troubleshooting. If you have 25 volts at terminal 2 on the IID but no spark on the pilot, check the pilot. If it looks good, test the orange high-tension cord. The easiest way to do this is to remove the end of the cord from the IID and hold it about ⅛-inch away from its terminal on the IID. Turn the heater on and if you get a spark jumping from the cord to the terminal, the cord is probably good. If you get no spark, it is faulty.

If you get a spark here but not at the pilot, the IID is faulty. If you get a spark but no gas to the pilot, the problem might be in the high tension cord (it might be sparking the pilot, but not sensing that the pilot has lit). If replacing the cord doesn't solve the problem, the fault is in the IID.

Finally, if you have a spark and a lighted pilot but no voltage at the main valve (MV) terminal of the IID, the IID is faulty. I always carry a spare high-tension cord and IID. If I suspect either component, it is easy to replace them to locate the problem. If it turns out that the component is okay, then my spare goes back in the toolbox and I put the original back in the heater.

IIDs, like combination gas valves, can be repaired at the factory but not by you. However, the cost of a new one ($70 to $100) is less than the labor, parts, and trouble of taking it to the factory for repair.

Electric-fueled heater reset buttons   The small red button found on electric-fueled heaters is a safety circuit breaker. Some are activated by an overheating condition, such as when the unit is allowed to run dry. Some are only sensitive to excess amperage, like the circuit breakers in your house. After they cool (in a minute or so), they can be pressed back in to reset them. However, as with other troubleshooting, you must determine the cause of the circuit breaking.

Usually the reset breaker is tripped by an electric heater that is overheating from low (or no) water flow. Check your circulation before resetting, and afterwards observe the unit in operation for several minutes to make sure it doesn't pop again. Of course, sometimes these little breakers simply wear out from old age and need to be replaced, but this is not common.

Preventive maintenance

The best preventive maintenance is to use the heater regularly. As noted previously, corrosion, insects, nesting rodents, and wind-blown dirt create many heater problems that can be eliminated by regular use. The heat helps to dry any airborne moisture that might otherwise rust the components. It discourages insects and rodents before they get too comfortable. It keeps electricity flowing through the circuits, preventing corrosion that creates resistance that might ultimately break the circuit completely. It burns off the odd leaf or debris that lands inside the top vents before a greater accumulation builds up, which when the heater comes on might start a fire or send flying embers into the air. In short, running at least a few minutes a day is great therapy for a heater.

Otherwise heaters need only be visually inspected from time to time. A look around will detect sooting, gas or water leaks, or other problems before they begin. Keep leaves and debris off the top of the heater. Look at the pilot and burner flames. Are they strong, blue, and burning straight up at least 2 to 4 inches? Open the drain plug on the heat exchanger and look for scale buildup.

SPARES KIT

The easiest way to work on heaters is to maintain a simple kit of spare parts. It helps in troubleshooting because you can replace a part you have doubts about as a test. It might also save you a trip to the supply house. Having spares saves you time and money and takes up little space in your toolbox. It also means you can complete more repairs on the spot and get paid by the customer right away.

Here's what I carry.
- Scripto lighter (BBQ-style throw-away)
- Honeywell IID
- Millivolt pilot assemblies (including generator)
  ~ Clip-in type (Teledyne Laars)
  ~ Screw-in type (Raypak)
- Electronic pilot assemblies
  ~ Teledyne Laars style with bolt-on electrode
  ~ Raypak style with clip-on electrode

- Orange high-tension cords
  - ~ Teledyne Laars style with bare extension wire and ceramic
  - ~ Raypak style (longer) with clip on the end
- Pressure switch with adapter for different models
- Teledyne Laars set of high-limit switches
- Raypak set of high-limit switches
- Electronic thermistor
- Transformer
- Wire of various lengths and colors (available with connectors in kits called *harnesses*)
- Teledyne Laars flange gaskets (1½- and 2-inch pairs)
- Raypak flange gaskets (1½- and 2-inch pairs)
- Small dental-type mirror for examining pilot flames

Generally, combination gas valves, thermostats, and switches are too expensive, numerous, and varied to carry as spares—you will keep several, but never have the one you need. Other components don't fail often enough to justify carrying spares.

In the past few years, several companies have noticed that, as is the case with automobiles, spare parts on heaters are unbelievably expensive. Building a heater from spare parts would cost three times as much as buying a factory-assembled unit. As a result, many companies now offer replacement parts that fit the popular brands of heaters. Generally these generic brands work just as well as factory replacements, although the factory won't necessarily agree.

One last word about replacement parts. To ensure an uninterrupted supply of parts, manufacturers use parts from various suppliers or different models of parts from the same supplier. A part might look different, but that doesn't necessarily mean it is different. Examine it and ask your supply house.

A REMINDER

Always remember that a heater is potentially the most dangerous component of pool and spa equipment. Whenever you make a repair, consider **safety first**. Refer back to the safety checklist at the beginning of this chapter before you start work on a heater.

7

Additional equipment

I have now examined the major components of water circulation, filtration, and heating. There are several pieces of equipment that complete the mechanical resources available to the pool, spa, or water feature.

This additional equipment is important to understand because many customer calls relate to these ancillary devices or supplies. The technician's ability to supply or maintain these items will impact the bottom line of any water-related business.

TIME CLOCKS

A key component of any water maintenance system is a time clock. Time clocks make sure that water is circulated through the filter each day, that it is heated sufficiently and/or at the correct time of day, that fountains or decorative lights come on at the right time, that automatic cleaners operate daily, and a host of other useful functions.

Electromechanical timers

Figure 7-1 shows an exploded view of a typical 120-volt time clock, and Fig. 7-2 shows the same product assembled. It allows various on or off settings throughout each 24-hour period. Some models also includes a seven-day feature, allowing you to determine which days of the week you want the clock to control the system. The only difference between a 110-volt and a 240-volt clock is that the 240-volt unit does not have a neutral line, but includes two incoming lines and two outgoing loads for the equipment. The neutral location is a ground. For explanation purposes however, this diagram shows the basic parts found in any time clock.

The small electric motor runs on the same push-pull concept of opposing magnets as the motors used for pumps. A winding is electrified that creates an electromagnetic field and sets up the rotation, which happens at a predictable 400 rpm.

Tripper

Reduction gears

Clock face

Motor

Drive gear

Dial spring

Visual motor check

Frame

Time pointer

Tripper

Terminal board

On/off switch

Fireman's switch

7-1 Typical electromechanical time clock. Intermatic, Inc., Spring Grove, Ill.

7-2
The electromechanical time clock and twist timer.

The 24-hour dial would not operate very well at that speed, so a series of reduction gears in the motor housing result in the actual drive gear turning at a speed that rotates the dial just once every 24 hours. In this way, you can tell the time of day by the clock face and use it to set on or off switching of the system.

Most ordinary watches or clocks have a face that remains fixed, while the hands of the clock rotate to give the time. The time clock operates on the opposite concept—it has only one hand, which remains stable, and a face that rotates.

Note the on and off trippers (the words on and off are etched on the trippers). By setting these where you want the system to go on or off, the clock controls the flow of electricity to your system accordingly. More than one set of trippers (also called *dogs*) can be affixed to the dial, so the system can go on and off several times during a 24-hour period. As you can see on the top tripper, the nipple on the bottom of the tripper is located almost midway along the edge, while the nipple on the lower tripper is located at the end. These nipples engage the cam on the on/off switch of the clock, turning it on or off as the clock face rotates.

To set the clock, you simply pull the face toward you and rotate it until the number on the dial, corresponding to the correct time of day, is under the time pointer. By pulling it forward, its drive gear is disengaged from the motor drive gear and it rotates freely. As noted earlier, the time pointer is the only hand on the clock. The dial is divided into each of the 24 hours of the day (differentiating between A.M. and P.M., some clocks say Day and Night) and each hour is further divided into quarter hours, so when setting the correct time of day you can be fairly precise. After setting the correct time, release the dial and it snaps back into position, reengaging its drive gear with the motor drive gear.

When the switch lever moves to the right, it lifts the contacts apart, breaking the electrical circuit. When the lever moves back to the left, the contact arm is allowed to drop back into contact with, the lower contact, completing the electrical circuit. Simple, eh? The lever also allows manual operation, so regardless of where the trippers are set you can manually operate the system.

The contacts are attached to screw terminals for attaching the wires of the appliance to be controlled. On a 120-volt model, the hot 120-volt wire is attached to the terminal marked Line and the neutral to the terminal marked Neutral. The hot wire for the load (the appliance) is connected to the load terminal and its neutral is connected to the neutral terminal. Some clocks provide separate neutral terminals for the line and load, but they both join to the same wire that returns to the circuit breaker. As noted, the 240-volt version includes two line terminals for the two hot lines coming in, and two load terminals for the supply to the appliance.

The wires feeding electricity to the clock are wrapped on the line and neutral terminals (with a 240-volt clock, the one lead would be attached to each line terminal). This way there is a constant supply of electricity to the clock.

A quick way to tell if power is getting to the clock (assuming the clock motor is in working condition) is to look through the opening provided marked Visual Inspection, or words to that effect. Look in here to see the motor gears spinning. Some clocks have the motor mounted on the front of the clock and the center hub of the drive gear is visible. Close inspection will reveal if the hub is spinning or not.

Waterproof boxes in metal or plastic are available to house time clocks. The unit shown in Figs. 7-1 and 7-2 simply snaps into clips built into the box. Other versions have built-in brackets that are designed to align with the screw holes on the clock plate, and these are held in place by two machine screws. The boxes are built with knock-out holes to accommodate wiring conduit of various sizes and have predrilled holes in the back for mounting the box to a wall.

Intermatic makes most of the pool and spa time clocks you will encounter. Paragon is another maker, and Len Gordon makes a miniature version included in its clock and control packages. As with other components of pool and spa equipment, if you understand the parts and operation of one, the others are easy to figure out.

When a second clock is used to run an automatic pool cleaner booster pump, make sure it is set to come on at least one hour after the circulation pump comes on and is set to go off at least one hour before the circulation pump goes off. These boosters rely on the circulation pump for water supply, and without it will burn out their components. When checking or resetting the system clock, be sure to follow this guideline with the booster pump clock.

Twist timers

Twist timers (Fig. 7-2) are used mostly with spa equipment such as booster motors or blowers when you want a limited amount of appliance operation as the user demands. A twist timer is built to fit in a typical light switch box and contains no user serviceable parts. This unit has a faceplate showing 15, 30, 45, or 60 minutes. The knob attached to the shaft that comes through the faceplate is twisted until its pointer or arrow aligns with the desired number of minutes. The circuit is completed in any position except off and the mechanical timer is spring-loaded to unwind for the number of minutes selected. When the spring is unwound, the circuit is broken and the appliance shuts off.

Twist timers are available in 120 and 240 volts and are used in place of a simple on/off switch, usually where users might forget to turn off such a switch. This also functions for safety. You might want to use a twist timer on a spa so a user cannot turn on the spa and remain in the hot water too long. When the system shuts down, the user is forced to leave the spa and reset it, perhaps making them realize that they've had enough. Another good use for a twist timer is for lights in a public area, where the user must set the desired time and the lights will go off even if the user forgets.

In the repairs section later, I don't discuss twist timers because, as noted, there are no user serviceable parts. If they fail, replace them. It is no more difficult than replacing a light switch and all you need is a screwdriver. When you buy the replacement, follow the instructions in the box if you don't find it self-explanatory. As with any electrical repair, be sure the electricity is disconnected at the breaker panel.

Electronic timers

Do you own a VCR? Are you one of those people who never sets the time of day on it, but lets it flash 12:00 all the time? Well, when you learn about electronic timers for water system controls, you will also know how to set your VCR.

Electronic timers are included mostly in spa control packages that also include electronic thermostats and other sophisticated controls. Because there are many makers and types of these controls, often requiring an advanced electronics degree to test and repair them, I cannot adequately describe them here. Generally, even electricians don't want to work on these high-tech controls. I have found they can be time and money quicksand, so if I suspect control problems with electronic devices, I call the manufacturer for the name of a local factory technician. Let them have the work.

Setting these units, however, is usually easy and sometimes self-directed (the prompts come up on the screen to tell you what to do). Again, there are so many types and manufacturers that I could not outline typical instructions. If you are servicing a unit with no obvious instructions, call the maker and ask for a duplicate set. It pays to do this because customer complaints of malfunction are often related to electronic timers that are simply not set correctly. For instance, sometimes the timer functions have been inadvertently set to override manual functions, so the customer finds his spa shutting down in the middle of use. I have also found that after a power failure some systems require resetting, and customer complaints of malfunctioning spa equipment can be traced to this simple procedure. It is the easiest service call fee you will ever earn, so when responding to spa repair calls, check the function and settings of the electronic timers if the system is so equipped.

The advantages of electronic controls are precision and low voltage. Some units are designed with such low voltage that you are permitted to install them next to a body of water (like a spa) and not worry about electrocuting your customer. Most have digital readouts so you can precisely set the time and temperature desired, as well as programming a host of other features.

Some allow you to connect the system to your touch-tone phone and control operations from any other phone in the world. The rock star Sting and Hollywood super-agent Michael Ovitz are among my customers who like to control their pool or spa by phone. On the way to your Malibu beach house in the limo, you can call your spa control system and, through touch-tone pulses, instruct it to turn on the heat and get the spa up to 104°F by the time you get there. Sounds like an everyday situation, right?

Lastly, most electronic devices have small backup batteries so that when main power is interrupted, they don't lose the time of day or previous programming. Again, system malfunctions can sometimes be traced to batteries that need to be replaced.

Repairs

There's not much that goes wrong with electromechanical time clocks and it's easy to service most of them.

Replacement

When a time clock rusts or the gears wear out and the unit needs to be replaced, it takes longer to buy the new unit than to perform the replacement. When buying the replacement, be sure to get the same voltage clock as you have in the existing installation. Also, buy the same make of clock, because a different manufacturer's clock probably won't fit in the existing box.

1. Turn off the power supply at the breaker.
2. Remove the line and load wires from the old clock and any ground wire. Remember (or mark) which wire is which.
3. Unscrew the holding screws or unclip the old clock to remove it from the box. Snap or screw the new one in place.
4. Reconnect the wires to the appropriate line or load terminals. Replacement clocks might not be arranged the same as the old one, so be sure you are getting the line wires (two hots or one hot and one neutral) attached to the line terminals and the two appliance wires attached to the load terminals. If it is not clearly marked, the simple way to tell which terminal is which is to notice where the clock motor is connected—the two clock wires are always attached to the line terminals.
5. Turn the power back on and test the clock operation by turning the manual on/off lever to On.

Most manufacturers sell just the clock motor for replacement. I have found that if the motor is old enough to burn out, the clock mechanics are probably wearing out too, so it pays to replace the entire mechanism. Besides the $20 or so price difference between the motor and the entire clock unit is not worth messing with motor replacement only. In fact, it takes longer to replace the motor only than it does the entire clock.

By the way, if your installation is a new one, follow the electrical guidelines discussed in the chapter on basic electricity and remember to supply power directly from a breaker. If the power supply goes to an on/off wall switch or other device first, the power might be interrupted, meaning the clock cannot keep accurate time. Keep in mind that the clock motor itself must always have a source of power.

Cleaning

Sometimes insects will nest in your time clock. I have found countless times when a clock fails to turn a system on that it is clogged with nesting ants. The cure is simple: turn off the power and remove and clean the contacts. When finished, apply a liberal amount of insecticide to the interior of the housing before reinstalling the mechanism. Don't spray the mechanism itself. You might create unintended electrical contact through the liquid.

Other than that, corrosion will occasionally bind up a clock. If you know it is getting power but cannot see the gears turning, take the tip of your screwdriver and gently force the gear (the one visible through the inspection hole) in a clockwise direction. Often that will get it started and it will continue to run fine after that. If it happens once, it will probably happen again unless you give the gears a good general lubrication.

To lubricate a time clock, turn off the power source and remove the clock from the box (wires still attached) to expose the rear of the clock. Spray WD-40 liberally around the gears. Do the same on the front to lube the gears behind the dial face. As noted, be careful not to wet the electrical contacts. Put the clock back in the box and turn the power back on. Turn the clock on and off a few times to work in the oil. If the clock fails again, replace it.

Mechanical failures

When you pull out the dial face of the clock to set the time, take care that when you release it you get a true reengagement of the dial with the motor drive gear. Try setting different times on the clock and you will note that sometimes the two gears don't mesh, but rather the dial gear sits on top of the motor gear. Obviously, in this case the clock won't work.

The answer is to wiggle the dial face as you release it. As you release it, twist the dial back and forth very slightly in your hand to make sure the gears mesh. With a few practice settings you'll feel the difference between a dial that has gone back into place completely and one that is slightly hung up. Often, clocks that don't work are a result of this setting problem, so make this one of the first things you check when you suspect time clock failure.

The second mechanical problem of time clocks is with the trippers. If the screw is not twisted tightly on the face of the clock, they come loose and rotate around the dial, pushed by the control lever instead of doing the pushing themselves. Check the trippers regularly because they can come loose over time or from system vibration.

Also, if you are trying to set a time that happens to be close to a tripper setting, the dial will not engage with the motor gears. You need to move the tripper a little to make room for the dial gear. Finally, trippers can wear out, so if the clock is keeping good time but not turning the system on or off, try a new pair of trippers.

Settings

When setting on/off trippers, you can place them on the dial face side by side to what appears to be about 30 minutes between them. I have found that when they are too close, they won't operate the lever. Generally, trippers must be at least an hour apart to operate.

Make frequent checks of your time clocks. Power outages, someone working at the house who shuts off all the power, the twice yearly daylight savings time changes, and any number of other household situations can interrupt power to the time clock. Each time this happens, the clock stops and needs to be reset when the power returns in order to reflect the correct time.

You might also find that you set the clock to run the filter eight hours a day and the customer resets it to run two hours a day. Very often a cloudy or green pool is the result of someone fooling with the time clock.

REMOTE CONTROLS

Used mostly with spa equipment, remote control devices allow you to operate pumps, heaters, lights, blowers, and other devices without actually going to the spa. Such devices are also available with switches at or in the spa, so you can control appliances without getting out.

Remote control devices fall into two categories: those operated by pneumatic (air) switches and those operated by electronic wireless switches.

Air switches

You have probably driven into a gas station over the black rubber hose and heard the bell in the station. Pool and spa air switches operate on the same idea—by compressing air in a hose, you force a switch to go on, off, or to a new operating position.

Fortunately a man named Len Gordon pioneered these devices with push buttons for pool and spa use so it isn't necessary to use a car each time you want to compress the air in the hose.

Figure 7-3 shows a typical air switch button. By depressing the plunger in the middle, air is forced into a flexible plastic tube (usually ¼-inch diameter or less). The force is transmitted along the air tube until it reaches a simple electromechanical switch at the other end, turning it on, off, or to another position. You can mount this type of button near or even in the spa and not worry about electricity near the water. Air switches operate up to 200 feet from the button.

 1 Switch box
 2 Switch assembly
 3 Box lid gasket
 4 Box lid
 5 Lid screws
 6 Double pole-double throw relay
 7 Air switch sensor
 8 Mounting plate
 9 Terminal plate
10 Ground bus bar
11 Spa/tub air switch button
12 Gunite air switch button

7-3 Air switch components. Len Gordon Company, Las Vegas, Nev.

The most basic of these are simple on/off switches. Some activate a rotating, ratcheted wheel that in turn activates up to four switches. You push the button once to turn on the circulation pump and heater, a second push to add a blower, a third push to add a booster, and a fourth push to turn it all off. Most of these four-function units allow combinations that vary the pattern, depending on your use preference or the additional equipment you have.

The electrical part of the air switch is usually located in a waterproof box, housing the wiring connections, the actual switches (usually microswitches), and the terminals for attaching line and load just like a time clock. In some units there is a small electromechanical time clock that can operate one or more appliances at preset times in addition to the activation by the air switch button.

As noted earlier, much of this application of technology is owed to Len Gordon and the company that produces the vast majority of these devices which bears his name. There is little that can go wrong with air switch units and they require virtually no service. Here are a few pointers, however.

Troubleshooting

To find out if the problem is the button, the hose, or the electrical switch, I carry a piece of hose with me (about 3 feet long). I disconnect the hose from the switch nipple (inside the electrical component box of the switch system) and place my hose on the nipple and blow through it. If the switch operates normally, I know the problem is not there. If it doesn't, I know the problem is not likely in the hose or button. Similarly, if I suspect a defective button, I can disconnect it and attach it to my test hose. If it activates the switch, I know that both the button and switch are okay, so the problem must be in the hose.

The microswitch itself can wear out, although this is rare, and can be bought separately and replaced (an operation that is self-explanatory after removing the faceplate of the switch box to reveal the actual microswitch location—usually only two screws and a couple of bayonet wire terminals are involved).

Air leaks

When you push the button and the equipment fails to activate, you might have an air leak in the hose. If you can reach both ends of the hose, tape the new hose to the old, at either end, and pull the old hose out. When you get to the taped joint, you have just snaked a new hose into the conduit. If you cannot access one of the hose ends, follow the replacement procedure detailed in the following paragraph.

Replacement

Sometimes air switches wear out, but they are easily replaced. Some are designed with a collar that is mounted into a deck or spa wall and have a removable button. Note in Fig. 7-3 that the button is six-sided so you can grip it easily with pliers or wrench to unscrew the center button portion for replacement. As can be seen in the drawing, this unit is fitted with a large nut on the underside so you can mount it on a deck and tighten the nut to hold the collar in place.

When you remove the center button portion, if there is no slack in the air hose, the hose might come off the end of the button and might be hard to reach for attachment to the new button. Try pushing the hose from the other end (the end attached to the electrical switch box) toward the button to force it back out of the conduit. If there just isn't enough hose, pull out the old hose and run an electrician's snake through the conduit. When the end comes out, tape the new hose to it and pull the snake back through. Now you can attach the hose ends to the button and the switch.

New installation

Original installation of an air switch system is not difficult, but you might want to hire an electrician and include it in your bill to the customer. If you try it yourself, follow the basic electrical instructions for wire sizes and codes provided in the basic electricity chapter, as well as the directions that come with whatever system you choose. To save time and money, you might want to buy the system, mount the electrical box, run the hose, and mount the button, then call in the electrician to do the electrical connections.

You will find that not all air hoses are run in conduit. Because they have no water or electricity in them, there are no standards or rules about running a length of hose. When making a new installation, I strongly recommend running the hose through a ½-inch PVC electrical conduit. This protects the hose from the elements and keeps it supple. It also keeps out rodents or dogs that might want to chew the hose.

Use electrical conduit and fittings rather than plumbing PVC. If you have to use a snake to put in a new line, the tape will get hung up on the sharp angles of the plumbing elbows and connections, whereas electrical connections are swept gradually into the elbow or angle to make that less likely.

Wireless remote control

In the past few years, wireless remotes (Fig. 7-4) have gained popularity. These are composed of a battery-powered wireless sending unit with anything from 4 to 24 buttons in a waterproof case (low voltage for safe water-side use). The buttons send a signal to a receiver that might be as much as 1500 feet away (usually mounted in the pool or spa equipment area). The receiver interprets which button has been pressed and activates a switch that turns on or off that piece of equipment. In essence, it works just like the air switch, except the signal is sent by radio signal instead of compressed air.

Some units do not have batteries, but plug into any household outlet. When you press the button, the signal is sent along the household wiring to the receiver, wherever it is located, which is powered by the same household current. Obviously, because these sending units are powered by 120-volt household current, they cannot be located near the body of water. The value of these is that they are not subject to battery failure or weak radio signals that sometimes fail to penetrate thick walls or long distances.

The 120-volt remote control is usually installed where the customer plugs the sending unit into an electrical outlet in the house so he is able to turn on and heat up a spa that might be located out in the yard. The disadvantage, obviously, is that to turn appliances on or off while using them you must get out of the spa.

There are too many makes and models of these wireless remotes to detail the wiring or technical data for installation or repair. As with a new air switch system, you might want to include your friendly neighborhood electrician in the price of installing a new system. Many are sold in modules that allow you to configure any system arrangement you wish, adding modules to existing wiring and equipment.

7-4 Wireless remote controls. Top row (left to right): 110-volt control transmitter, battery-powered control transmitter; bottom row: 110-volt receiver module (front and rear view), remote receiver module.

As mentioned, the main failure of these units is that the battery becomes weak or the radio signal cannot penetrate thick walls or long distances between the spa and equipment location. Actor Stacy Keach has a system like this and I am forever replacing the batteries. His electrician also installed manual override switches in the house, so Stacy can fire up the pump and heat before going out to the spa, or turn the pool and spa lights on at night from the house. Unfortunately, if the overrides were left off, no power would go to the switches in the equipment area and the entire system would not work. It took several tries and some new batteries in the remote unit before I figured that out. Be aware that the troubles in systems like this are not always in the remote control system itself.

FLOW METERS

As the name implies, flow meters record the flow rate in a particular pipe or piece of equipment. Most look like clear Plexiglas thermometers with a graded scale (expressed in gallons per minute) over a liquid-filled tube (Fig. 7-5). Inside, a small lead weight rises and falls depending on the flow of water in the line. One end of the flow meter is male threaded for easy installation into a female threaded opening in a pipe or, for example, on the drain plug opening of a heater.

Flow meters are generally used in commercial installations where health department or building inspectors test the system by the rate and volume of water flow. Some codes require that one flow meter be installed somewhere in the system. Only one is required because the flow rate is the same anywhere in the system. Remember, pressure might change where restrictions occur, but the rate of flow (in gallons per minute) is the same anywhere in the system.

When installing a flow meter, locate it away from the inflow end of a 90-degree elbow. Most codes say installation must be at least ten times the diameter of the pipe away from the inflowing side of a 90, and two times the diameter of the pipe on the outflow side. To illustrate this point, assume you have a 2-inch diameter pipe. Your flow meter would have to be 20 inches before the next 90-degree elbow in the path of the water (10 × 2 inches = 20 inches) and 4 inches past the last 90-degree elbow (2 × 2 inches = 4 inches).

To install the flow meter, cut the pipe and plumb in a T fitting to reconnect the pipe. Two ends of the T are used in reconnecting the pipe, and the third opening of the T is plumbed with a female threaded reducer to make the opening the same size as the male threaded end of the meter, which is then screwed into place. Apply Teflon tape to the male end to prevent leaks. As with pressure gauges, I don't use pipe dope for these installations because any excess dope can clog the opening of the meter.

Another method is to drill a hole in the pipe that is slightly smaller than the diameter of the male threaded end of the flow meter. Because the PVC is softer than the metal end of the meter, when you apply pressure and screw it into the pipe it will self-tap. These installations usually leak, so I don't recommend this method un-

7-5
Flow meters, top and side mount.

less the plumbing or equipment area is just too tight to cut the pipe and add the T fitting. You can use this method with copper plumbing because although the copper is soft, it is less likely to strip when you screw in the flow meter.

Don't always believe a flow meter. I know commercial pool technicians who remove the lead weight from the flow meter, cut off a small amount, and reinstall it. Obviously if the weight is less, it will go higher in the tube than it should, reflecting a higher flow rate than is actually the case. This is done to get a passing grade from the health department inspector even though the filters are dirty or some other problem needs fixing. It is a cheap way to get passing marks on an inspection and avoid the trouble of being cited or even closed down, but of course, it is illegal. The point I make here is that flow meters are not always accurate and should be used as a reference in conjunction with other observations you make about a system if you are called in for troubleshooting, maintenance, or repairs.

DIVING BOARDS, SLIDES, LADDERS, AND RAILS

A good source of added income for a pool technician is installation of new diving boards, slides, or ladders. Here are a few guidelines.

Diving boards

Diving boards are not as popular as they once were because pools today are built smaller and shallower as people try to save water and energy and because more people use pools for swimming exercise laps rather than for jumping into deep water. Still, there are many boards out there, and as they age they need repair or replacement. Diving boards are wood, for flexibility, covered with fiberglass, for waterproofing, with a nonskid tread applied on the top.

Diving boards come in simple fulcrum models and spring-assisted jump models for extra bouncing action. Figure 7-6 shows several designs. Residential boards are from 6 to 12 feet in length and 18 inches wide. Commercial boards are usually 20 inches wide and up to 16 feet in length.

Some building and safety codes specify pool dimensions for a diving board to ensure that there is adequate water area and depth for safe diving. Generally, to accommodate the smallest board, a pool must be at least 7½ feet deep and at least 15 feet wide by 28 feet long. These suggestions go up for longer boards or those with more springing action. Check local codes before adding a diving board to a pool. Remember, the customer depends on you for this type of advice and you might be held liable for installing a product they can't use safely.

Repair and replacement

Most diving boards you encounter will need repair or replacement, not new installation. I have repaired or replaced hundreds of boards, but in the past 10 years I have not been asked to install one new board. Also, although replacement boards are not cheap ($200 to $500 wholesale, including hardware), I suggest replacement rather than repainting or repairing.

Because of the wood interior, if the fiberglass has cracked or started to delaminate, the wood underneath might have wet or dry rot. If you cosmetically repair a board by patching cracks with fiberglass resin and adding new nonskid, then the board snaps in half two weeks after you worked on it, you are asking for a lawsuit.

Figure 7-6 shows the hardware used to attach a board. Here's how you proceed.

1. Looking underneath the board, remove the nuts, lockwashers, flat washers, and/or boltplate from the mounting hardware that holds the board to the stand. As you remove the last one, remember that the board might tip forward into the pool, so have someone hold it or put some weight on the end of the board, away from the pool side, so you can control the removal.

2. Pull out (or push up from the bottom) the bolts that hold the board on the stand. There are two bolts at the very end and possibly two more in the middle at the fulcrum. Remove the bolt plate and gaskets. Remove the board from the stand. Be careful and have some help nearby—boards can be very heavy, especially if the wood inside the board has become waterlogged.

Spring board

Jump stand

Cantilever
jump stand

Fiberglass
jump stand

Fiberglass
diving stand

Deck level
diving stand

Bolt

1/4" steel plate
(above and
beneath board)

Lock
washer

Nut

7-6 Diving board styles.

3. Lay the new board on the deck. Lay the old board on top of the new one and use the old holes as a pattern for drilling the new holes. Drill the needed holes. On some stands, the fulcrum has a rubber cushion that might need to be replaced. A new cushion comes with most replacement boards. Lift off the old cushion and push the new one in place. Most stands also have a rubber gasket or cushion on the bolted end as well. Apply the new gasket there.

4. Lay the new board in place. Again, remember that it might tip into the pool if you don't have someone help you hold it. Lay the bolt plate on the new board and insert the bolts. Replacement boards come with new bolts, but even if the old ones look good, replace them. They might have hidden cracks or rust and after years of pressure are bound to be metal-fatigued. It is no good to have a strong new board installed with bolts that might snap off at any minute.

5. Working underneath, replace the nuts with whatever lockwashers, flat washers, and/or bolt plate that are included for the underside. If the replacement doesn't come with lockwashers, add them so that the nuts can't vibrate loose over time, causing failure of the entire board. Never add nuts without washers or an underside bolt plate. The stress on the bottom of the board is enormous each time a jump is made and, over time, the nut by itself will tear through the board. The washers or bolt plate distribute that stress evenly.

6. When the new board is bolted on, walk out on the length of the board and jump up and down. I have always felt it was better to have something fail under me, when I'm expecting it, than under the customer who is not. Be sure your bolts are tight, but don't be crazy and torque them so far that the board cracks or weakens at that point.

For legal and safety reasons, do not replace a board with a longer one. Use the same type and length.

Having said earlier that I wouldn't upgrade a board cosmetically, the one exception might be to replace the tread. Sometimes, on a board that is only a few years old, heavy use and sunlight can destroy the nonskid material on the top. This can be bought in self-adhesive rolls (called *tread kits*) at the supply house with instructions for application. Replacement springs are sold for each make of jump-type board with instructions for replacement, and again, although fiberglass patch kits are available, I wouldn't use them.

Slides

Slides are generally made of fiberglass with metal frames and steps. A straight slide will be 8 to 13 feet long, requiring a lot of deck space. If deck space is limited, left-handed or right-handed curved slides are available (Fig. 7-7). As with diving boards, safety is a key consideration—the higher the slide, the deeper the pool needs to be.

Installation

Slide installation is simple and anchor kits are available at your supply house with instructions. Some installations provide brackets that you bolt to the deck with special concrete fasteners. Others require drilling 1½- or 2-inch holes for each of four

7-7 Slides, ladders, and rails.

legs and securing the installation with quick-set cement. Following the instructions in the kit, anyone can install a slide.

Some slides are plumbed to provide a sheet of water along the slide area to make them more slippery. These can be plumbed into a water supply line, but prolonged use will lead to an overflowing pool. A better installation is to plumb the slide to a return line from the pool's circulation system. This might require running a long and complicated pipe from the equipment area to the slide, but the supply pipe is usually ¼-inch diameter, so concealing it in gardens or along building edges is not difficult. Always fit the line with a shutoff valve so the slide can be turned off when it is not in use. The water running over the large surface area of the slide, which is usually quite hot, will evaporate rapidly, slowly draining the pool.

Maintenance

Because of constant exposure to the elements, slides often discolor. Your supply house sells a glaze/polish kit that can restore the appearance to almost new. I don't recommend painting slides, even with epoxy paints, because the constant sliding friction and exposure to sun and temperature extremes will quickly make the painted surface look worse than the original problem. Painted surfaces often oxidize and your swimmers will end up with powdery streaks over the backs and bottoms of their swimsuits.

The only other maintenance concern with slides is safety. Make it a habit when you service the pool to shake the slide to detect loosening bolts. Because water on the deck can lead to rusted bolts, make a visual inspection and replace bolts with signs of rust. Finally, check the water supply line (if the system has one) for leaks that might deplete the pool water level, leading to equipment that runs dry and expensive repairs.

Ladders and rails

Ladders and rails are available in a wide variety of configurations to meet the design needs of any pool or spa. Some are anchored in the deck only, simply resting on the bottom or side of the pool, while others are anchored in the deck and pool.

Installation

Installation of a rail or ladder is easy, in fact easier than the manufacturers try to make it. Most installation kits have you drill holes in the concrete deck and/or pool floor and fit the holes with bronze cups that are equipped with bolts and wedges. You mount the cup in the hole, secured by some quick-set concrete, then put the legs of the rail or ladder into the cups and tighten the wedges. I do not use this method because the bolts and wedges ultimately come loose, allowing the ladder or rail to fail under someone's weight. These cups are designed to allow easy replacement of the rail or ladder in the future, but I have found that if the material of the rail or ladder has discolored or rusted, so has the metal cup and bolt system. My installation is more secure and if you need to replace the item, you can cut the old unit out with a hacksaw and drill out the deck holes with a concrete drill, just like creating holes for a new installation. In the long run, safety is the key and my method is much safer.

1. Fit the pool or spa for the desired rail or ladder by deciding where you want the item and measuring the available space. In the case of a ladder (Fig. 7-7), it needs 36 inches of deck width and 24 inches depth below the waterline. With a rail, you measure the horizontal distance from the deck to the end of the steps. With these measurements, your supply house can provide the appropriate rail or ladder.

2. Rent a concrete drill from your local tool rental yard with a 2-inch bit. Most rails and ladders are made from 1½-inch diameter stainless steel pipe.

3. Position the item on the deck as needed and mark the spots where the legs touch the deck and pool. In the case of the rail, which also needs to be anchored in the pool bottom, the pool must be empty for this procedure. Drill out the holes in the deck and pool bottom to a depth of 3 to 6 inches, depending on how high you want the final installation to be.

4. Dry fit your ladder or rail. Slide the escutcheon plates into place (the decorative rings that conceal the holes in the deck or pool). Be sure to put them on facing the correct direction. After the item is set in the concrete, you can't remove the escutcheon plate. Set the item into the holes and be sure you have the desired fit and height. Be sure the item is level—that it doesn't lean to one side or the other. If you drilled your holes to the same depth, the item should also be horizontally level. Now remove the rail or ladder.

5. Mix a batch of quick-set concrete (available at any hardware store). Fill the holes with the mixture and set the ladder or rail into the holes as deeply as it will go. The concrete should fill and surround the hollow pipe. Keep it in position until the concrete sets enough to hold the rail or ladder by itself.

   Let the concrete dry overnight before refilling the pool or applying any pressure to the item. Slide the escutcheon plates into place to conceal the installation holes and test the rail or ladder. Apply pressures and forces consistent with normal pool or spa use to test your installation.

SAFETY EQUIPMENT

Commercial installations require certain safety equipment that is a good idea for residential sites as well, because pools and spas quickly become the most popular meeting place in the neighborhood.

Signs

Most building or health department codes require certain signage to be prominently displayed near the pool or spa. Check your local codes for specifics and required proximity to the water's edge. Many pool and spa technicians ignore these signs, but I advise you to be more careful. First of all, providing and posting these signs is a job that can generate profits—you mark up the signs like any other piece of equipment you sell. Also, you charge for the labor to install the signs.

More important however, if you are the service technician for a commercial pool and someone is injured, the first thing their lawyer will check is if the appropriate warning signage was posted. If not, you might be held liable for negligence. At the least, if the customer gets sued, you will probably lose the account, so it pays to do this job right.

The basic signs (Figs. 7-8 and 7-9) you need (some or all, depending on your local codes) are

- Maximum bather load

    ~This sign usually says, in simple bold, block letters Occupant Capacity ——
    —— or Maximum Occupants ————. You fill in the blanks, depending
    on the local codes. Many jurisdictions allow 15 square feet of surface area

7-8
First-aid procedures.

for each bather, so a 15-by-30-foot pool would allow a maximum of 30 occupants at a time (450 square feet of surface area ÷ 15 = 30). Other rules guide spa occupancy or wading pools for kids, so check local requirements before writing the number in the blank provided. Use paint or an indelible marker so the number cannot be changed by pranksters.

- Artificial respiration

~This is a sign that diagrams artificial respiration (Fig. 7-8) and first-aid techniques in case of emergency. It's a good idea to familiarize yourself with these techniques by at least studying the sign (or better yet, take a first-aid class), because you will be around pools and spas all day long and sooner or later you will run into someone who needs help.

- Warning: No lifeguard/no diving

~This sign advises bathers that no lifeguard is on duty at this pool or spa and, if appropriate, no diving is allowed.

- Emergency phone numbers

~This sign prominently displays 911 for general emergency calling and also provides spaces to include the phone numbers of a local emergency room or doctor, fire department, and police. Since most communities now have the 911 emergency number, some jurisdictions only require this to be posted.

- Pool/spa rules

~This sign is available with residential or commercial rules and includes hours for use, age restrictions of users, common-sense rules such as No glass or No running, etc. These rules might be generic common-sense rules or might have specific language required by the state. Arizona and Florida, for example, have their own specific requirements for this language.

- Emergency shutoff

~This sign is posted next to the electrical switch that cuts all power to all equipment, in case of emergency. It simply makes users aware that the shutoff exists in case someone is injured in the pool or spa or, more often, when clothes get sucked into powerful pump suction openings in the spa.

- Danger: pool closed

~This sign is a good one to keep in your truck in case you need to close down a pool. I carry some glow-in-the-dark plastic ribbon, available at any hardware store, to tape off a pool or spa area that must be closed and also mount the sign prominently. You use this in case you have to drain a pool or spa or when chemical or repair procedures require closing a pool. The sign and tape prevents accidents and keeps you from being liable if someone jumps into an empty pool.

Toss Rings

A toss ring (Fig. 7-9), like a life ring on a ship, is a foam plastic ring (17 to 24 inches in diameter) that has a rope attached. It should be mounted prominently near the pool for rescue purposes if someone is drowning. Most jurisdictions require the toss ring for commercial pools and spas only, but I strongly urge you to sell one to your residential customers as well.

7-9 Signage, toss rings, and life hooks.

Life hooks

Like the toss ring, the life hook (Fig. 7-9) (a pole with a large metal hook on the end) is required to be mounted near the pool or spa to help someone out of the water if needed. The pole is 6 to 9 feet long with a broad, looped hook on the end (made so the hook doesn't do more harm to the victim than the drowning).

Thermometers

Some commercial pools require an in-water thermometer be available at each pool and spa and most residential customers want one too. For commercial installations, I recommend a unit that is built into the skimmer cover (Fig. 7-10) and takes the temperature at the skimmer. Other models usually disappear or are tampered with. You can also simply tie a thermometer to a rail or ladder, or there are tube models with a float on the top that float on the surface of the water. These usually float into the skimmer basket, which is a good place for them to keep them out of sight. You and the health department inspector need to check the thermometer, the 200 residents of the apartment building don't.

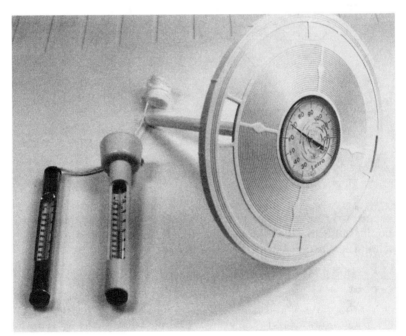

7-10 Styles of thermometers.

In-line thermometers, designed like in-line flow meters, are also available for installation into the equipment plumbing or directly to the heater manifold. For you high-tech techies, digital readout, battery-operated thermometers and pH testers are also available in floating models or with test probes that you put in the water while reading the information on a small hand-held device about the size of a calculator.

AUTOMATIC POOL CLEANERS

There are numerous designs of automatic pool cleaner available today (Fig. 7-11). Some of them are no longer manufactured but you might encounter them in existing installations.

7-11 Styles of automatic pool cleaners.

There are two categories of automatic pool cleaner in common use today and three other less effective technologies that might still be lurking around a pool you are servicing. Let's dispense with those three first.

Electric robot

The electric robot type of automatic pool cleaner sounds futuristic, but is nothing more than a battery-powered (some are 120 volt, some transformed down to 12 volt) vacuum cleaner with a bag that catches debris as the unit patrols the pool bottom. As you might guess, these are expensive and are found most often on large commercial pools.

Boosterless water pressure designs

As you will see in a moment, some automatic cleaners work like your leaf vacuum (see the chapter on cleaning and servicing) by sending a pressurized stream of water up into a catch bag, creating a vacuum for the unit as it patrols the bottom. The pressure is created by taking return water and boosting the pressure with a separate pump and motor. There is nothing wrong with this boostered type of automatic cleaner.

A variation on this theme, however, is the unit that uses the circulation pump itself as the booster, by connecting to the return line after the pump but before the filter. This style is called boosterless because it uses no separate water pressure boosting device. These units require an automated valve and control system so that the heater doesn't try to operate while the cleaning unit is in operation.

Herein lies the great failing of these systems. If the cleaner and heater are on simultaneously, the cleaner receives the return water before the heater, thus starving the heater. Low water circulation in the heater will cause it to shut off or overheat. Boosterless cleaners are not popular units because you can't filter and heat the water at the same time you vacuum debris, and because expensive additional plumbing and controls are needed.

Self-cleaning systems

The idea of the self-cleaning pool was clever, but I have yet to see a design that realized the dreams of the promise.

The bottom of the pool is fitted with a series of jets, connected to a diverter at the circulation equipment area. As the water leaves the heater destined for the pool, it passes through the diverter which sends it to the floor jets on the shallow end first, then the next deeper jets, and the next, and so on.

The theory was that the jets would push the dirt from the shallow end to the deep end, each jet sweeping the dirt toward the deepest part of the pool where the main drain would suck it into the filter system. Of course this presumed that the bottom had only fine dirt, not sand or leaves or other heavy debris that would not move, or, if it did, would not clog the main drain when it arrived. Further, it assumed dirt would not stick to the bottom of a pool, which it does. Last, it assumed that these jets, some of which are made like pop-up sprinkler heads that rotate 360 degrees, would equally cover all areas of the bottom. Obviously, the jet sweeping action is greatest near the source, then gets progressively weaker as the jetstream moves outward, resulting in uneven cleaning. Finally, these systems assume that return line water pressure will always be strong enough to power such a system. But dirty filters, solar heaters, and clogged skimmer baskets and pump strainers all conspire to make these systems unreliable.

The other problem with this design is the stress it creates on the entire plumbing and equipment system. If you are able to observe one of these systems in operation, watch the pressure gauge on the filter for a few minutes. Each time the diverter switches flow from one jet (or pair of jets) to the next, it is like turning a diverter valve. For a moment, the circulation is severely restricted or closed completely until the diverter lines up with the next jet. The pressure in the system spikes off the scale before releasing into the next jet, creating backpressure and stress on the entire system three or four times a minute.

You might get the feeling I don't like these self-cleaning designs. To improve circulation and remove unnecessary stress on the system, I disassemble the diverter in the distribution plumbing so that return water flows evenly back to the pool through all of the jet outlets. I suggest you advise your customers of the shortcomings of these systems and do the same. The distributor is usually a domed-shape plastic helmet with six or eight pipes coming out the bottom (one goes to each return jet). Take the distributor apart by removing the retainer clamp (like a filter) or the bolts and remove the helmet. The diverter design will be obvious once you get inside, and all you have to do is remove the diverter wheel.

Now I have discussed the three systems I don't think you'll see much of, so let's move on to the two that you will.

Booster pump systems

As just described, booster pump systems take water after the filter and heater, which is already on its way back to the pool, turbocharge or pressurize it by running it through a separate pump and motor, then send this high-pressure water stream through flexible hoses into a cleaner that patrols the pool bottom.

In one style, called a *vacuum head* type, the cleaning device has its own catch bag for collecting debris, much like a vacuum cleaner. In another variation, the *sweep head* type floats on top of the water with long flexible arms that swirl along the walls and bottom, stirring up the debris. A special basket is fitted over the main drain so that the stirred-up debris is caught in either the main drain or the skimmer and any fine dirt is filtered out normally. Let's review the details of each type.

Vacuum head type

Although there are several manufacturers, Polaris Vac Sweep is the leading maker of this style and will be the example. As with other pool and spa equipment, if you understand the leading manufacturer's equipment, you will easily comprehend the operating concepts of the others.

Figure 7-12 shows a vacuum unit with catch bag. (Model 180 is the most common type already in service, although newer models are now available.) Pressurized water from the booster pump enters the unit through the stalk and some is immediately jetted out the tail. This water pressure causes the tail to sweep back and forth behind the unit to brush loose any fine dirt on the bottom that is then filtered out by the pool circulation system. The remainder of the water powers a turbine (item 20) that has a horizontal shaft with gear teeth to engage comparable gear teeth on the inside of the single left-side wheel (item 21) and the front right-side wheel. A small right-side drive wheel (item 2) transfers power to the trailing right-side wheel as the unit moves forward.

Some jetted water is diverted to the thrust jet (item 26) which can be adjusted up or down to help keep the unit from moving nose-up. The head float (item 31) also serves this function and keeps the unit upright.

I have found the Vac Sweep to be an effective and valuable pool cleaning tool, saving me countless hours. They can also be a valuable source of installation, repair, and maintenance income to the pool technician.

Installation Vac Sweeps are available as preplumbed units (where the supply pipe from the equipment to the pool area is plumbed into the original pool plumbing) or as over-deck models, which requires a garden hose be run from the equipment area over the deck to the pool's edge. The booster pump and vacuum unit are identical, only the plumbing between the two are different with these two models. A complete installation guide is provided with either unit when purchased, so I will outline here only the general steps to give you an overview of what you will be doing.

Polaris180
VAC • SWEEP

Install
dished side
out

Install
cap side in

Serial number

1 Lock washer	20 Drive shaft assembly
2 Small wheel	21 Large wheel
3 Ball bearing assembly	22 Turbine cover
4 Backup valve (not shown)	23 Flow elbow
5 Spring	24 Sand and silt bag
6 Axle screw	25 Turbine ball bearing
7 Body screw	26 Jet assembly
8 Swing axle kit	27 Wheel washer
9 Bottom housing	28 Sweep hose weight
10 Wheel screw	29 Frame
11 Component assembly screw	30 Turbine shaft tube
12 Tire	31 Head float
13 Hose clamp	32 Self tapping assembly screw
14 Hose	33 Axle
15 Housing top	34 Turbine shaft shield
16 Flow adjustment screw	35 Orifice tip and guard
17 Large axle (rear wheel)	36 Sweep hose clamp
18 Axle plate	37 Wear ring
19 Axle bolt/nut assembly	

7-12 The vacuum head type of automatic pool cleaner. Polaris Pool Systems, Inc., San Marcos, Ca.

1. Install the booster pump. The Polaris booster pump is a ¾-hp pump and 120/240-volt motor. The electrical connections are the same as described in the pumps and motors chapter, normally powered through a time clock.

   The pump's suction plumbing is taken from the return line of the pool after any other equipment (after the heater and solar panels) but before an automatic chlorinator, if any. Where the circulation system supplies both pool and spa and is diverted to one or the other by a three-way valve, you must install the Polaris plumbing after the three-way valve on the return line to the pool only. If you added it before the three-way valve, when you circulate the spa water, some water will be returned to the pool through the booster pump line, slowly draining the spa.

2. Follow procedures outlined in the pumps and motors chapter for setting, plumbing, and electrifying the pump and motor.

3. Plumb from the booster pump to the pool. If an automatic cleaner line was included in the pool construction, as it probably was if the pool was built in the past 15 years, this step has already been completed. If you are adding over-deck plumbing, either hard-pipe or use a garden hose to run the water supply from the pump to the pool edge. Garden hoses tend to fail, so I strongly recommend ¾-inch PVC pipe for this installation, and in either case, run the line through garden areas or along building edges as much as possible to avoid the line being run over, kinked, or kicked. Make the shortest open-space run possible. Bring the line to the middle of the pool's longest side.

4. Add the feeder hose. A long plastic hose (30 feet provided with the unit, and additional 10-foot lengths available) is supplied to run from the booster pump supply pipe to the vacuum unit itself. Every 10 feet a swivel connection is provided so the hose doesn't coil up during use. Lay the hose out to its full length, preferably on a hot deck. Let it lay in the sun for an hour or so. The plastic has a memory and wants to recoil itself back into the form it was as packed in the box. This layout procedure helps the plastic hose relax.

5. One end of the hose is fitted with a female snap connector to attach to the male end that is plumbed to the end of the supply pipe from the booster pump. Connect this fitting together. The other end fits over the top of the stalk on the vacuum unit itself. A securing nut is provided (with reverse threads) to attach the hose to the stalk. When these break or wear out, I use a 1-inch stainless steel hose clamp instead.

6. Walk the supply hose and vacuum unit around the pool deck. If you can reach all around the pool, you have enough hose. Allow enough for the depth of the pool at the deepest end. Now cut off any excess hose length or add more sections if it's too short. After taking the vacuum head unit out of the box, you might need to attach the tail, fitting the open end over the tail jet on the stalk and securing it with the hose clamp provided. Add the catch bag provided, secured by a Velcro strip. Set the unit on the deck and turn on the pool circulation pump.

7. After the air has purged from the lines, you should see water squirting from the tail, the thrust jet, and out through the catch bag on top. Lower the unit into the pool and turn on the booster pump (it is educational to try this once while the unit is still on the deck so you can observe the difference between running the water with and without the booster in operation). The unit will now patrol the bottom.

8. Adjustments. Each pool has design quirks that make adjustments to the vacuum head unit necessary.
   • Is the tail sweeping vigorously, but not so hard that it flies off the pool bottom? The tail jet flow can be adjusted by tightening or loosening the screw (item 16) that acts like a valve to open or close the flow to the tail.
   • Is the vacuum unit running nose up? Does the unit go in left or right circles? If so, loosen the holding screws (item 11) of the thrust jet and

adjust the aim of the thrust jet to compensate for the misdirection. A few experiments with the thrust jet position will teach you this technique.

- Is the unit hanging up on steps, ladders, or deep corners? You might need to shorten or lengthen the supply hose to allow the unit to turn around more easily when it reaches a corner. If the hose is too long, the unit might just circle in one area.

After each adjustment, watch the unit for several minutes. Some problems work themselves out and others might not be immediately apparent, so take your time with this procedure to get the best results. Follow additional directions in the manufacturer's instruction booklet.

Operation A few operational guidelines will help you keep the vacuum head automatic cleaner cleaning the pool efficiently. Never operate the booster pump without the circulation pump also working. The booster is not self-priming, it relies on the system circulation pump to provide water. If it runs dry, the plastic pump will overheat and warp the seal plate and burn out the seal.

Set the booster time clock to come on at least one hour after the circulation pump and to go off at least one hour before the circulation pump does. This allows for slight time differences between the clocks. The vacuum head will probably cover as much of the pool as it's going to cover in about three hours. More than that and you are just wearing out components. In windy or very dirty locations, you might want to set it to operate for two hours in the morning and two hours in the afternoon.

Before putting the catch bag in place, slip the end of an old stocking over the opening and secure it with a rubber band. Then install the catch bag. The fine mesh of the stocking will capture fine dirt and sand, saving you from having to vacuum the pool so often. Some supply houses sell small stockings made for this purpose, which might be easier than constantly raiding a nearby panty drawer.

Empty the catch bag and/or sock as needed, but don't be obsessive about it. The bag actually works better with some leaves or dirt in it because the debris helps filter as well, catching fine particles in the leaves. Make sure the openings on the bottom and through the center of the unit are not clogged with large leaves so there is always a clear path for the debris to get into the bag.

Watch the unit perform for a few minutes each time you service the pool. If it runs listlessly or fails to run in a pattern that will clean the entire pool, follow the adjustment procedures outlined in the installation section.

Repairs Perhaps the simplest way to explain the few repairs needed by these cleaners is to list the symptoms of the problems you might encounter.

- Water is not flowing out one or more of the jets in the vacuum unit. Because the jets inside the unit are small, grains of sand can clog them. A fine mesh strainer is provided that is installed at the connection of the plumbing to the feeder hose to catch any small particles or DE that might get past the filter system. Dirt or sand can, however, be picked up by the unit and clog any of the internal jets. If this happens, there is probably sand or dirt in other parts of the unit as well, creating excessive wear and abrasion on the drive system, tires, and other components.

Using Fig. 7-12, disassemble the unit and clean each part thoroughly. As with every other disassembly and repair procedure in pool and spa work, carefully note how the unit comes apart so that you will know exactly how to put it back together. Use a thin wire to clear out the jets. Follow the path of the water and simply clean it all out.

- Wheels are not turning. If the jets are all clear, there are two other reasons for sluggish performance. First, if the booster pump is not getting enough water because of restrictions in the main circulation system, the vacuum unit will not perform well. Clean the filter or address whatever other problem is creating poor circulation in the entire system and you will usually find that the automatic pool cleaner works better as well.

  Second, over time the metal drive gear wears out the plastic drive gear inside the wheels. Check to make sure the gears are meshing and that there are enough teeth on the inside of each wheel. If the wheels are sloppy, they will also fail to properly engage with the drive gear.

  Replace the wheel bearings (item 3), which simply pop in place like a pump seal. Last, as the tires (item 12) wear out, they do not properly engage on the drive wheel, so replace them. They simply stretch in place over the wheel like a large rubber band. Polaris provides a tire gauge to help you judge the thickness of the tire all around its circumference to determine when you should replace them.

- Vacuum unit falls over. Remove the head float by gently but firmly pulling it off of the stalk (be careful—the stalk it is mounted on can easily break). If it is full of water, it is not floating the unit upright. Replace it.

- Stalk breaks. Polaris sells a mast repair kit. Using a hacksaw, you cut off the mast just above the tail jet. The replacement glues in place with PVC glue and includes a new headfloat stem. Simple, detailed instructions are included with the kit.

- The wheel screw breaks. The screws that secure the wheels are plastic and overtightening will snap them off. If this happens, you can gently tap a flat-blade screwdriver into the broken stem to unscrew, remove, and replace the screw.

- Tail wear. The tail assembly will be the first thing to wear out because it is constantly sweeping the pool bottom and sides. Water will squirt out of parts of the hose where it shouldn't, making the tail swing wildly. To help prevent this, the tail is fitted with rubber rings that absorb the wear, so as you see these donuts wearing down, replace them before the tail goes.

- Wheels seize up. Sometimes the drive wheel (item 2) gets hung up and actually prevents the wheels from turning. Tension for the drive wheel is spring loaded (item 8 assembly), and as the spring wears out the tension will be too much or too little. (The new model 360 uses an internal belt-driven propulsion instead of this cumbersome assembly. The folks at Polaris might not agree, but I have actually removed this drive wheel completely from the units I service and had great operational results. The turbine powers the front wheel on the right side and the single wheel on the left side with the rear right-side wheel just trailing behind. The unit works fine and the wheels never seize up. Try it.

- Unit gets caught in ladder, corner, or steps. The irregular-shaped pools that are popular today are the automatic pool cleaner's nightmare. If all adjustments and hose lengths are correct but you still have problems, a backup valve is the answer.

Installed about halfway down the feeder hose, this valve shuts off the water supply to the vacuum unit about every five minutes, shooting the water out of the valve to act as a jet to pull the unit backwards. It sounds complex, but it is actually amazingly simple. Read the directions that come with the backup valve for installation and servicing instructions. They work very well.

• Unit runs too fast, just skipping over the dirt. On some pool systems where the return water pressure is very strong, the vacuum head pressure is too great for normal operation. Polaris makes a pressure tester, available for your own assembly from a kit. Follow the simple instructions provided and test the pressure at poolside to determine if pressure-reducing washers are needed. If so, this simple reduction technique employs a washer with a smaller diameter than the plumbing, thus restricting the amount of water that can flow to the vacuum head.

This pressure tester is a valuable tool to use when you suspect inadequate pressure might be the cause of sluggish operation. Pressure values and test techniques are explained in the installation booklet or test kit instructions.

Polaris (and most other manufacturers) offers terrific seminars all over the country (often with dinner included) that give you hands-on training in teardown and rebuilding, troubleshooting tips, adjustment procedures, etc. These seminars will not only give you a good foundation in automatic pool cleaning technology, but they will keep you up to date on new features and products. For these reasons, it is worth attending one each year, but also because the first time you might not comprehend the subtleties or ask the important questions. After some field experience with automatic cleaners, you will get more out of each seminar. Attend one.

Sweep head type

The other type of cleaner that uses a booster pump is a floating head that controls long sweeping arms which brush around the pool. Arneson is the leading maker of this technology and will be the example here.

Figure 7-13 illustrates this technology. The booster pump, plumbing to the pool, and feeder hose to the head are essentially the same as the Polaris booster system described previously. The head of the Arneson Pool Sweep floats on the water surface and is driven around the pool by a propeller underneath. Some water drives this propulsion system and the remainder fills the sweep tails which stir up dirt and debris from the bottom. A leaf basket goes over the pool main drain so that suction will pull the stirred-up debris into the basket for later removal.

Although some customers swear by these units, I am not a big fan of them. You can easily swim around the Polaris-style units, but the big sweeping arms of the Arneson are forever getting in your way when you must share your pool with one. Moreover, the technology relies on an efficient main drain to catch debris, which is rarely true because so much suction in modern pools is taken at the skimmer. The tails get tangled in ladders or their own connector fittings at the pool wall, even when using the ladder guards provided. It has no way to remove coarse sand and dirt, relying again on the pool circulation system. In short, I can brush a pool in about three minutes and accomplish what these units do all day.

If you service a pool with one of these units, they are not hard to maintain or figure out. Installation, operation, and troubleshooting guidelines are very much as described previously. Because of the obvious drawbacks, few new systems of this

Head
assembly

Head float

Supply hose

Quick
disconnect
Hose nut

Sweep hose

Deck
mount

Ladder
guard

Main drain
debris catch
basket

7-13 The sweep head type of automatic pool cleaner. Arneson Products Inc., Fairfield, Ca.

type are installed, so as time goes by you will encounter fewer of these in the field and most will be old units operated by customers who have put too much money into the system to throw it out.

You can, however, point out to a customer that the expensive part of the system is the booster pump, electrical system, and plumbing. All of these components can be used with a vacuum head system, so the customer only needs to replace the sweep head with a vacuum head to have a more efficient system.

Suction-side systems

Suction-side automatic pool cleaners work off of the suction at the pool's skimmer. It is like vacuuming the pool without a pole or pool person. In this design, a standard vacuum hose of 1½-inch diameter is connected between the skimmer suction opening at one end and a vacuum head that patrols the pool bottom at the other end (Figs. 7-11 and 7-14). Arneson makes a very good unit in this design (Pool Vac).

The problem with these systems is that the skimmer was designed to skim dirt and debris from the surface of the pool before it sinks to the bottom. By using the skimmer's suction for another purpose, everything sinks to the bottom. Special valves can be installed on the skimmer to allow some suction to continue the skim-

7-14

A suction-side automatic pool cleaner.
Kreepy Krauly USA, Inc.

1 Wall climb adjuster
2 Float restrictor
3/4 Dive float assembly
5 Weight holder
6 Main body
7 Wing
8 Pleated vacuum seal
9 Foot pad
10 Swivel assembly
11/12 Roller bumper strap
13 Drive tubes
14 Hose connector
15 Valve cap
16 Regulator valve
17 Valve plate
18 Extension ring
19 Locking bar
20 Compression adapter

ming action while the remainder operates the vacuum head, but few pumps have enough suction to do both efficiently, so you end up with two systems that work poorly. Some valves alternate the suction between the skimmer and the vacuum head, but then you are back to the original problem that when the vacuum head is running, the skimmer is not.

The second major problem with these systems is that the debris collected is sucked directly into the pump strainer, often overloading the basket and cutting off circulation, making the entire system of filtration and heating inefficient. Canisters have been designed to float in-line along the suction hose, but when these clog, you are still slowing the circulation of your pool which means it is not filtering or heating effectively.

As you might notice, I am not a fan of suction-side systems. I think the additional cost of purchasing and operating a booster pump system is well worth the added efficiency. By the time you add the leaf canister and skimmer suction valves, the price differential is not that great anyway.

These systems might be good for homeowner use if someone is willing to check on them and clean them out daily, but the typical pool serviceperson comes once or twice a week and will be blamed if the pool looks cloudy and the heat doesn't work because the system has gotten full between service calls.

Because of these design flaws, there are far fewer suction-side systems in the field. Kreepy Krauly is the leader in this technology. There is not much that goes wrong with these units, because they are essentially hoses with a vacuum head. Troubleshooting will usually find leaves and debris clogged somewhere in the system or the inability of the circulation pump to generate enough suction to make the vacuum effective.

To buy or not to buy

How do you advise your customer about whether to purchase an automatic pool cleaner in the first place? Well, here are some factors to consider.

Price
Obviously the cost of buying, installing, maintaining, and running the equipment is the first question (don't forget it costs money to replace parts as they wear out and costs more to run the cleaner itself, which is a good reason to use it no more than necessary and to keep the system pressure adjusted to specifications). Many customers are turned off by spending $1000 to $2000 on top of weekly pool service.

Expectation
Make sure your customer understands what the various types of systems can (and cannot) do. Many customers buy a system and then expect the pool to be spotless every minute of every day thereafter. It just doesn't work that way, so make realistic service predictions if you are describing or trying to sell one to a customer.

Existing pool and equipment
Will the pool actually benefit from an automatic system? Is it in such a heavily landscaped area that any system cannot keep up? Is the circulation system itself old and unable to do its part with an automatic system added? Is the design of the pool conducive to an automatic system or will odd contours within the pool, decorative rocks, tight corners, etc., cause problems? Look at any pool with an automatic

cleaner and you will notice dirt on the steps, loveseats, in corners, and up the walls. Is that satisfactory to your customers needs?

The pool service technician

Yes, you too are a factor in the decision. I have had customers buy automatic cleaners from me, then cancel the service because the system worked well and they were willing to do the chemicals themselves. Compare the profit of the installation job to the potential loss of your customer. Some systems just get in the way and add time to the service call, because you clean the pool and still have to clean and adjust the automatic cleaner.

Okay, that's the downside, but the upside is that you might make profits from the initial sale and more in future years because the system will need replacement parts and maintenance. Think about it.

To be honest, I don't think most pools benefit from automatic cleaners if they have regular service already. If a customer doesn't look at the pool all week, but uses it on the weekend, you set up your weekly service call for Thursday or Friday and the pool is fine during the time of its highest use. Most of my customers take twice or three times a week service, so the pool generally doesn't get dirty enough between calls to justify the expense. Then again, Olympic decathlon champion Bruce Jenner and his family named their vacuum head unit and treated it like a household pet.

LIGHTING

External lighting of your pool, spa, or water feature is possible with a variety of high- and low-voltage systems. For installation and maintenance of external systems, I suggest you consult your local electrician. For lighting the water from within, however, there are also a variety of options. This is where the water technician needs to be an expert.

Standard 120/240-volt lighting

Figure 7-15 shows a typical pool or spa light housed in a stainless steel conical-shaped fixture, about 8 inches in diameter by 6 to 10 inches deep. The fixture is mounted in the wall of the pool or spa in a container called a *niche*.

Like the lights in your house, pool and spa light fixtures have a standard, screw-in socket for a bulb (in fact, your household 100-watt bulb will work in a 120-volt pool or spa fixture). The cord that supplies electricity to the fixture is waterproof and enters the unit through a waterproof seal. There are no user serviceable parts in the cord, seal, or fixture except for changing the bulb. Although the correct term is *lamp*, for clear understanding I will use the more common expression *bulb*. Fixtures and bulbs are available in 120 or 240 volts, but 120 volts is most common for residential use. Bulbs generally run from 300 to 500 watts.

Fixtures are sold with cords of 10 to 100 feet, so you need to know the distance from the light niche to the junction box to determine what you need when purchasing a replacement fixture. A simple rule of thumb is to buy the next longer length than you think you need, because you can always use the cord for something else and the few extra bucks is cheaper than getting one a little too short and making another trip to the supply house.

7-15 The typical pool or spa light installation.

Smaller versions of these fixtures are available for spas. They come in 120 or 240 volts, but usually employ smaller-based specialty bulbs that might screw in or have bayonet-type sockets. A quartz or halogen bulb will give you the most light from a small low-wattage bulb. These new technology bulbs are very expensive but save energy and need replacement less often than hotter, higher wattage standard bulbs.

You might encounter very old fixtures that are basically designed the same as modern ones, but they have a *mogul* base. The socket for these are also female threaded, but they are about twice the diameter to accommodate bulbs with that mogul-style base. Replacement bulbs are still available for these fixtures, but I don't know anyone who still sells the fixtures (or would want to).

Light fixtures are sealed to be completely watertight, so the air inside reaches extreme temperatures unless it is cooled by contact with the water. Never turn these lights on if there is no water in the pool.

Most fixtures have lens overlays available in a variety of colors. These are plastic lenses that snap over the glass lenses to make the light blue, green, or whatever color. Being very thin plastic, these too will quickly melt if the light is operated out of the water.

Fixtures are held in the wall of the pool or spa in a light niche. A niche is a metal can large enough to contain the fixture, cemented watertight to the side of the pool. A waterproof conduit leads away from the niche to above the water level out of the ground to a junction box.

Installation

Since you probably won't be installing too many light niches on new pools, I will outline here the procedure for replacing a fixture when it rusts out or otherwise fails.

1. Turn off the power source at the breaker. Find the junction box. On older pools, this will be in the deck directly above the light niche, often under a 4-inch diameter stainless steel (or bronze) cover plate that is held in place by three screws that probably have rusted or stripped heads, making them impossible to remove.

   Later building codes required the junction box to be at least 5 feet from the edge of the water and 18 inches above the surface of the water, so on more recently constructed pools, look in the garden directly behind the light niche and you will often find it sticking up there.

2. Figure 7-15 shows a typical junction box (simply called a J-box). Remove the four screws and take off the cover. Three wires come into the box from the breaker or switch and three go out of the box to the light fixture. The three to the light fixture will be individually insulated, colored white, black, and green, and are bound together with a single rubber sheath that waterproofs the package. This is the cord of the light fixture as described previously. Disconnect the three wires. Unscrew the cord clamp (Fig. 7-16, item 9).

3. Lean into the pool (yes, you have to get at least an arm wet) and remove the face rim lockscrew (retaining screw) from the faceplate that holds the fixture to the niche. The top of the fixture will float outward, the bottom is hooked into the niche and can be lifted out.

4. Uncoil the excess cord to give you enough slack to raise the fixture out of the water onto the deck. Go back to your J-box and tug at the cord of the fixture. If you see the cord move at the fixture end, you know the replacement is an easy one. If it's tight, follow the steps outlined later (see tight cords).

5. Cut the cord where it enters the old fixture. Strip off the rubber sheath, back about 6 inches. Remove the string or paper threads that run alongside the wires (these were put in the cord to add strength for when you pull on it). Remove the insulation from each single wire back about 6 inches.

6. Remove the new fixture from the box and float it in the pool. I have laid the new fixture on the diving board or deck and it always gets tugged off or kicked, shattering the lens, so just get in the habit of laying it in the water and you won't face this problem. Take the wires of the new fixture and strip the cord back as described in step 5. Now bind the wires of the new cord to the ones of the old, folding each wire over as if to make a hook (Fig. 7-15), then connecting the two (like two fishing hooks) and twisting the loose end around the base wire. Use electrical tape to cover the exposed wire, wrapping it tightly and thoroughly.

Don't tape so much that you make a connection thicker than the cord itself. It won't pass through the conduit easily. The idea is to make a union of the wires that will not separate when you pull.

Lay the cord out freely into the water. Pull the old cord at the J-box until you have pulled the connection and the new cord through. Some niches have the conduit connector in the top of the niche pointing upward at an angle, so when pulling the cord, it might snag at your connection.

To solve this problem, have someone lean into the pool and angle the cable down into the water, at the same angle as the conduit connector itself. That person can feed the line as you pull from the other end. Keep pulling until you have enough cord left with the fixture for it to easily reach up to the deck for future bulb-changing access. Now untape your connection and remove the old wire.

7. Reach into the pool again and coil the excess line around the fixture, then reset the fixture in the niche. Replace the lockscrew on top.

Sometimes, particularly if the fixture is way below the waterline, you might have to get into the pool to loosen or reset the fixture. If you're on the deck and having trouble lining up the face rim lockscrew with the hole of the niche, use a piece of coat hanger wire. Slide it through the hole on the faceplate and into the hole of the niche. Slide the fixture into place and the holes will line up for the screw. Remove the wire and insert the screw. Cut off any excess cord at the J-box, leaving enough to make a good connection with the supply wires. Reconnect the wiring and close the J-box. Test the fixture.

Tight cords

If the cord won't budge from either direction, the cord might have swelled from age, heat (electricity creates heat), or moisture. Try pouring tile soap down the conduit from the J-box. You might have to let it sit overnight to slither all the way along the conduit and be effective. Some service technicians use oil, but whatever you use will eventually end up in the pool, and I'd rather have a little nonfoaming tile soap than an oil slick.

Another trick is to use leverage. Place a long 2-by-4-inch board down into the pool and run the cord up along its length. Using the edge of the pool coping as a fulcrum, try to pry the cord away from the fixture or get your local electrician to help you with his wire pulling levers (but if you can't get the cord out with a 2-by-4, he probably can't either). Another trick is to cut the cord as low as possible at the J-box side of the conduit. Using your drain flush (blow bag) and the garden hose, try forcing water from the J-box to the pool through the conduit to loosen the cord. Compressed air might also work. Be careful not to exert too much pressure with these techniques, however, because you might crack the conduit, which might be brittle PVC or thin, rusting copper.

Sometimes a combination of air pressure and soap will do it. Be patient and persistent. If all techniques fail, the conduit will have to be dug out and replaced or the light abandoned.

Sometimes the cable comes out, but the new one tears off as you are pulling it. Remove the new cord and use an electrician's snake to run through the conduit, starting at the J-box and working toward the pool. When it comes out the other end, hook the wires of the new fixture to it and try to pull them through again.

Dick Clark's spa had an old copper conduit that had broken and the snake wouldn't go through because it kept hitting the break in the conduit and pushing outside of it. When I tried to retrieve the snake, it got caught in the broken conduit and wouldn't come out. I had to cut off 10 feet of my snake, and because Dick didn't want to cut up his deck to replace the conduit, I put the old fixture back in the niche for cosmetic purposes and to this day it has no light. Also to this day 10 feet of my snake is in his spa conduit.

Replace the bulb

The most important factor in bulb replacement is to maintain the waterproof integrity of the fixture. Follow each step carefully to avoid leaks that will not only damage the new bulb and fixture, but might also lead to electrical shock of the next swimmer.

1. Turn off the power, then get the fixture up on the deck as described previously.
2. Remove the lens (Fig. 7-16, item 2) that is held in place by a clamp or a set of screws. Gently pry the lens away from the fixture, taking care not to gouge the lens gasket (item 3).

1 Bulb
2 Lens
3 Lens gasket
4 Washer
5 Stainless steel face plate
6 Assembly screw
7 Clamp assembly (including 9, 10)
8 Colored lens cover

7-16 Light fixture assembly. Triton Pool Systems, Sanford, N.C.

3. Replace the bulb. Inside some fixtures you will find a bare coiled spring wire. This is nonelectrical but is designed to break a circuit. Notice that without a bulb in place, the spring lays to one side of the fixture. Hold it up against the opposite side and screw in the new bulb. The spring lays on the bulb itself.

If the bulb bursts when in use, the spring sweeps across the filament, cutting the electricity in the circuit. In this way, if water has gotten into the fixture, a live electrical circuit won't stay in contact with the water, ultimately electrocuting someone in the pool.

4. Now place the fixture on the deck and turn on the light to make sure your new bulb works. Never operate a closed fixture out of the water, but with the lens off, the heat can escape without problem.

5. Reassemble the fixture. For the $2 extra, I always use a new gasket. After long use, heat, and harsh chemicals, the old one is probably compressed, and if it doesn't fail immediately, it might fail before the next bulb change.

6. In reassembling the faceplate, if it has the series of screws, tighten them on opposite sides until you have done them all. This applies even pressure and prevents gaps in the gasket that will ultimately leak.

7. Lay the fixture back in the water. Hold it underwater for several minutes to make sure it doesn't leak. A few bubbles might rise from air trapped under the lip of the faceplate, but a steady stream means the fixture is filling with water. Take it apart again and dry it thoroughly. Go back to step 4 and be more careful.

8. If it passes the leak test, turn it on for a few seconds before putting the fixture back in the niche. During reassembly or testing you might have banged the unit around enough to break the bulb or its delicate filament, or you might have a bad bulb. I hate getting wet and wasting the time getting the fixture back in the niche, only to find out that the bulb was damaged during the process. Remember, the closed fixture depends on water for cooling, so conduct this test for literally no more than two seconds. Now reset the fixture in the niche as outlined previously.

Safety

A few words of wisdom about lights and pools. Electricity and water mix only too well, often with deadly results. Therefore, since you can be injured or your work might make you liable for someone else's injury, take note.

- If a fixture, lens, or gasket looks suspicious, replace it. It's not worth the time, hazard, or money to "try that old one, one more time."
- Use the bulb, gasket, or replacement lens that fits the fixture. Try to get the same manufacturer's replacement parts or use the generic brand designed for that make and model. You might be able to force a bulb, gasket, or lens into place and it might stay watertight for a few days, but what happens when it ultimately overheats and leaks, bringing water, swimmer, and electricity into contact?
- Stubby bulbs are popular replacements. They are squat, short bulbs so they fit almost any fixture. The problem is that by being short, the heat is brought closer to the end of the fixture than the originally designed long-stem bulb. Also, bulbs with a reflector material painted on the underside of the bulb will reflect more light into the pool, so you get the appearance of a higher wattage bulb. The reflective surface is also designed to direct heat away from the rear of the fixture. This is significant because the resin at the rear of the fixture, that makes it waterproof, can melt.

- For the same reason, don't put higher wattage bulbs in a fixture than it was designed to take. If it isn't marked and you can't read the old bulb, don't use greater than 400 watts in your replacement.
- When changing a bulb, as with any repair, you are required to work to local building and health department codes. This might mean bringing an older installation up to code by adding a ground fault interrupter (GFI) as outlined in the chapter on basic electricity.

Troubleshooting

Replacing a bulb or fixture as outlined previously is about all you will need to do with lights and water. But how do you know if it's the bulb, cord, fixture, or breaker? If the light won't come on, try this procedure.

1. Make a visual inspection. Look at the lens of the fixture (remove any color overlays first). If you can see water or black soot (the result of a bulb bursting in the fixture), you know it will have to come apart. If the water has filled the fixture, it might only look cloudy, so it pays to know what a dry, well-operating fixture looks like underwater.
2. Turn off the power supply at the circuit breaker. Open the J-box and expose the wiring connections. On old J-boxes mounted in the deck, the gasket often fails and allows water to fill the box. The water shorts out the wiring. Clean and dry out the box and replace the gasket and lid firmly.
3. If that is not the obvious problem, turn the breaker and light switch on and use your multimeter to verify that power is getting to the light fixture cord. If not, trace back the problem to the breaker or switch. Sometimes there is more than one switch, such as a remote control, and the problem might just be a second switch being off.
4. If the breaker trips off when you turn on the light, disconnect the light cord from the power wiring at the J-box and reset the breaker. If it still trips off, the problem is the breaker or GFI, or more rarely, the wires between the breaker and J-box have shorted together. If the breaker stays on, the problem is in the fixture or its wiring.
5. Set your multimeter to test circuit continuity. Touch one lead to the white wire of the fixture cord and one to the black. If you have continuity, the bulb and wiring are good.
6. Touch one lead to the green and the other to the white, then the black. If continuity exists between white and green or black and green, the wiring is bad. You can test bulbs that way too—touch one lead to the threaded part of the bulb base and one lead to the tip of the base. A good bulb will have a complete circuit (continuity), a bad one will not.

Low-voltage lights

Designed like their 120/240-volt cousins, low-voltage fixtures employ a transformer to drop the voltage to 24 volts. As a result, the highest wattage fixture does not deliver much light, so they are used in spas or fountains where lower wattage is sufficient to light a smaller area. Installation, repair, and replacement techniques are identical to those described previously.

Fiberoptics

The most significant recent development in water lighting is the use of fiberoptics. The hardware looks the same, but instead of running electricity to a fixture, the cord in the conduit contains thin plastic fibers that conduct light, not electricity. They terminate in a lens in the water to shed the light into the pool or spa. The light source and electricity are located safely away from the body of water. Most of these are used in smaller bodies of water because manufacturers have not yet developed fiberoptics with enough power to light large pools. But they will.

Troubleshooting and maintenance of low-voltage and fiberoptic systems are similar to those described previously. If you understand what has already been discussed, you will have no trouble figuring out these as well.

COVERS

Pool and spa covers have come into increasing use over the past ten years in an effort to save heating energy costs and water through evaporation. As water evaporates, the minerals in it are left behind in the remaining pool or spa water. Chlorine, which is made from salt, leaves minerals in the water as well. If you live in say, Malibu, where the water comes from wells or other very hard sources, this is a very real problem. Covers slow down the rate of evaporation and, therefore, reduce chemical use. Covers keep out leaves and dirt which also absorb chemicals. Keeping out the dirt means you vacuum less and clean your filter less.

Shielding the water from the sun further cuts down on chemical use. The result is the water in the pool or spa doesn't get so hard, so quickly. That saves your customer money in water bills, chemical costs, expensive draining and acid washing, and, as if that weren't enough, the heater does not scale up inside from hard water.

Covers also help retain heat. Some estimates say a cover alone will heat the pool or spa 10° to 15°F, meaning money is saved on gas or electricity. When you do turn on the heater, the water starts out warmer and heats up faster—a real benefit to a spa when you have a hot date on a cold night.

Selling covers can be a good profit center, because in addition to the first sale, sooner or later the cover will need replacing and you can sell the customer another. Cleaning the pool or spa will be easier and the customer will be happier because of the ecological, financial, and appearance savings they are making on their pool or spa. Okay, I'm sold. What are the choices?

Bubble solar covers

Have you ever unpacked something sent in the mail that is wrapped in that plastic bubble wrap? If you're like me, you squeeze the little bubbles between your thumb and finger to pop the bubbles, right? Well someone realized that large sheets of this stuff would make good pool and spa covers.

Figure 7-17 shows a typical bubble cover (also called a *sealed air* or *solar* blanket). In profile, the cover has one flat side and one bumpy side. In fact, the cover is made from two sheets of blue plastic (usually available in 8 or 12 mil thickness), heat welded together with air bubbles in between.

7-17 Solar blanket cover. Cantar Corporation, Youngstown, Ohio

The sun warms the air bubbles which transfer the heat to the water (thus the term solar cover). Similarly, the trapped air acts as an insulator for the heat coming up from the water. Always lay a bubble cover on the water with the bubble side down. In this way, the spaces between the bubbles also act as pockets for trapped air, further insulating the water.

Because they are thin, lightweight, and flexible, bubble covers can be cut to any size and are sold in large sheets from 5 feet by 5 feet for spas up to 30 feet by 50 feet for pools, with many intermediate sizes to fit any job. They are easily cut with scissors or a razor knife. Bubble covers are cheap, costing about 12 cents per square foot wholesale. They last two to four years depending on water chemistry, weather conditions, and user wear and tear.

The disadvantage of a bubble cover is that in heavy winds they can blow off or away into the next county. Also, they don't really keep out dirt or debris, because as you remove the cover the dirt falls into the pool. Much of the dirt does stay on the cover, meaning you have to spread it out somewhere and clean the cover as well as the pool.

On a large pool (especially if the cover has no roller), taking the cover off and putting it back on can be a real chore. If you drag it off, where do you put it? If you lay it in the grass or on a nearby deck, it might pick up more dirt that goes into the pool when you put the cover back on, or it might tear as you drag it back into place.

As the bubble cover ages, sunlight and chemicals make the plastic brittle, causing the bubbles to collapse and sending little bits of blue plastic into the pool and circulation system. In short, bubble covers are only good for their thermal properties, which are valuable, especially if you heat the pool or spa a lot.

Installation

As you might guess, installation of bubble covers is very easy.

1. Measure the pool or spa and buy a cover that will overfit the water surface. In other words, get a size larger than you actually need.
2. Lay the cover out on the water surface and leave it for two or three days (Some manufacturers recommend as many as 10 days, but unless it is very cold, I have not found any difference after two or three days). You can remove it to use the pool or spa during this period, but the idea is to give the material time to relax to its full size in the sun and to shrink to any degree that it might (I have found shrinkage is not more than 5 percent in any direction).
3. Using shears, scissors, or a razor knife, cut the cover to the water surface size of the pool or spa. As you walk around the cover, cut slightly less than you think you should—you can always go around again and trim off a bit more, but it sure is hard to add any back on if you cut too much. There are no frayed edges or seams to worry about.

Rollers

As noted, handling large bubble covers can be inconvenient. You can try to fold the cover into large folds before taking it off, which is easier with two people of course. You can also cut the cover into two smaller pieces, which might be easier to handle one at a time. Or you can buy a roller (also called a reel system). Figure 7-18 shows a typical roller. The concept is to attach the cover to the barrel of the roller with straps so that when you rotate the barrel, the straps roll up first, pulling the cover along and rolling it on the barrel.

7-18 Cover roller (detail). Odyssey Reel Systems Ltd., San Clemente, Ca.

Even these can be cumbersome without two people, but they do make storage, cleaning, and handling the cover much easier. Replacement parts are available for rollers and wheels can be added to make it easier to move the entire assembly out of the way when using the pool. Plastic covers are also sold to cover the cover. If left on the roller for extended periods, this cheap plastic cover keeps sunlight and dirt off the bubble cover, extending its life. Rollers cost about $250 wholesale.

Foam

Much like bubble covers, foam covers are sheets of lightweight compressed foam (⅛-inch thick) that float on the surface of the pool or spa. Because foam is much more expensive (about 75 cents per square foot wholesale) than bubble plastic, these are primarily used for spas. Installation is the same as with bubble covers, described previously.

Sheet vinyl

Particularly in cold climates where it is desirable to cover a pool in winter, sheet vinyl is sold like a bubble cover without bubbles. Sheet vinyl has little insulating value and doesn't even float very well, but it is very cheap. Sandbags or plastic ballast bags that you fill with water are sold with sheet vinyl covers to anchor the cover on the deck as it lays over the water (Fig. 7-19, page 239).

Some building and safety codes require a pool to have a cover, on the theory that someone falling into a covered pool will not sink to the bottom as fast. These codes don't require you to use the cover, of course, so most of the vinyl covers I have sold were to satisfy the building inspector then never taken out of the box. These covers are generally $10 to $20 for a 20-foot by 40-foot cover and can be purchased at any discount department store as cheaply as at your supply house.

A higher grade variation of the vinyl cover is a heavy-duty material reinforced with wire to make them kid-proof. These are hooked onto fittings in the deck and are a real pain to take off and put back on. They are a beast to handle. The hooking system usually requires dozens of fastening points which means lots of stretching, tugging, and fighting.

When I see a customer with a mesh or other hooking cover like this, they either agree to have the cover off the pool on service days or I charge an additional $10 per service call for the labor and time of cover removal and replacement. **Caution:** Before you accept a customer with one of these covers, try removing and replacing it yourself, no matter what the added fee. Large ones simply cannot be handled by one person.

Electric covers

Somewhat a misnomer because you don't cover the pool with electricity, I call a sheet vinyl cover on a track, rolled up on a roller that is operated by a motor an *electric cover* (Fig. 7-19, page 238).

I deal with these separately from sheet vinyl covers as discussed earlier because the vinyl used in the electric cover is usually heavier or reinforced with a fabric or thread pattern to make it substantially stronger. Also, you can buy and supply to your customer the sheet vinyl cover, but not the electric cover.

Electric cover systems, with the cover material stretched between tracks or rollers on the edge of the coping and a motor-driven pulley system to roll it up on a barrel roller (which is often concealed under the deck or in a bench-box at one end), are sold by several major companies like Poolsaver. You refer your customer to them and they make the deal, do the installation, make an obscene profit, and send you a $50 commission check. They don't work for you, they only work directly with the homeowner.

Anthony Pools, Montebello, Ca.

Loop-Loc Safety Swimming Pool Cover, Hauppauge, NY

7-19 Reinforced vinyl cover (electric roller system), a strength test for a deck hooked style cover, and a winter pool cover.

7-19 Continued. Cantar Corporation, Youngstown, Ohio

This, however, is not a bad thing. Electric covers are large, complex pieces of machinery. The cover can easily tear if it gets caught or bound up, because the motor doesn't know when to stop it. The pulley system and ropes can break or get bound up and the cover won't open or close, often burning out the motor. Switches and controls fail. Manufacturers sell small submersible pumps with their cover systems, because the owner is responsible for pumping the water off the cover after rainstorms. In short, within the first year, most of these systems need a lot of maintenance and warranty work, so you don't really want to be involved anyway.

Electric covers are, however, just what I had in mind when writing about all of the positive features of covers at the beginning of this section. They protect, save, and insulate, and, best of all, they keep dirt and debris out of the water. Of course, they can be a pain if a lot of debris gets on top of them because you have to clean the cover before removing it. They also provide protection against someone falling into the pool. These systems are generally $5000 or more installed.

Fiberglass systems

A system you don't want to install and probably won't see many of, but should at least have a passing knowledge of, is a fiberglass panel that lifts off the pool like a garage door.

Saudi Arabian billionaire Adnan Kashoggi bought a house near Pepperdine University so his kid could attend the school. He had a $30,000 cover system added to the pool of his $5 million house. Large fiberglass panels, fitted in an aluminum frame, rest on the deck over the pool.

At the touch of a button, motors driving long arms lift the panels 8 feet into the air and they can be adjusted to any angle. When they are not acting as a cover, the panels provide shade over the pool as needed. As uncommon as these are, they're probably the best cover you could hope for—a simple hosing to clean them off, no debris entering the pool, all the saving benefits of other covers, and true safety when in place.

Mesh covers

Where security is a real concern, you might use a plastic mesh fabric (Fig. 7-19, page 238) that stretches over the entire pool and anchors on the deck all around (like the reinforced vinyl cover described earlier and with all the related problems of handling). The middle part of the cover sags slightly into the water and the material is made waterproof for this reason. The actual installation is a series of hooks and eyes on the cover and deck that secure the unit in place.

These covers are strictly for security. They are a close woven mesh, like the mesh strips used on lawn chairs, reinforced with steel or other wires running in crisscross patterns within the fabric. Water can evaporate through these covers, although some water is trapped back into the pool.

Specialty spa covers

Some spa covers are made of the same material as pool covers; however, because of the relatively small size of spas, clever inventors are constantly coming up with creative designs and materials for spa covers (Fig. 7-20). Here are a few.

Wood

Any good carpenter can make a wooden cover for a spa that sits on the deck over the top of the spa. These are particularly popular with wooden hot tubs and where child safety might be a factor. They are sturdy and lock hardware can easily be added. Most wood covers are made in two sections for easier handling and storage. The sections can be hinged for convenience or left separate for lighter handling weights.

If you contract with a carpenter to make a custom spa cover, make sure he uses stainless or bronze nails and screws that won't rust. Check the size of the job and get an estimate from him of the weight of the materials—when a wood cover gets over 6 feet in diameter, it might be too heavy for your customer to handle. Also, make sure redwood, cedar, or other hardwoods are used. Pine or fir will rot quickly unless they are varnished. Don't paint wood covers—the heat and chemicals of the spa water will peel paint very rapidly. I pay around $300 wholesale for a redwood cover of up to 8 feet in diameter.

Reinforced upholstered foam

A popular alternative to heavy wood is a rigid cover (that also sits on the deck rather than on the surface of the water) made of Styrofoam, framed for strength with plastic or aluminum, and covered with plastic or Naugahyde. These covers are usually made in two sections with a fabric hinge to connect them. Because of the materials and labor involved, expect to pay $250 wholesale for a cover up to 8 feet in diameter.

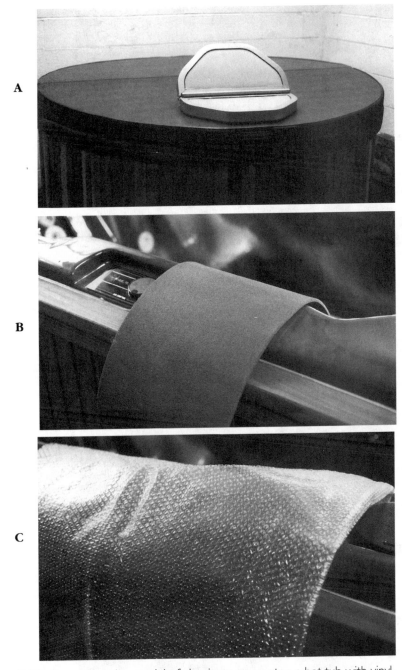

7-20 Spa covers. A—model of aluminum cover atop a hot tub with vinyl-covered reinforced foam cover (detail seen in H); B—section of floating foam cover; C—sealed-air "bubble" cover; D—rolling cover (foam with redwood reinforcement strips); E and F—TubCap™, folding-style cover; G—EZ Lifter™ cover support system; H: construction of hot tub cover.

D

E

F

7-20 Continued.

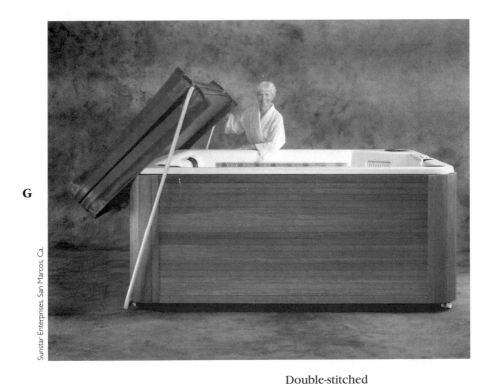

G

Sunstar Enterprises, San Marcos, Ca.

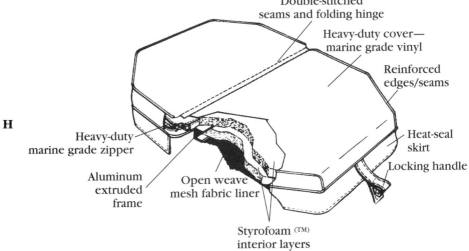

H

Double-stitched
seams and folding hinge

Heavy-duty cover—
marine grade vinyl

Reinforced
edges/seams

Heat-seal
skirt

Locking handle

Heavy-duty
marine grade zipper

Aluminum
extruded
frame

Open weave
mesh fabric liner

Styrofoam (TM)
interior layers

7-20 Continued. York Manufacturing/United Leisure Industries, St. Petersburg, Fla.

Ping-Pong balls

One of my favorite covers, which I have seen used extensively in Europe but not the United States, is to cover the water surface with floating Ping-Pong balls. Sold in basic white or a variety of colors, Ping-Pong balls insulate with the same trapped-air concept as the bubble cover, but you don't have to remove them to get in the spa.

In frigid northern countries with outdoor saunas and spas, this is a great advantage and lots of fun.

The obvious drawback is the skimmer. If your system uses a strong skimmer, the balls will be constantly sucked into it, but a piece of aluminum chicken wire or screen wire makes an effective strainer. To remove them, use your cleaning net to scoop them out and store them in a plastic garbage can or bag. In the Netherlands I swam in a large commercial pool that used the Ping-Pong ball cover idea. If your customer is interested, you'll have to do some research to locate a wholesale supplier of the balls in your area and divide the water surface square footage by the diameter of each ball to estimate the number you will need. Good luck.

Wood and foam

A variation on the covers I have already discussed is the popular foam cover with redwood slats, as shown in Fig. 7-20. The foam connects the wood slats and provides insulation. The wood slats provide rigid safety, but the combination is not as heavy as a solid wood cover. An added convenience is that you can roll it up for removal and storage. These hybrids sell for about $400 wholesale for a 6-foot-diameter cover.

8
Water chemistry

Having previously warned you that various topics of water maintenance create disagreements among experts, I must now tell you that you have reached the granddaddy of them all: water chemistry. Put five pool people in a room and ask them to write this chapter and you'll have at least six different opinions about the correct analysis, maintenance, and corrective actions concerning pool, spa, and fountain water treatment. Worse, each of the resulting opinions will offer confusing terms, formulas, technical names, and a wide variety of other information designed to make you give up completely any attempt at understanding. The result is, like many pool people, you throw in a gallon of chlorine and maybe a quart of acid for good measure and hope it all works.

Well, the good news is that this chapter will demystify the subject of water chemistry by starting with plain, simple concepts and approaches. Then I'll add the more technical information for those who want to understand more. In either case, I recommend that you skim over this chapter to get an overview, then read it through twice—once for concepts and the second time for detail. No matter how it's presented (or by whom) it is a lot to absorb at one time.

INTRODUCTION

Let's begin with the easy part, the part that all pool and spa people, health departments, manufacturers, and builders can generally agree upon. Here are the components of healthy water (each of which will be explained in detail, so don't fret if it looks Greek to you now).

- Chlorine residual   1.0–3.0 ppm
- Total alkalinity   80–150 ppm
- pH   7.4–7.6
- Hardness   200–400 ppm
- Total dissolved solids   Less than 2000 ppm
- Cyanuric acid   30–80 ppm

Oh sure, you'll get slight variations from one expert to the next even on these basics, but no one would completely disagree with these recommendations and no health department would close you down within these ranges. In fact, if your pool or spa tested out exactly within these parameters, you'd probably get a medal and your picture in a pool service magazine.

So if everyone generally agrees on the guidelines listed, where does the disagreement enter into it? The heated debate begins when you discuss how to achieve and then maintain these parameters; how to correct a severe imbalance of one or more of the components without throwing other components out of alignment; how to kill an unexpected algae bloom; how to eliminate blue staining from the plaster, etc., etc., etc. Well here's my approach.

DEMAND AND BALANCE

If you get nothing else from this chapter, learn this section on demand and balance—you will be ahead of many water technicians who think they know a lot about water chemistry.

Water chemistry is a process of balance. Change one component, even to bring it into a correct range, and you might adversely affect another component, thus adversely affecting the entire pool or spa. Imagine that the water quality parameters are stones, each of equal size and weight, evenly distributed around the edge of a dinner plate. Now imagine balancing that plate on one finger. You can do it if you find the exact center of the plate, where each stone balances the others. But now imagine that one stone is doubled in weight or removed. It changes the balance of the plate—the other stones will slide into new positions or off the plate completely, eventually making the entire plate fall from your finger and crash to the ground.

So it is with water chemistry. A balancing act, with each component working on the others. Moral of the story: learn about each component of chemistry, learn how to achieve ideal conditions for that component, but think before you pour chemicals. Learn about the consequences of each action (or lack of action).

This introduces the concept of *balance*, but what is the *demand* part? Water is a solvent. It will dissolve and absorb animal, vegetable, and mineral until it can no longer hold what it dissolves (the *saturation point*). After this, it will dump the excess of what it has dissolved (the *precipitate*). With this process in mind, understand that water makes demands on anything it comes in contact with until those demands are satisfied.

As you will see in the discussion of pH, for example, if the water is very acidic, it will demand to be balanced with something alkaline. If such water is in your plastered pool, the alkaline lime in the plaster will be dissolved into the water until that balance is achieved. When the water is no longer acidic, it will start dumping excess alkaline material, depositing it on tiles and inside pool equipment, as well as back on the plaster as rough, uneven calcium deposits. So demand and balance are inextricably related as will be increasingly apparent with each section of this chapter.

COMPONENTS OF WATER CHEMISTRY

Our original ideal parameters are worth repeating before I discuss each one individually.

- Chlorine residual   1.0–3.0 ppm
- Total alkalinity   80–150 ppm
- pH   7.4–7.6

- Hardness   200–400 ppm
- Total dissolved solids   Less than 2000 ppm
- Cyanuric acid   30–80 ppm
- Temperature, wind, evaporation, and bather load

I've added one more component to the list. Obviously not a chemical component, temperature has an impact on some of the other components, either directly or indirectly, as do evaporation, wind, and dirt. More on that later, but for now just add it to your list.

Sanitizers

To discuss the first item on the list, chlorine residual, I will review the concept of sanitizers in general. So far in this book, when mentioning a sanitizing agent, I've only mentioned chlorine, but there are four forms of chlorine, each with its own properties and problems. There are four alternate sanitizing methods in common use, and there are a great number of "wonder" products for killing off special algae or solving stain problems. Chlorine is used as a benchmark because it is the most common sanitizer.

The purpose of any sanitizer is to kill bacteria in the water. Bacteria carry disease and stimulate algae growth. Sanitizers accomplish this by *oxidizing* the bacteria and other waste in the water. Rust is oxidization in progress. Oxygen is, in essence, dissolving the material with which it comes in contact. Sanitizers, oxidizing in your pool or spa are essentially "rusting away" the bacteria and other waste material in the water.

The most basic form of chlorine is as a gas, which does not occur in nature. It is man-made by passing electricity through a saline solution (saltwater). The result is a liquid, hydrochloric acid (also called muriatic acid), and a gas, hypochlorous acid (also called chlorine gas). Liquid chlorine is then made by passing this chlorine gas through water.

Dry chlorine is subsequently made by removing the water from such a solution. Before I examine the properties of each, let's look at some properties common to all, then some evaluation criteria for each form. The chemistry that follows is about all you need to know for this entire subject to make sense, so don't be confused or discouraged by the notations and formulas—it's not really much to learn and I promise there is no more.

The chemical processes of chlorine

Chlorine, in any form, mixed with water forms hydrochloric acid (HCl, commonly called muriatic acid) and hypochlorous acid ($HOCl$), the killing form of chlorine. In scientific notation, it looks like this:

$$Cl_2 + H_2O = HCl + HOCl$$

The $HOCl$ part of the equation (the hypochlorous acid or killing part of the chlorine) further breaks down when it goes in your pool or spa. It breaks down into a positive ion of hydrogen and a negative ion of hypochlorite, which is a noneffective oxidizer of bacteria. The process looks like this:

$$HOCl = H^+ + OCl^-$$

If these resulting products are ineffective, then obviously the goal is to stop the chemical reaction at HOCl without further breakdown. How is that done? By maintaining a slightly lower than neutral pH in the water. As pH increases, chlorine becomes less effective because the process described previously is allowed to continue. Simply stated, available chlorine in the water does this:

- at pH 7.2—80 percent killing HOCl + 20 percent useless OCl⁻
- at pH 8.0—20 percent killing HOCl + 80 percent useless OCl⁻

See what I meant about balance, how one aspect of water chemistry effects another?

Chlorine demand

As I mentioned in introducing the ideas of water chemistry, water demands a lot. *Chlorine demand* can be defined as the amount of any chlorine product (in any form) needed to kill all the bacteria present in a body of water.

Chlorine residual

Chlorine residual is the amount of chlorine (expressed in parts per million) that is left over after demand is satisfied.

Chlorine evaluation criteria

Here are a few criteria for comparing various forms of chlorine and, for that matter, each of the other sanitizers I'll deal with. In each case, I will note any special properties of the sanitizer (for example, if the product is reasonably stable I will make no comment, but if it is very unstable I will say so).

Chlorine availability What does *available* mean—that you can ask it out for a date at any time and it will say yes? Well, sort of. Chlorine in its various forms combines with other elements present in the water, meaning some of the chlorine is locked up or unavailable for oxidizing bacteria, while some is available for that sanitizing purpose.

In considering the effectiveness of any sanitizing program in a body of water, this availability question must be considered to know what is really going on. In other words, you might add 5 gallons of liquid chlorine to your pool, but have less effective, active, sanitizing *available* chlorine than if you had added 2 pounds of chlorine gas. I will explain in a moment.

Stability Chlorine is by nature very unstable, which is to say that it is easily destroyed by contact with ultraviolet (UV) light (sun). Efforts are made, as you will see, to make it more stable using various methods.

pH Since you add some form of sanitizer to your water with such regularity, you must consider the pH of the product, because it will alter the pH of the water if enough is added.

Convenience How easy is it to administer the product?

Cost This is self-explanatory.

By-products Does the form of chlorine you are using add anything to the water besides chlorine?

Before continuing, if this is your first time through this chapter, you might want to skip ahead to the section on pH and get into the chemistry detail the next time through when such concepts are not all new to you. If you already have a general knowledge of what's ahead, press on.

Chlorine gas

Chlorine gas, *hypochlorous acid*, is about two and one-half times heavier than air and light green in color. Chlorine gas lowers the pH of water when introduced into it. The advantages of chlorine gas are that it is cheap (about one-fifth the cost of liquid chlorine, the next cheapest form) and that it is 100 percent available. Because it is pure chlorine, there are no by-products added to the water when using it. So if chlorine gas is 100 percent available and cheap, why would there be any other form, why doesn't everyone use it? Because it is lethal. Yes, chlorine gas is deadly and very dangerous.

One day in my pool supply store I was in one of those cleanup moods and thought I would combine two half-filled bottles of chlorine into one. What I didn't know was that one bottle contained muriatic acid and one contained chlorine. When I poured one into the other, holding them close in front of my face, the resulting explosion of chlorine gas (released by contact with the muriatic acid) choked me, burned my eyes and lungs, and I thought I was going to die. Well fortunately I didn't, but the lesson was one I hope I can learn for you—don't make the same mistake.

By the way, liquid pool and spa chemicals come in color-coded plastic bottles so, in theory, you can't make that kind of mistake. The color varies by manufacturer, but acid usually comes in red or green bottles while chlorine comes in yellow, light blue, or white bottles. Whatever the color scheme, don't ever do as someone in my shop did that nearly fatal day—don't ever put a different chemical in a bottle meant for something else.

Chlorine gas can kill in other ways as well, as a case in Florida a few years ago demonstrates. A technician administered chlorine gas to a public pool one morning using a cylinder of chlorine and a wand. The wand is placed just below the water surface and a certain volume of gas is injected into the water. As he removed the wand, some gas residue escaped from the wand or hose and floated on top of the water. Remember, chlorine gas is heavier than air, but not heavier than water. The technician allowed kids to jump in the pool as soon as he was finished. One kid dove in and surfaced exactly where the chlorine gas was floating on the surface of the water. He took a deep breath as he came up, inhaling the gas. It killed him.

That's why chlorine gas is not universally loved in the water maintenance business. It's cumbersome (pressurized tanks of gas, valves, wands, or expensive injection devices attached to the circulation equipment), and it's deadly if mishandled.

Liquid chlorine

Liquid chlorine, *sodium hypochlorite*, is simply chlorine gas already injected into water, creating a liquid solution. Caustic soda is added to the mix to help keep the chlorine from escaping from the solution. This gives liquid chlorine a very high pH, between 13 and 14. Chemically, liquid chlorine is $NaOCl$.

Remember, chlorine gas was formed by electrifying saline (salt) water, so liquid chlorine is very salty. In fact, you are adding 1.5 pounds of salt to the water for every gallon of liquid chlorine you add. This adds to water hardness.

Liquid chlorine is produced at about 16 percent strength and by the time you use it, it is about 12.5 percent available. When you add sugar to your coffee, it tastes sweet. If you add twice as much, it tastes twice as sweet. If the object were to put as much sugar in solution in the coffee as possible, you would keep adding sugar until the coffee could simply not contain anymore.

Why not do the same thing with liquid chlorine? Why not have, say, 65 percent chlorine? Then you would only need to carry 1 gallon of the product for every 5 gallons you now carry—it would be a more concentrated, convenient product. Unfortunately, because chlorine is so unstable, producing it at such strength is pointless. Within hours, that 65 percent solution would deteriorate to less than 15 percent anyway. So that's what you get from the factory to start with, ending up with about a 12 percent solution after some normal deterioration.

This brings me to another important point about liquid chlorine. Air, sunlight, and age accelerate the deterioration process, so keep your supplies fresh and covered. By the way, household laundry bleach, which is nothing more than liquid chlorine, is produced to around 3 percent strength. You realize how profitable the household bleach business must be when you note that a gallon costs about the same as a gallon of pool or spa liquid chlorine, yet the pool chlorine is four times stronger.

The advantages of liquid chlorine are that it is easy to use (just pour it into the body of water being treated) and it goes into solution immediately, because it's already a liquid. When pouring liquid chlorine, pour it close to the surface of the water to prevent splashing (bleaching your shoes) and to prevent unnecessary dissipation of strength (by contact with air).

Dry (granular or tablet) chlorine

There are essentially two types of dry chlorine sanitizers. The more popular of these two are called *cyanurates*—they contain stabilizer (cyanuric acid) to help prevent breakdown, generally about 1 pound of cyanurate to each 4 pounds of chlorine product. But more on that later. First, let's review the other type of dry chlorine.

Calcium hypochlorite Commonly sold under the brand name HTH, this product is popular for it's relatively low cost and convenience of use. It is available in granular or tablet form.

Calcium hypochlorite is unstable, slow dissolving, and leaves substantial sediment after the chlorine portion of the product enters into solution in the water. It is 65 percent available chlorine, so it is rather potent (remember, liquid chlorine is only 12 percent available). Its pH is 11.5, so it tends to raise the pH of the water.

Calcium hypochlorite is often used as a shocking agent (see the discussion of shocking later). When shocking a pool, your main concern is available chlorine, delivered for immediate sanitizing use. Thus, a product that contains no stabilizer is preferred, because there is no point in paying the extra cost for chemicals that you don't need for this particular application. For regular sanitizing and for maintaining the chlorine residual, one of the cyanurates might be better because they are longer lasting with fewer by products.

Cyanurates There are two types of cyanurates, which are chlorine sanitizers containing stabilizer (cyanuric acid). The first form is *dichlor* (sodium dichloro-s-triazinetrione). Okay, I know I promised not to muck up this chapter with unintelligible chemical jargon and this product sure sounds like just that. But this is really simple. Most technicians refer to this product as dichlor. Triazinetrione just means stabilized—the product contains a stabilizing agent.

So what are the properties of dichlor?
- 56–63 percent available chlorine,
- No sediment or by-products as with HTH, and
- A pH of 6.8, so it is slightly acidic.

When you buy dichlor products they won't be called dichlor. They will be called by various brand names, but read the label.

The second form of dry chlorine is *trichlor* (trichloro-triazinetrione). Trichlor is the most concentrated (and therefore the most expensive) form of chlorine produced—90 percent available chlorine. To achieve both stability and strength, trichlor is produced mostly as tablets that slowly dissolve in a floater, although dry granular trichlor is produced as a super-killing agent such as Algae-Out for problem algae blooms.

Trichlor's pH is 2.8 to 3.2, very acidic. So make sure the floater (floating duck, dispenser, etc.) that you use in the pool is well away from any metal and/or skimmer or intake. Never put trichlor tablets in the skimmer as I have seen many pool technicians do. The acidic properties will have the same results as dumping acid directly into the system—the metal will dissolve. When tablets are left in a skimmer when the pump is off, the tablet dissolves creating a few very acidic gallons of water. When the pump comes back on, this acid water is sucked directly into the metal plumbing and components of the circulation system. This daily acid bath will quickly dissolve metals, destroying plumbing and equipment and depositing metal on the plaster walls of the pool or spa.

One major advantage of trichlor is that it dissolves slowly, so if a pool is not serviced frequently, chlorine will still be released into the water daily. A disadvantage of both dichlor and trichlor is that they are 50 to 58 percent cyanurate, so continued heavy use leads to a buildup of stabilizer, requiring at least a partial draining of the pool.

There is one more form of dry chlorine, available in powdered form, called lithium (lithium hypochlorite). Lithium is very soluble; in fact, when broadcast over the surface of the pool it will dissolve before it reaches the bottom. This makes it good for vinyl-lined or dark plaster pools where you might otherwise be concerned about discoloration.

Lithium is not in common use, however, and so I won't discuss it in detail. The main reason for its lack of popularity is price. It will cost you more than double to maintain your pool with lithium as opposed to liquid chlorine, and substantially more than all other sanitizers.

Chloramines and ammonia

As you might have guessed from the name, *chloramine* is a combination of chlorine and ammonia. Remember I said that chlorine likes to combine with other elements in the water. It is this fact that makes it an effective oxidizer of bacteria, as it tries to combine with the bacteria, thus killing it. But when ammonia is present in the water, the two will combine together to form chloramines.

You don't need any chemical formulas here. Just know these facts about chloramines:

- Chloramines are very weak cleaners (weak oxidizers of bacteria) because the chlorine is locked up with the ammonia and is not available to kill bacteria. The colloquial term is *chlorine lock.*
- Chloramines are formed when ammonia is present in the water from human sweat or urine (chemically very similar) and when there is insufficient chlorine present to combine with all the available ammonia and to oxidize bacteria. One active swimmer produces one quart of sweat per hour.
- Chlorine odors in a pool (and irritated eyes) are caused by chloramines, not too much chlorine.

- *Breakpoint chlorination* is the point at which you have added enough chlorine to neutralize all chloramines, after which the available chlorine goes back to oxidizing bacteria and algae instead of combining with ammonia in the water.
- My method of achieving breakpoint chlorination is to add 1 gallon of anhydrous ammonia and 8 gallons of liquid chlorine to every 20,000 gallons of pool water and turn off the equipment for 24 hours. After that time, add another 8 gallons of liquid chlorine, turn on the equipment, and allow swimmers back into the pool when the chlorine residual reads normal.

Now this might seem odd—you're trying to use chlorine to eliminate ammonia, so why add more ammonia to the pool? Ammonia and chlorine combine to form chloramines. When ammonia decomposes, it forms nitrogen compounds, which are nutrients for algae. When the algae ingest the nitrogen, they also ingest chloramines, which kills them from the inside out.

Superchlorination

As the name suggests, adding a lot of chlorine to a pool is superchlorination. Why would you do it?

Even if you maintain a sufficient residual chlorine in your pool at all times, the ammonia and other foreign matter in the pool might keep your chlorine from being 100 percent available. That's why algae grows in a pool even though it has a high chlorine residual when you test the water. Superchlorination is, therefore, adding lots and lots of whatever chlorine product you use so that there is plenty in your water for the foreign matter to absorb, leaving enough to oxidize bacteria (kill algae).

Superchlorination is recommended by various experts as something to do monthly, three times a year, etc.—in other words, every pool person has a different opinion based on experience. The truth is that there is no one answer for every pool. You only want to superchlorinate when you have a substantial volume of ammonia or other foreign matter in your pool to soak up. Since you can't really know how much ammonia is in the water that might be locking up your chlorine, you can guess based on how dirty the pool gets every week and how much it is used. Of course, the more swimmers, the more sweat and urine are likely to be present. Therefore, it's a judgment call. Each chlorine product requires different amounts to achieve superchlorination. Remember, you have to close the pool to swimmers until the residual comes back into normal levels. So it pays to use enough, but no more. Follow the directions on the package for superchlorination. With liquid chlorine, I use 8 gallons of chlorine per 20,000 gallons of water. Keep the water circulating 24 hours a day until the residual reads normal.

Another time to superchlorinate is when the pool has been allowed to get exceptionally dirty or full of algae. A quick way to kill off a bad algae bloom (growth) is the same method I described to eliminate chloramines from your pool. Add 1 gallon of anhydrous ammonia and 8 gallons of liquid chlorine for every 20,000 gallons of pool water. In the chloramine discussion, I said that chloramines are not an effective cleaner. That was only half the story. Chloramines will, in fact, kill and dissipate quickly. They have a reputation as a poor cleaner because chloramines are totally unstable—they leave no residual, they kill and are gone.

So for especially bad algae problems, I use that method with the pump running 24 hours a day and continue to add chlorine daily (usually 8 gallons per day per 20,000 gallons) until the chloramines are absorbed and a normal residual is restored. The white powder you see in the pool the next day is the bleached, dead algae.

Summary scorecard of chlorine

To give you an overview of the chlorine products just discussed, here's a simple summary of their properties. They are arranged in order of relative cost, with gas being the cheapest. Note the relationship between cost and stability—the more stable you try to make this inherently unstable product, the more you must add to it and therefore the more expensive it becomes.

Product	pH	Available chlorine	Common form	Stability
Cl_2 gas	Low	100 percent	Gas	Very unstable
Sodium hypochlorite	13+	12.5 percent	Liquid	Unstable
Calcium hypochlorite	11.5	65 percent	Dry granular	Stable
Dichlor	6.8	60 percent	Dry granular	Very stable
Trichlor	3.0	90 percent	Tablet (or granular)	Very, very stable

In terms of relative effectiveness, this comparsion is useful: 1 lb. chlorine gas = 1 gal. sodium hypo = 1 lb., 8.5 oz. calcium hypo = 2 lbs., 13.5 oz. lithium hypo = 1 lb. 12.5 oz. dichlor = 1 lb. 2 oz. trichlor

Chlorine alternatives

There are many alternatives to chlorine, although none are as widely used or as easy to use.

Bromine Bromine is a member of the chemical family known as *halogens*. In fact, so is chlorine, iodine, and fluorine. All are oxidizers. So if bromine is a cousin of chlorine, why isn't it used more? First of all, it cannot be stabilized and is therefore expensive to use. Having said that, it must be noted that bromine is more stable than unstabilized chlorine *at high temperatures*, and that's why many people use it instead of chlorine in their spas and hot tubs.

Nevertheless, a stabilized chlorine product is more effective and cheaper than bromine, so the only contribution made by bromine to the world of pools and spas is for people who use their spas a lot and object to the chlorine odor. As noted in the section on chloramines, there won't be a chlorine odor if the correct amount is maintained and if the user doesn't urinate or sweat in the water.

The ultimate irony about bromine, however, is the fact that for it to work it must have a catalyst, which is usually chlorine. The small amount of chlorine added to bromine products is not detectable, so most technicians don't know that. You might have guessed, I'm not a fan of bromine.

Iodine Iodine is cheap, tasteless, odorless, causes no eye irritation, and is easy to administer and test for. Wow! So why don't we all use iodine as a sanitizer? Because despite this glowing assessment, iodine in sufficient quantity to oxidize bacteria in pool water will turn that water green. Moreover, iodine doesn't kill algae. That's why you won't find any at your pool supply house. But now if your customers ask why you don't use iodine, you can tell them.

Fluorine Just like iodine, this isn't one you will refer to very often, but you will want to know about it for your well-rounded water education. Fluorine is one of the most effective sanitizers known, but it is prohibitively expensive, so you won't find this one on the shelf anywhere either.

Ultraviolet (UV) light When I tell you that I am not a fan of ultraviolet light as a means of killing bacteria in your pool or spa, some technicians will say, "This guy is too opinionated and tied to the old-fashioned ways of liquid chlorine." I will gladly embrace any high-tech twenty-first century means of pool sanitizing . . . if someone would invent an effective one. You know what ultraviolet light is: light waves produced at the highest end of the spectrum—the rays that when they come from the sun turn your skin to Kentucky Fried Chicken. While very effective at producing skin cancer, the problem with ultraviolet light as a sanitizer is that it is reactive, not preventive.

After the water has passed through your pump, filter, heater, and other equipment, it passes through a chamber where it is exposed to a beam of ultraviolet light. Any bacteria in the water passing through this light is killed. Do you begin to see the problem? The bacteria must be present in the water to be killed, whereas with a residual level of any chemical sanitizer already in the pool water, the bacteria is prevented from growing in the first place. Also, algae growing in the pool, which is not suspended in the water passing through the circulation system, are totally unaffected because the ultraviolet light has no way to kill the algae inside the pool itself. So ultraviolet light is not effective by itself. You still need to use chlorine or some other chemical agent to establish a residual in the water. For this reason, in many health department jurisdictions, ultraviolet by itself is not even legal.

So if you have to use chemicals anyway, why buy the ultraviolet gadget? Some will say that it allows you to use less chemicals and therefore helps eliminate chlorine odors, etc. Well, remember, it's not the chlorine that causes the odor, it's the chloramines that are caused by ammonia in the water. In short, ultraviolet is just a gimmick, not a truly effective sanitizer.

Ozone Ozone is almost as useless as ultraviolet. Ozone is actually supercharged oxygen (O_3) (wow, three molecules of oxygen). It is produced when the water is circulated through an ozone generator, where bubbles of oxygen are actually pumped into the water. When you see a steady stream of tiny bubbles coming from the return lines of a pool, you can guess leak or ozone generator.

Ozone is produced in two ways for pool and spa use. One method, called the *corona discharge method*, is to shoot an electrical charge through oxygen and water, creating ozone. The other method is to electrify Freon (the gas used in air conditioners and refrigerators). In both cases, the production happens in an *ozonator* (ozone generator) built into the circulation system.

Unlike ultraviolet, ozone actually gets into the pool, and although it is very unstable, there is a small residual quality to it and it will kill bacteria and algae that might be hiding in the pool. But because the residual is so small, as with ultraviolet, ozone cannot be used alone. It too must have a booster chemical.

Potassium monopersulfate Not very common, yet commonly available, PM is a very efficient oxidizer used most often in products sold for superchlorination. It is also an effective catalyst for bromine and is sometimes found in small quantities in bromine tablets (or sticks). But you can't go into your supply house and ask for PM in the way you might ask for chlorine. You'll simply find it when you read the label of various sanitizing products.

pH

pH is a way to assess the relative acidity or alkalinity of water (or anything else for that matter). It is a comparative logarithmic scale (meaning for every point up the scale, the value increases tenfold) of 1 to 14, where 1 is extremely acidic and 14 is extremely alkaline. In the middle, 7.0, is neutral, but for our purposes, 7.4 is considered neutral, neither acidic nor alkaline. In fact, the Los Angeles health department requires a pool's pH to be between 7.2 and 7.8.

pH has an amazing effect on water. If it is allowed to become very acidic (from adding too much acid, from too many swimmers sweating or urinating, from too much acidic dirt and leaves in the water) the water becomes corrosive, dissolving metal it comes in contact with. If it is allowed to become too alkaline, calcium deposits (scale) can form in plumbing, equipment, and on pool walls or tiles.

More than that, pH determines the effectiveness of the sanitizer. For example, you might have a chlorine residual of 3 ppm in a pool with a pH of 7.4. But increase the pH just four-tenths to 7.8 and you will still have a chlorine residual of 3 ppm, but the ability of the chlorine or bromine to oxidize bacteria and algae will decrease by as much as one-third. Therefore, maintaining a proper pH is not just a factor of interesting chemistry or saving your equipment, it is a matter of effective maintenance.

pH is adjusted with various forms of acid (to bring down a pH that is too high) or alkaline substances (to raise a pH that is too low). With plaster pools and spas, the calcium in the plaster creates a situation where the water is always in contact with alkaline material, slowly dissolving it (remember, water is the universal solvent), causing the water to be slightly alkaline. Therefore, in plaster pools and spas you will always be adding small amounts of acid to keep the pH down.

Acid is most readily available in liquid (a 31-percent solution of muriatic acid) or dry granular forms. As noted elsewhere in the book, never mix acid and chlorine—the result is **deadly** chlorine gas.

Because each product is different, I won't try to guide your use of acid; just follow the instructions on the package or in your test kit for the amount to add. Keep this general guideline in mind, however. If you add acid in small amounts, you can always add a bit more to get the desired results. But if you add too much, the water becomes corrosive, which means you must now add alkaline material to bring the pH back up.

When you add acid, particularly liquid acid, pour close to the surface of the water so the acid doesn't splash on the deck or your shoes where it will do damage. Keep it away from your face and hands as well. If you get some on you, it might not burn right away, but don't be deceived—it will start eating your skin. Wash the area immediately with fresh water and seek emergency first aid if you are exposed to a lot or if you get it in your eyes.

Also, when you add acid to the water, distribute it as much as possible, walking around the pool or adding it near a strong return outlet while the pump is running. The idea is to dilute it as soon as possible. Never add acid near a main drain, skimmer, or other suction inlet. The idea is not to have full-strength acid sucked directly into your pump and other equipment which could be damaged.

When adding soda ash or other alkaline material, dissolve it in a bucket of water first and add it slowly. Adding too much or adding it too quickly might result in the water turning milky blue. Follow the package directions. By the way, soda ash has a pH of 12 and bicarb has a pH of 8. These products are described later, but for now, use soda ash to raise pH and total alkalinity. Use bicarb to raise total alkalinity alone.

In both cases, when adding acid or alkaline, allow several hours for the pH to stabilize before testing again—the next day is even better. If you test right after adding chemicals, you will get a false reading. Similarly, don't test pH after adding chlorine or other sanitizers. Remember, each has a significant pH itself, so if you test right after adding sanitizer, you might be testing water saturated with that product and not truly reflective of the water's pH.

pH is tested using a chemical called phenol red. For easy reference, all discussion of testing techniques is in the section on chemical testing.

Total alkalinity

Total alkalinity is one of those concepts that frequently gets confused with pH or hardness. While the total alkalinity of a pool or spa has an impact on those and other factors, it is not the same thing.

Imagine a waiter brings you a cup of coffee already sweetened with sugar. You sip the coffee and rate the flavor as slightly sweet, moderately sweet, very sweet, very, very sweet, and so on. But does that rating tell you how much sugar was actually put into the cup? To know that, you would ask the waiter and he might tell you, for example, it was one tablespoon.

Total alkalinity is to water chemistry what that tablespoon of sugar is to the cup of coffee; while pH is to water chemistry what the flavor rating is to that coffee. A pH test will tell you the relative acidity or alkalinity of the water, while the total alkalinity will tell you the quantity of alkaline material in the water. It is a measurement of the soluble minerals present in the water (like the measurement of sugar in the coffee).

To use another analogy, you might say that a pH reading is similar to a temperature reading on a thermometer, which simply states the present temperature; while a total alkalinity measurement represents the volume of heat that brought you to that temperature. The usefulness of such information is that you can then predict the added quantity of heat required to reach a desired temperature. Therefore, total alkalinity readings will tell you the amount of acidic material required to reach a desired pH.

In other words, *total alkalinity* is a measurement of the alkaline nature of the water itself and, therefore, the ability of that water to resist abrupt changes in pH. In fact, adjusting water chemistry to a proper total alkalinity acts as a buffer against fast and extreme changes in pH (called *spiking*).

The actual test method is discussed later, but it is important to know at this point that an appropriate reading for your water's total alkalinity is 80 to 150 ppm. Less than 80 ppm means that too much acid has been added even if the pH reading is high. Therefore, you always adjust the total alkalinity level first, then the pH.

Proper maintenance of the total alkalinity of a pool will pay great dividends. You will use less chlorine, less acid, and see fewer algae problems.

Hardness

Hardness (or calcium hardness) is a component of total alkalinity. It is a measurement of the amount of one alkaline, soluble mineral (calcium) out of the many that might be present.

So why single out this one alkaline mineral for special measurement, especially if you have measured and balanced total alkalinity? Because, when in sufficient quantity, calcium readily precipitates out of solution and forms salty deposits on the pool, tile, and in equipment. This deposit is called *scale* and is the white discoloration you see on rocks or tile at the waterline of a pool or spa. If you live in an area with hard water (water that contains large amounts of minerals) you might also have seen it in your home—in drinking glasses or flower pots that are left with standing water for long periods.

The acceptable range when testing for hardness is 100 to 600 ppm. Over 600 ppm and you will see scale. The only cure for water that hard is to partially or totally drain the pool and add fresh water.

Total dissolved solids (TDS)

Now that I've discussed two measurements of minerals in the water, there is one overall category that helps you keep track of the big picture. *Total dissolved solids* is, as the name implies, a measurement of everything that has gone into the water and remained (not been filtered out), intentionally or not.

The total of minerals (including calcium), cyanurates, chlorides, suntan lotion, dirt, etc., etc., equals TDS. The main contributor to TDS increases in a pool or spa is evaporation. When the water evaporates, it leaves its contents behind. You add more water, and the solids keep building up as it evaporates. As noted earlier, the liquid chlorine you add to the pool is made from salt, so when the liquid evaporates the mineral remains behind, adding to the total amount of dissolved solid material in the water.

In most places, water from the tap already has a TDS reading of 400 ppm. You add about 500 ppm more each year with chemicals. Depending on where you live, the evaporation rate might add another 500 ppm per year. In California, for example, between wind and temperature you can expect to lose the equivalent of the entire volume of your pool each year to evaporation.

When a body of water reaches a reading of 2000 to 2500 ppm, it's time to empty it. There is no other way to effectively remove all those solids floating around in your water.

If you can't see, taste, or smell TDS, why make such a big deal about it? Because TDS acts like a sponge, absorbing chlorine and other chemicals you put in the water, rendering them ineffective or requiring you to use far more of each chemical to attain the same results. That becomes expensive and creates further problems (remember, balance).

Cyanuric acid

Cyanuric acid simply extends the life of chlorine in water by shielding it from the ultraviolet rays of the sun, which would otherwise make the chlorine decompose. Cyanuric acid is a powder that is added once or twice a year in most pools, maintaining a level of 30 to 80 ppm. Above 100 ppm serves no function and creates other chemistry problems, so this is prohibited by most health department codes.

Because cyanuric acid helps prevent chlorine decomposition, it is also frequently called *stabilizer*. Because it extends the life of the chlorine, the result is that you use less chlorine and save money. You will thereby be adding less mineral to the water, slowing the rate of increase of TDS.

Cyanuric acid itself does not decompose or burn out in pool or spa water as is commonly believed. The only way to dissipate it is to remove it through water replacement. Much is also lost when bathers splash and leave the pool and when removing leaves, dirt, or other debris during cleaning.

As with the other components of water chemistry, the testing methods are detailed later. The method of administering cyanuric acid to water is detailed in the chapter on cleaning and servicing.

Weather

Perhaps the least discussed and yet most important aspect of water chemistry is the impact of varying weather conditions. Although you can't do much about these factors, understanding the role they play in your pool or spa will help you make decisions about adding water and chemicals and whether or not to use a cover.

Sun

As mentioned, ultraviolet rays in sunlight speed the decomposition of chlorine. Without stabilizer, 95 percent of chlorine can be lost in two hours on a sunny day. In addition to protecting chlorine with stabilizer, a pool cover might pay for itself in the chemical savings over many years. Similarly, wind, low relative humidity, and high temperatures will speed evaporation and, as I have just discussed, that increases the frequency of draining your pool.

Temperature

In Malibu, pool water will stay in the 70s during the summer and below 60°F in winter. Although unscientific, I have found that below 65°F you will have little or no problem with algae growth. Above that temperature you are always fighting it.

Wind and evaporation

Wind not only speeds evaporation, but it also carries dirt into the pool, adding to the TDS (not to mention adding to your cleaning problems). Evaporation becomes a real problem when customers see their water bills going up and begin to question if there is a leak in the pool.

A simple way to determine if the evaporation is normal for the weather conditions or if there is a leak is to leave a bucket of water next to the pool for a week. Mark the level in the pool and in the bucket and don't add water to either during the experiment period. After a week, mark the level of water in each and measure the number of inches of water lost. Each container should evaporate the same number of inches. If they do, there is no leak and you have just provided your customer with a dramatic example of how weather influences the operating costs of pool ownership. Be sure the bucket and pool conditions are comparable. A bucket in the sun or on a hot deck will have faster evaporation than the pool.

In conducting this experiment, however, take into consideration the use of the pool or spa. If many bathers have been in and out, they will take significant amounts of water with them or splash it out of the pool. Some slight variation between the bucket and pool can be explained by this. Of course, when conducting this experiment, don't backwash or conduct any other maintenance that might artificially change the level of the pool water.

Dirt

The dirt, leaves, and other debris carried into a pool by wind also impact on chlorine use. Such debris will absorb chlorine, effectively removing it from the water. You will need to add two to three times as much chlorine to a dirty pool than you will to a clean one to reach the same results. One more reason to keep the pool clean, as well as the skimmer basket, strainer basket, and filter.

Rain

Rain is another weather problem. In some areas, such as Southern California, the rain comes in enough volume during certain months that you have to pump water out of the pool to prevent overflowing and flooding. In doing so, your carefully balanced chemistry is quickly unbalanced. The extra water saves the customer's water bill, but plays havoc with maintaining chlorine residual and other levels, requiring far more tweaking of water chemistry during rainy months than dry.

In summary, it is important to realize that varying weather conditions and bather loads have varying impacts on your water chemistry. (Two people in a 400-gallon spa destroy 40 ppm per hour of chlorine residual.) Simply being aware of that probability will keep you testing and adjusting your water frequently enough to stay ahead of rapid changes that can be caused by weather influences.

ALGAE

Algae are one-celled plants, of which there are over 20,000 known varieties. Algae include microscopic ocean plankton, giant kelp that grows 2 feet a day, and virtually everything in between. The word algae is derived from the Latin word for seaweed and is already plural (the singular being alga).

Sunlight speeds algae growth, appearing in the pool as a green, brown, yellow, or black slime often resembling fur. It thrives in corners and on steps where circulation might not be as thorough as elsewhere in the pool.

When you brush algae from pool surfaces, some will immediately flake off while much will remain tenaciously attached to the plaster, appearing as a stain. It is fairly easy to brush algae off of smooth surfaces, such as fiberglass, and very difficult to brush it off of rough plaster where it can burrow in the cracks and crevices. Black algae (actually a blue-green), for example, is completely impervious to brushing, forming a hard protective shell over itself.

Although green and black algae tend to flourish in water that lacks chlorine or other residual sanitizer, algae can grow in the presence of both chlorine and well-balanced water. Some forms become increasingly resistant to the amounts of chlorine you maintain in the water. This is another reason to superchlorinate (described previously) on a regular basis.

How does the algae get into pool and spa water in the first place? Despite treatment of domestic water supplies, algae is often found in the water you use to fill the pool. Algae spores are also carried by the wind, dropping into open bodies of water and commencing growth if conditions are favorable.

Favorable conditions include temperature (the higher the better for algae), lack of sanitizers, light (for photosynthesis), and nutrients (among other things, algae consume the minerals found in pool water). Because water tends to be cooler at depth and less light will penetrate deeper water, algae also prefer shallow areas of the pool or spa.

Algae can have an impact on the water chemistry components themselves. For example, algae can break down bicarbonates in water, raising the pH significantly and adding to the alkalinity at the same time.

Forms of algae

There are three forms of algae that you will encounter in pools and spas.

Green algae (chlorophyta)

The most common algae is green in color and grows as a broad slime on pool and spa surfaces. The slime can be removed by brushing, but that doesn't kill the plant. Super-chlorination, combined with other procedures, will solve most green algae problems.

Green algae will first be seen on steps or in pool corners in very small patches. That is the time to take action, because any algae is a sign of a pool or spa that is not being properly maintained. I have seen pools, especially if they are heated to about 80°F, that have gone from those first few patchy signs of algae to entire green coverage in 24 hours. In other words, it can grow and get out of control very quickly.

Yellow algae (phaeophyta)

Yellow algae, which at times appears brown or muddy in color, is also called mustard algae. It does not grow as rapidly as green algae, but it is more difficult to kill. It grows with the same broad, fur or mold-like pattern as green algae. Brushing has little visible effect, although it will remove the outer slimy layer, exposing the algae underneath to chlorine. As with green algae, the best killer is superchlorination combined with remediation techniques and general good maintenance.

Black algae (cyanophyta)

Black algae is actually dark blue-green in color and is the pool technician's worst nightmare. I don't know any pool person, no matter how fastidious about their maintenance procedures, who has not encountered and dealt successfully with green or yellow algae. But at the first sign of black algae, you need to consider the pool or spa as a patient that is in critical condition.

Black algae grows first in small dots, appearing to be specs of dirt on the bottom. It seems to be less influenced by depth than green or yellow algae, growing sometimes from the deepest part of the pool first. As time passes, these specs begin to enlarge and then appear all over the pool. The growth rate is slow at first, but then exponential. In other words, it starts slowly, but once black algae takes hold it quickly makes up for lost time. I was called to a pool in a rural canyon area because the homeowner thought she had a problem with green algae. Sure enough, when I arrived, I found a nice black plaster pool with early green, slimy algae growth on the steps and shallow corners. When I began brushing, I noticed patches of discoloration in the black plaster which led me to try a few other tests. I reached into the water with a screwdriver and scraped a small area of the plaster, discovering that I was actually dealing with a white plaster pool covered entirely by black algae, over which green algae had started to form.

The homeowner failed to mention that the pool was white plaster and that about a year ago it slowly started turning to black. In other words, it took awhile, but the black algae had completely taken control of the pool. Of course, the homeowner had failed to apply virtually any basic maintenance during that period and the only remedy was draining the pool and replastering. That's right, we were unable to acid wash or otherwise remove the algae because black algae grows so deeply into the plaster and concrete below that the only solution is to remove the plaster and start over.

Having seen that kind of plant power, I have the greatest respect for black algae and never want to have to explain to my customer why I allowed it to grow in his pool while he was paying me for weekly maintenance. Believe me, you don't either.

Part of the reason for black algae's virulence is that instead of creating a slimy substance as a protective barrier like green or yellow algae, black algae covers itself with a hard substance that resists even vigorous brushing. Only a stainless steel brush will break open the shell, allowing sanitizers or algicides to penetrate the plant. A more detailed method of black algae removal is discussed in the next section.

Algae elimination techniques

If there is any one reason for preventive maintenance in pools, spas, and water features, algae growth is that reason. Maintaining proper water chemistry, balance, and a clean body of water are the best safeguards you have that you will not need to deal with remedies instead.

Nevertheless, as mentioned previously, even the most careful pool technician will encounter algae problems at one time or another. The mere fact that you service a pool only once or twice a week means that you might not be able to maintain that perfect balance or cleanliness every single day. Often the cause of an algae bloom can be attributed to a homeowner trying to save money on the electric bill and shortening the circulation time of the system. Weather extremes, especially between service calls, can unbalance your carefully planned chemical and cleaning plans. A heavy bather load in the pool or spa can also add nutrients for algae and hasten decomposition of sanitizers.

Finally, simple things can foster algae growth, such as customers who don't like too much chlorine in their pool and don't add (or forget to add) the amount you have left for them to add between service calls, especially during hot summer months. In short, the best laid plans of mice and pool technicians can easily go astray, so don't feel badly if you arrive at a customer's pool and find algae growing.

But the best cure for algae is keeping a clean pool, filter, and baskets. Keep your chemical components balanced and brush the pool frequently, even if there is no apparent algae or dirt. The brushing exposes microscopic algae growth to sanitizer before it has a chance to really blossom, preventing a major problem.

If you do encounter a situation of advanced algae growth, there are two approaches to deal with it. One is a general elimination program that will work with most algae blooms. The other is to combine that program with a special algicide. Please note, however, that there are no miracle cures. Despite the claims of some manufacturers, there are no chemicals you can simply add to the water and walk away, expecting the algae to disappear by the next service call. Every elimination technique requires hard, repetitive work—another good reason to take every preventive measure in the first place.

General elimination guidelines

The following steps will help with any type or spread of algae growth and should be undertaken before adding algicides.

1. Clean the pool. Dirt and leaves will absorb sanitizers, defeating the actions you are about to take. Be sure you have a clean skimmer and strainer basket and break down the filter as well, giving it a thorough cleaning. As the algae dies or is brushed from the pool surfaces, it will clog the filter, so it is important to start clean. You might need to repeat this process several times during a major algae treatment.
2. Check the pH. Adjust it if necessary (see water treatment section).
3. Add 1 pound of trichlor for every 3000 gallons of water. Brush the entire pool thoroughly, stirring up the trichlor for even dissolution and distribution. Never use trichlor or other granular sanitizers at such strength on dark-colored plaster, painted, or vinyl-lined pools. These surfaces will discolor. Instead, use liquid chlorine as outlined in the section on chloramines.
4. Run the circulation for 72 hours, allowing the trichlor to attack the algae at full strength, brushing the pool at least once each day. Adjust the pH as needed and continue to add liquid chlorine to maintain a chlorine residual of at least 6 ppm.
5. As noted earlier, the dead algae might clog the filter, requiring teardown and cleaning several times. Keep an eye on the pressure. When the chlorine residual returns to 3 ppm, resume normal maintenance. You will need to vacuum the pool frequently during this period to remove the dead algae and the inert ingredients of the trichlor, both of which will appear as white dust when you brush or otherwise disturb the bottom.

    You will also need to brush frequently, not only at first, but for at least one week after you can no longer see any trace of algae. Believe me, it's still there and will rebloom if you let up.

The second general procedure to understand is for particularly stubborn algae, such as yellow (occasionally) or black (always). This method applies even when you can only see a few small patches, because there are other contaminated areas that you can't see yet.

1. Follow the routine outlined in the general elimination guideline.
2. Use a stainless steel wire brush and vigorously brush the surface of the patches. Brush the remainder of the pool normally. In addition to the chemical application already recommended, pour some trichlor directly over the top of the algae spots. You might need to turn off the pump to accomplish this, so that the currents in the pool don't redistribute the trichlor you are trying to get on top of the algae.

    Another trick to accomplish this is to use a length of 2-inch PVC pipe. Insert the pipe into the water, with one end over the algae. Pour a cup of trichlor into the other end of the pipe and allow it to settle over the spot. You can also maneuver a chlorine tablet to lay over the algae.

    On deeper locations or where the algae is on a vertical location, fill a stocking with trichlor or tablets and hang it on the pool wall so that the stocking makes contact with the spots. It isn't necessary to cover every inch of every algae location. The act of locating such a concentrated dose of chemical within a few inches will have the desired effect.
3. Continue daily brushing of the spot with the wire brush and reapply chemicals until the algae is gone. On areas of especially difficult access, I

have often put on a mask and fins and gone into the pool to make a more effective treatment of chemical or brushing.

If these measures are not effective on your particular situation, or for more certain, rapid results, you can also use an algicide.

Algicides

Algicides kill algae, while algistats inhibit their growth. In fact, most products can be used in varying strengths for each purpose. There are countless companies producing countless products designed to prevent and/or kill algae in pools, spas, or water features. Brand names vary and because many of them use a chemical as part of their name, they might be confusing.

Algicides fall into two basic categories, depending on their intended use. There are those made for pools and those made for spas. Since all algicides are regulated by the U.S. Environmental Protection Agency as pesticides (because they are designed to kill living organisms), they are required to be registered for pool or spa use, and it is a felony to use the product for purposes other than intended. In short, a great deal of testing and engineering has gone into each of these products, so follow the product instructions. The various types of algicides, based on their active ingredient(s), follow.

Copper sulfate (liquid copper) Effective on all types of algae, but recommended for use in decorative ponds or dark-colored pools because of staining.

Copper sheets were nailed to the bottom of sailing vessels for centuries to eliminate marine organism growth, such as barnacles. It remains a leading ingredient in paints applied to boat hulls. In the last hundred years, copper sulfate (also called *bluestone*) has been used as an algicide in lakes and decorative ponds.

The copper actually destroys the algae's ability to breath and consume food, killing it in the process. It also tends to coat the pool surfaces, acting as a preventive long after the original problem it was applied for has disappeared.

The only drawback to this product is that it stains the porous surfaces of plaster pools, along with swimmer's hair, that distinctive blue-green copper color you see on buildings that have copper roofs or trim. Modern products containing copper sulfate also contain other chemicals (called chelating agents) to prevent the staining, but in the simultaneous presence of strong oxidizers, like chlorine, the staining returns. Products containing copper sulfate are extremely effective, but I only use them on decorative ponds or dark-colored pools where the potential stains will go unnoticed.

Colloidal (suspended) silver Effective against all types of algae in all situations.

Silver works the same way as copper, without the side effects. I have heard of extreme cases where silver, which is sensitive to light, has turned black in a pool if administered in sufficient quantity and catalyzed by the presence of excessive amounts of acid or stabilizer. In practice, I have never seen this occur and find products such as Silver Algaedyne to be remarkably effective.

The first time I used Algaedyne I followed the label instructions and poured a quart into the skimmer of actor Dick van Dyke's pool. In a minute, the return lines of the pool were spouting forth a brown liquid akin to motor oil. I was horrified because the label failed to warn me about this temporary effect. Fortunately, it dissipated in a few minutes and actually allowed me to dramatically observe the excellent, total circulation pattern in his pool. The moral of the story is to follow label instructions, but be aware that they might not tell you the whole story. When you buy a new product, ask the salesperson or another pool technician who has used it what to expect.

Polymers Effective on all types of algae to some degree, but works best on green and yellow algae.

Polymeric algicides might be any of a number of chemical compounds, preceded by the prefix *poly*, meaning many parts which duplicate each other. In short, they work by repeating the job over and over.

The way polymers accomplish the job is to invade the outer membrane of the algae and effectively smother it. Polymeric compounds contain a strong, positive electrical charge and are attracted to algae, which is naturally negatively charged. Herein lies the drawback. They are also attracted to dirt in the water or filter, so their strength might be diluted. Also, because they work by attacking the exterior of the plant, they're not very effective against black algae, which has a very strong outer shell, or yellow algae, which has a stronger exterior than green, unless you have done a complete job of wire brushing (which might be impossible in some pools). Also, polymers don't coat pool surfaces and build much preventive effect. In other words, use them on green algae problems or where you're sure you have brushed thoroughly, but not with any hope of preventive maintenance. When you do use them, apply 20 percent more than the label suggests to allow for dirt in the pool.

Quaternary ammonium compounds (quats) Good on green algae in early stages.

Quats work like polymers, but the positive electrical charge is not as strong. Generally I have been unimpressed with the results of these products. Further, many of them will foam up if agitated, so don't use them in spas because they will create foam when you turn the jets and blower back on. Filtration will ultimately remove the foam.

Chlorine boosters Great against stubborn yellow or large green algae blooms.

The most effective chlorine booster is anhydrous (or aqua) ammonia. As described previously, it turbocharges the chlorine so it is more effective and kills algae from literally the inside out. It is good and inexpensive, especially for large pools.

Miscellaneous products Mostly for specialty applications such as lakes or decorative ponds. Also referring to mechanical devices that don't work in the real world.

If you have unique installations or requirements, such as a large water feature or pond, you might wish to consult your supply house for products such as simazine. Simazine is essentially a polymer (usually combined with polymers), and although it is effective it will hasten the decomposition of chlorine, so it is not popular in pool and spa applications. It is important for you to know that there are special products being developed all the time, so if you have unique requirements on a job, ask.

The other type of algicide worth mention are the electronic devices that are plumbed in the pool or spa equipment area and which claim, through magnetic or other mysterious means, to kill algae. Of course this assumes that the algae enter the circulation system, presenting themselves to these mechanical marvels for execution. In short, such devices don't work; avoid them like, well, a dose of black algae.

With all algicides, follow the instructions on the label. Some general guidelines and specific tips will be presented in the following sections on water treatment.

COMMON WATER PROBLEMS AND CURES

I have reviewed the most common problems associated with pool and spa water—algae—and now let's take a look at the symptoms you might see with other common problems. Because the concentration of chemicals, heat, and bodies in a spa create unique problems, I will discuss the problems common to all water followed by those that are more specific to spa installations.

Cloudy water

Cloudy water is most often caused by inadequate filtration, either signifying that the daily filter run is not long enough or the filter itself is dirty. Solutions to those two problems are obvious. DE leaking through a filter back into the pool can also cloud the water, so check filter grids for holes or seams that have separated.

Finally, check the total alkalinity and pH. Sometimes high readings result in cloudy water. In spas, this might reflect high TDS levels in a relatively confined, small body of water. Adjust as described later.

Colored water

If the problem is not algae, color in pool water means metals. It also means problems with plaster staining, so make corrections promptly.

Green

Green or blue-green means copper. The water is probably etching (very acidic), so balance the pH. The copper might have come from plumbing or heater components, stripped away by acidic water and deposited on soft plaster walls. It can become very extreme and actually appear as black, sooty-looking deposits. When this happens, immediately shut down the circulation equipment until you can ascertain the problem and solution. Apply the corrective measures and distribute the chemicals by brushing the pool. After several hours, take another pH reading. If it is now normal, turn the equipment back on, otherwise, keep balancing. The copper might also come from a copper-based algicide, so examine the other products you might have used to eliminate any sources.

Finally, apply a metal chelating agent to the water. These products, available in various brand names, will attract and combine metals so they can be filtered out. Follow the label instructions on the product you choose and in all cases, run the filter for 72 hours once you have balanced the chemistry.

Brown towards red

This is caused by iron. Check and adjust the total alkalinity first, then the hardness, and finally the pH. You might need to drain part or all of the water and add fresh. Iron precipitates out of the minerals in the water, built up over time, from corroding light fixtures, ladders, and rails or from plumbing corroded by acidic water. Add a chelating agent to the water.

Brown towards black

This is caused by manganese or other heavy metals (including silver). Follow the same instructions as for iron. Remember, the precipitated metal might have come from another chemical product you were using for algae control or some other purpose, so examine what has recently gone into the pool. Usually, however, the cause of all metal staining is water that is too acidic.

Corrosion of light fixtures, rails, and ladders

You might notice discoloration of the pool fixtures before you see deposits on the pool walls, so take advantage of the early warning. Corroding fixtures mean the pH is too low (acidic), the total alkalinity is too low, or the calcium hardness is too low. In each case, the water is demanding neutralization by oxidizing the metals.

If you have perfectly balanced water and still find fixtures turning black or rusting, you might have a problem with electrolysis. Discussed in the chapter on advanced plumbing (automated controls), this is more likely to be a problem in the equipment area, but if you see no other solution, look for signs of electrolysis. Remember, for metals to dissolve through electrolysis you need two dissimilar metals, electrical current (perhaps a slight water leak in a light fixture or J-box), and water with enough minerals (salts) to conduct the weak current.

Scale

As described previously, scale is a buildup of calcium carbonate precipitated out of water by evaporation or heat. Of course, excessive amounts of calcium need to be in the water in the first place for this formation to occur.

The solutions are simple. Check the hardness. If it is near or exceeds the standard of 200 to 400 ppm, drain and replace some or all of the water. If the hardness is within acceptable limits, the problem might be high pH or total alkalinity. Check both.

Scale can be a frequent problem in spas, where heat and high bather loads cause heavy use of chemicals and high rates of evaporation. The water quickly becomes hard, especially in commercial installations. Be prepared to change spa water completely on a regular basis.

Eye or skin irritation

Irritation of the eyes or skin can be caused by too much chlorine, although that is rare except perhaps if you are too heavy-handed with the trichlor in a spa. Most such problems can be traced to low pH and/or too many chloramines. When the free, available chlorine is insufficient to oxidize the chloramines, you will have the characteristic chlorine odor and eye or skin irritation complaints. The solution is to adjust the pH and shock the pool or spa.

Colored hair, nails, or skin

Caused by the same problems as eye or skin irritation, discolored hair, nails, or skin can be a real problem for the service technician. It is bad enough for the customer to find some discoloration on the plaster, but when it turns his hair green, watch out. The most susceptible are light-skinned, blonde persons.

Odors

There are several odors you might use as clues to pool and spa problems.

Chlorine odor

See the section earlier on eye and skin irritation.

Musty odor

Algae growth or high bacteria in a small spa can lead to a musty, moldy smell. The solution is to eliminate the algae as described earlier, shock treat the water to kill any bacteria, or simply drain and refill it.

A smell that is closer to mildew might, in fact, be mildew on pool or spa covers or in deck crevices where water has been standing for long periods. The cure is to follow your nose to the source and take corrective measures.

Spas

Spas will exhibit the same symptoms described previously when problems are present, and they might be even more intense because of the relatively concentrated, small body of water. For example, discolorations will be more intense and spread more quickly, and spas have a few additional symptoms:

Foaming

A buildup of body oil, suntan oil, or organic dirt can create foam when you run the spa jets or blower. Soap of almost any kind, introduced by a bather or in a cleaning process, will also cause foaming. If the foam is smelly and off-white to tan in color, it is the former. If the foam is white, smells of soap and bubbles like soap, then it is the latter. In either case, the best remedy is to drain and clean the spa and refill it with fresh water. Don't forget to clean the filter or you will reintroduce the problem when you restart the equipment. Soap can be especially difficult to remove from a cartridge filter, so you will need to soak the cartridge overnight or replace it.

Skin rash

Although mentioned previously, skin rashes and other irritations are more common in spas where bacteria is allowed to build up, exacerbated by heat and turbulent water. The cure is to drain and refill the spa or at least shock the existing water.

Bromine odor

Bromine has a definite caustic odor. A strong bromine odor usually suggests a low pH. Check and correct it.

WATER TESTING

Having reviewed various potential problems of pool and spa water and learned that diagnosis is based on testing the water, I will now examine the methods of doing so. Each component of water chemistry has unique test requirements, but the most popular basic methodology is similar.

Test methods

There are three basic approaches to water testing: using chemical reactions and comparing the resulting colors, estimating values with electronic devices, and making observations of the relative cloudiness (turbidity) of a water sample.

Colorimetric

The most common approach to testing water is to collect a small sample in a clean tube and add some chemicals to the sample. These chemicals, called *reagents*, let you evaluate the sample by its changing color. By comparing the intensity of the color with a color chart of known values, you can determine the relative degree of each water chemistry parameter. To be legal for use in some health department jurisdictions (and to be practical for accurate reading), a color chart must have at least four different shades of the color in question. For example, in one such test, a sample turning yellow denotes the presence of chlorine. The stronger the intensity of the color, the more chlorine is in the sample.

Titration

Also a color-based test, *titration* is the process of adding an indicator reagent (a dye) followed by a second reagent (called the *titrant*) in measured amounts, usually a drop at a time, until a color change occurs. By counting the number of drops of reagent required to effect the color change, you can estimate the value in question. Some manufacturers supply tablets in place of liquids for titration testing. The amount of acid demanded by a body of water can be calculated with this method.

Test strips

Paper strips impregnated with certain test chemicals are also used as colorimetric test media. With this method, you dip a test strip into the pool or spa and move it around in the water for 30 seconds. When you remove the strip, you compare its color to the color chart of known values to determine the test results. Test strips are produced for individual tests or have several on one strip. Because they are more costly to use, test strips are not yet widely used, but because of their simplicity and accuracy they are likely to be more popular in the future.

Electrometric

Testing using electronic probes attached to calibrated digital or analog readouts is the most accurate method of chemical testing. Because of their high price, these methods are not yet widely used, although they are becoming more popular every year. Because colorimetric tests are more prevalent in the pool and spa industry, I will deal with those in more detail.

Turbidity

A test that uses the cloudiness of a water sample to detect a substance is also commonly used. Called a *turbidity* (cloudiness) test, this is used, for example, to measure the amount of stabilizer in a pool. In one type of turbidity test, a vial is provided with a dot on the bottom. A water sample is taken and combined with a reagent called melamine, which clouds the sample. When viewing the sample from above, you compare the appearance of the dot with a chart of known values to determine the results. In a similar test, a vial is provided with a plastic ladle. The ladle has a dot on it. You raise the ladle up through the sample until you can see the dot, then note the value on the scale that is printed on the vial to obtain the test result.

Although many manufacturers produce pool and spa test kits, if they employ these methods they are essentially identical. In each case, follow the exact directions included with your test kit. In the typical multitest pool/spa test kit, you will find five small squeeze bottles, a color chart, instructions, and a sample beaker (Fig. 8-1).

8-1 The typical pool and spa combination test kit.

Let's examine each parameter of water quality and identify which test is used for each.

Chlorine

Several reagents are used for testing not only the presence of chlorine, but also the free, available chlorine within the total.

OTO

The first bottle in most reagent test kits is OTO, or more exactly, orthotolidine. When combined with a sample that contains chlorine, OTO (which contains muriatic acid) turns color, from light pink to yellow to deep red, depending on the strength of the chlorine. You compare the color of the sample with the color chart to determine the exact parts per million of chlorine in the water.

The only drawback to this simple procedure is that OTO only measures the total chlorine content. It does not distinguish between free, available chlorine and that which is present but combined to form chloramines. It is, therefore, a good indicator test, but not a complete result.

The accuracy of OTO testing also makes this test more valuable as an indicator than as a precise, final answer. The pH, alkalinity, turbidity, temperature (OTO works best on samples at 34°F), and overall amount of chlorine in the sample can affect the accuracy of the result. For example, chlorine content in excess of 6 ppm cannot be measured and appears as a deep red color. Also, OTO measures total oxidizer in the sample, so if any other oxidizing agent is present the reading will be false.

Having mentioned these limitations of OTO testing, it is the most widely used test in the pool and spa industry for chlorine monitoring because of its simplicity and low cost. As a regular, weekly test it is quite adequate to indicate the presence and approximate residual of total chlorine in the water. When unexplained water problems appear and as a periodic evaluation, it is wise to supplement regular OTO testing with DPD or other more precise methods.

OTO can also measure bromine in the same way, using a slightly different color chart and with similar limitations. (Multiply the chlorine reading by 2.25 to get the bromine ppm.) Bromine results will turn a darker yellow.

DPD

A second colorimetric test, which determines the actual free chlorine in a sample, is conducted with a reagent called DPD (diethyl phenylene diamene). By subtracting the results of the DPD test from the OTO test, you will know the amount of combined chlorine—chloramines—in the sample.

DPD produces colors between pink and red for both chlorine and bromine. DPD reagents are available in liquid or tablet forms, with several manufacturers providing additional reagents for determining results of total, combined chlorine and then singling out free, available chlorine (or bromine). Each manufacturer provides detailed instructions for their particular product.

DPD has another advantage over OTO. DPD reagents are produced with pH buffers that make the test less sensitive to false readings at varying pH levels. Temperature and the presence of heavy metals or other oxidizers can also affect the accuracy of DPD tests. As with all sampling, however, it is wise to test samples between 60° and 85°F. Also, if the sample has a particularly high hardness value, the DPD reagent might precipitate mineral out of solution. If this occurs, it is easily corrected by placing the reagent in the sample vial first, then adding the water. More sampling parameters are discussed later.

One final note about DPD reagents. They contain the chemical aniline, which is extremely toxic and can be absorbed through the skin. Although you won't use much reagent with each test, if you use DPD on many samples daily, repeated contact can be harmful. Common-sense handling will solve this potential problem—avoiding direct skin contact and washing any spills from your skin immediately. Also, avoid touching your eyes if DPD is spilled on your hands.

pH

pH is also tested with reagents and a color comparison chart. The reagent for this test is phenol (phenosulfonephthalein) red. Before adding phenol red to a sample you must neutralize any chlorine in the sample to avoid a false reading. Various forms of chlorine exist at various pH levels and their presence in the sample can give a false reading. When adding phenol red to a sample with extremely high levels of chlorine or bromine, a chemical compound is created (chlorophenl or bromophenol) that turns purple. This color tells you that the sample is unreadable and no inference should be taken from such a test result. Before adding phenol red to a sample, you add a predetermined amount of the neutralizer sodium thiosulfate.

Phenol red is accurate in the pH range of 6.9 to 8.2 and is the most accurate around 7.4, where your water sample should be anyway. Therefore, at the extremes, use the phenol red test as an indicator only, trusting the test results only when they indicate your water is returning to the middle range.

All test reagents should be fresh to ensure accurate results, and phenol red is the one that will quickly go bad with age. As a water technician, you will be using test chemicals in sufficient quantity that they will always be fresh, but you need to remember this when a customer tells you his readings are different from yours and the test kit he is using has been sitting in the garage for the past several years.

Perhaps the best use of electronic testing devices is in determining pH. They are accurate, compensate for extreme temperatures (useful when testing spa water), and are not subject to problems from high chlorine or bromine levels. Some technicians say electronic testers are influenced by high sanitizer levels, but my experience is to the contrary. If you choose to invest in an electronic pH meter, try it on a sample of water known to have a high chlorine content. Then try it again after neutralizing the chlorine. I think you will find the results identical and your confidence in your pH meter complete. Follow the pH test with a base-demand or acid-demand test.

Total alkalinity

The most common chemical test for total alkalinity of a water sample is the titration method described earlier. There are no significant limitations in total alkalinity titration tests such as those described for other reagent tests.

Test strips are available for total alkalinity testing. With both pH and total alkalinity testing, there is one test to determine the relative value (what is the pH and what is the total alkalinity value?), followed by another test to determine the acid or base demand. In short, a second titration test determines the amount of acid or alkaline (base) material required to bring the water back to desired levels.

Hardness

Hardness is tested using a colorimetric and titration test combined or test strips as described previously. With combined testing, a buffer solution is added to the sample to increase the pH to about 10.0 to facilitate the accuracy of the subsequently added dye solution. The dye reacts with calcium and magnesium in the sample to produce a red color. A reagent called EDTA (ethylenediamine tetra-acetic acid) is then added a drop at a time until the solution turns blue. The number of drops is compared to a chart to evaluate the hardness of the sample.

Older kits using this method express the result in grains per gallon (gpg) rather than the more common parts per million (ppm). To convert grains per gallon to parts per million, multiply the gpg by 17.1. For example, if your older kit gives you a reading of 10 gpg, that is equal to 171 ppm.

Total dissolved solids (TDS)

There is no simple, inexpensive way to test for total dissolved solids in a water sample, so many technicians take a sample to a lab or local pool retailer that is equipped with electronic equipment to analyze for TDS. With the recent advent of less costly electronics, however, TDS meters have dropped below the $100 mark and are an excellent investment for the professional water technician. TDS is evaluated by the conductivity of the water when two test leads (electrodes) are placed in a sample and a small current is passed between them.

Heavy metals

As noted previously, there are times when you might suspect that there is iron or copper in your pool or spa, particularly when stains appear on the plaster or when fixtures corrode.

I have rarely tested for heavy metals, however, because if you keep the other chemistry components in line it is not likely that you will have problems with metals. The metals enter the pool or spa from improper chemistry that creates etching conditions, corroding equipment and fixtures. If you have this problem, the causes are not difficult to determine and correct, so knowing the volume of metals in the water is not meaningful. If you are interested in pursuing testing for metals, it is conducted by standard colorimetric testing.

Cyanuric acid

Cyanuric acid is evaluated with the turbidity test described earlier. There is an electronic device available, but it is expensive. Also, you can use test strips that have recently been added to the water testing marketplace.

Test procedures

Regardless of the type of testing you are conducting, some basic precautions and procedures are necessary.

Use only fresh reagents or test strips. Keep them out of prolonged direct sunlight and store them at moderate temperatures. Look at reagents before use. If they have changed color, appear cloudy, or have precipitate on the bottom of the bottle, discard them.

Thoroughly rinse the sample vials and any other testing equipment using the water you are about to test. Never clean equipment with detergents because chemical residue from these products can deliver false readings.

Consider where and when you are sampling to ensure truly representative results. Samples should be collected several hours after any chemicals have been added and thoroughly circulated through the body of water to ensure that their effect has been completely registered. The water should be circulated for at least 15 minutes before sampling. Sample away from return outlets and away from dead zones in the pool to make sure you are collecting a representative sample. There are times, however, when you might want to know more about various zones in the pool, to evaluate circulation for example, in which case you might specifically test these areas. Finally, collect samples from at least 18 inches below the surface to avoid inaccuracies caused by evaporation, direct sunlight contact, dirt, etc.

Make color comparisons in bright white light against simple white backgrounds to ensure accurate colorimetric comparisons. Most test kits come with a white card to place against the back of the test vial.

Observe and record the results immediately; never take samples for later evaluation. The chemistry of the water can change when a sample sits around.

Handle samples and reagents carefully. As noted earlier, some reagents can impact health, so avoid direct contact and never pour them into the pool or spa, even after the test is complete. Avoid contact with the sample because acids and oils from your hands can contaminate a sample, leading to false readings.

Take your time. Hurried testing leads to inaccurate results, which leads to improper application of chemicals. Follow the directions in the kit. Some might seem redundant or unnecessary, but believe me, they are stated for a reason, so follow them. Also, allow time for hot samples to cool for more accurate results.

Never *flash* test. Old-time pool and spa technicians often carry a bottle of OTO and one of phenol red in a leather case strapped to their belt. They arrive at a pool and dash a couple of drops of each into the pool, saying they can analyze enough of the chemistry from that. If you have followed the preceding discussion of the many factors influencing the accuracy of various test methods, you will know that this is simply false. On top of the many reasons for more careful analysis techniques, there is a very different color appearance when viewing a dispersing reagent in the pool as compared with a captured result in a vial held at eye level.

Conduct all tests before adding any chemicals to modify the water chemistry. If you test for chlorine residual and add significant amounts of chlorine in any form, then test the pH, your pH reading will reflect the value of the chemicals just added, not the water. After adding chemicals to the water, allow adequate time for distribution, then test to see if your actions were adequate. Allow at least 15 minutes for liquid chlorine to circulate and at least 12 hours for the pH to adjust before testing again.

Because test kit reagents can be replaced, you will probably use your kit for many years. The color chart might, however, fade over time. Compare your chart against a new one from time to time, or simply buy a new test kit annually (they're not that expensive).

This brings me to an important related point. As a water professional, your most valuable tool is your test kit. All of your water quality actions as a technician will be determined by this kit and, as a result, the accuracy of your evaluations is the basis for your recommendations. Inaccurate results can be more than embarrassing or inconvenient. Dealing improperly with water chemistry problems can result in very expensive repairs, health considerations, and even legal liability. Your reputation and livelihood depend on your test kit. The point is that because so much is dependent on that test kit, buy the best equipment you can afford, make necessary replacements regularly, and don't skimp.

Langlier index

In the 1930s, a researcher developed a water quality index that now bears his name and which reflects the balance of water chemistry components. Although not literally useful in the pool and spa industry (the index was designed to evaluate closed water systems, not open pools and spas), many certification courses require you to know the Langlier index. The basic principle is to assign a value to each component of water quality, resulting in a positive or negative numerical result. A positive value indicates water that is saturated by alkaline material and likely to precipitate scale; a negative value indicates corrosive water that is likely to etch.

As noted earlier, water is the universal solvent. Over time it will dissolve anything. At some point however, a body of water becomes saturated with the materials being dissolved. After that point, these materials precipitate out of the solution, back to their natural state. For example, you can continue to add sugar to your coffee, dissolving each spoonful. When the coffee can no longer hold more sugar, any additional sugar will not dissolve. Further, if you change other parameters, such as the temperature or pH, some of the previously dissolved sugar will come out of solution and precipitate back into granular form, depositing on the bottom of the cup.

When the coffee (or water) can no longer dissolve materials, it is said to have reached the *saturation point*. In the pool, the water takes the place of coffee and calcium takes the place of sugar. Water might be undersaturated (not yet full of dissolved calcium), saturated (dissolved as much as it can, called *equilibrium*), or oversaturated (returning calcium to the mineral state in the form of scale precipitating back into the pool).

Langlier's index is based on the water's state of saturation. He noted that extremely undersaturated water will be hungry, trying to dissolve minerals that it contacts (corrosive); balanced, neither dissolving mineral nor depositing any back into the pool; or extremely oversaturated, precipitating scale on pool surfaces and in the equipment.

Langlier's formula to determine where the water stands on this scale is simple (Fig. 8-2). After sampling and testing each water quality component, you assign a numerical value called a *factor*. The saturation index shows the formula for application of these factors.

I have seen several variations on the original Langlier index. Some forms include TDS in the equation and combine hardness and total alkalinity into one factor.

TEMPERATURE		CALCIUM HARDNESS		ALKALINITY	
°F	Factor	ppm	Factor	ppm	Factor
40	2.7	10	4.0	10	3.7
50	2.6	30	3.5	30	3.2
60	2.5	50	3.3	50	3.0
65	2.4	75	3.1	75	2.8
70	2.3	100	3.0	100	2.7
80	2.2	150	2.8	150	2.5
90	2.1	200	2.7	200	2.4
100	2.0	300	2.5	300	2.2
110	1.9	500	2.3	500	2.0
120	1.8	700	2.2	700	1.9
130	1.7	1,000	2.0	1,000	1.7

Formula:

pH – temperature factor – calcium hardness factor – alkalinity factor = saturation index value

Interpretation:

Positive value (+) = scaling condition in water

Negative value (–) = etching condition in water

Zero value = perfectly balanced water (theoretical)

(Provided by the Los Angeles County Health Department)

8-2 Langlier index.

Whichever you use, the result is still a simple equation such as the one in Fig. 8-2, with the result being positive, neutral, or negative.

Obviously, you can correct an extremely high positive or negative result by changing any of the factors. Herein lies the impracticality of the index as a water quality tool. For example, raising the temperature of the water by 90°F results in a change factor of only 1.0, and of course, you are unlikely to balance your water components by raising the temperature. Similarly, significant lowering of hardness (or in the alternate index changing the TDS) requires replacing the pool water. Therefore, the corrective measures available to you are pH control (which you want about 7.4) and alkalinity (which should be 80 to 150 ppm). Again, to bring the index into line you would not take either pH or alkalinity to great extremes.

In short, the Langlier (or similar) index is a pleasant parlor game, to add up the values of your water testing and see how balanced it is. If you have maintained proper chemistry and/or taken appropriate corrective measures, the result should always be neutral, so this is really not a useful analysis tool in the real world.

WATER TREATMENT

I have reviewed the various testing methods to determine the chemical needs of a body of water, but what about administering the actual chemicals that will deliver the desired results? There are as many products for sanitizing and balancing the chemistry of water as there are for testing it. It pays to follow the directions on product labels, but some general guidelines apply.

Liquids

Liquid chlorine and muriatic acid are generally sold in 1-gallon plastic bottles, four to a case. Some manufacturers provide heavy-duty, reusable bottles and plastic cases and charge a deposit to ensure return. Others supply thinner, disposable bottles in cardboard cases for a slightly higher price, but you don't have to mess with returns or the expense of lost bottles and cases.

When adding liquid chlorine or acid to a pool, pour as close as possible to the water's surface to avoid splashing it on your shoes or the deck, discoloring both. Air and sun contact will diminish the concentration of chlorine, so keeping it close to the surface of the water also minimizes those impacts. Add the chlorine slowly near a return line while the circulation is running to maximize even distribution. Pour some near steps or in any known dead spots in the pool, or directly over any appearance of algae. Never pour liquid chlorine into the skimmer. Avoid skin or eye contact with chemicals.

The amounts of various products needed to raise or lower chlorine, alkalinity, hardness, and stabilizer levels are detailed in Fig. 8-3. Because of variations in products however, it is always important to read the product label to ensure the desired results.

Granulars

Sanitizers, acids, and alkaline materials are available in granular form. As with liquids, study of the product label will give you exact application methods and amounts.

Amount of product required to obtain a change in 1000 gallons of water:

Item	Raise or lower 10 ppm	Product	Amount required
Calcium hardness	Raise	Calcium chloride	2 oz. dry weight
Total alkalinity	Raise	Bicarb of soda	2.5 oz. dry weight
Total alkalinity	Lower	Sodium bisulfate (dry acid)	2.5 oz. dry weight
Total alkalinity	Lower	Muriatic acid	¼ cup liquid
Conditioner	Raise	Cyanuric acid	1.5 oz. dry weight

Example: 20,000 gallon pool needs 30 ppm increase in total alkalinity using bicarb.
20 (thousands of gallons) × 3 (30 ppm divided by 10 ppm) × 2.5 oz. = 150 oz. required (9 pounds, 6 oz.)

Amount of chlorine needed to raise residual in 1000 gallons of water by 1 ppm (multiply results by 30 for superchlorination procedures):

Available chlorine of product used	Product required to raise 1 ppm
12%	⅛ cup liquid
50%	¼ oz. dry weight
80%	⅙ oz. dry weight

Example: 20,000 gallon pool needs 2 ppm residual increase using liquid chlorine.
20 (thousands of gallons) × 2 (2 ppm) × ⅛ cup (per ppm) = 5 cups

Note:
—16 oz. dry weight = 1 pound dry weight
—64 oz. liquid = 1 gallon
—8 oz. liquid = 1 cup liquid
—Use accurate measurements, such as a measuring cup or the cap of a product container after measuring the volume of the cap.

8-3 Chlorination, conditioner, alkalinity, and calcium hardness chart.

Granular products tend to settle to the bottom of the pool or spa before dissolving, so if the vessel is vinyl or dark plaster, it is wise to brush immediately after application. Another method is to create a solution of the product by dissolving it in a bucket of water before pouring it into the pool.

None of these products should be poured directly into the skimmer and all should be applied with the circulation running. Since granular products tend to be extremely concentrated, they can cause skin irritation or breathing problems if direct contact is made. Handle them with care.

Granular products tend to have a longer shelf life than liquids, but time or prolonged exposure to sunlight will diminish their efficacy as well. Store and treat granular products as you would a liquid.

Tabs and floaters

Chlorine tablets are sold to place in floating devices (sometimes shaped like ducks) which allow a slow dissolving process. Tablets are valuable when you can't service the pool or spa for some time and need a constant source of sanitizer. They are expensive and unregulated, which is to say you can't control the rate of dissolution.

Some floaters have a valve that allows more or less water to flow in them, theoretically controlling the amount of chemical entering the water, but the results are trial and error. Like granular, tablets left on the bottom of a pool or spa will bleach out any color, so use a floater unless the surface is already white. Finally, never leave a tablet in the skimmer. The low pH of tablets means you are assaulting your circulation equipment and any related metal plumbing with acid, and I have already discussed the vagaries of that.

Mechanical delivery devices

Chlorine is the only pool and spa chemical generally added by a mechanical device, although large commercial pools might use them for other chemicals as well. Mechanical delivery systems generally fall into two categories: erosion systems and pumps.

Erosion systems

As the name implies, these systems use the water passing over the dry chemical to erode or dissolve it into solution in the water. Like the floater, the erosion system is usually controlled by restricting the volume of water allowed to pass through the system and, therefore, the amount of erosion that can take place of the tablet or granular chlorine (or bromine) inside. Figure 8-4 shows a typical erosion chlorinator system.

Erosion systems are typically made of PVC plastic and are located in the equipment area, plumbed directly into the circulation lines. These chlorinators are easily installed following the directions supplied, and require only basic plumbing techniques. They must always be located after any other equipment in the system. If you

1 Cap wrench
2 3" threaded cap
3 O-ring
4 Chlorinator body (tablet holder)
5 Lock screw
6 Check valve
7 Flow base T assembly
8 $1/2$" nipple
9 Control valve
10-11 Feeder tube assembly
12 Reducer bushing
13 Small tablet holder insert
14 View of assembled unit
   (with optional top feed assembly)

8-4 Typical in-line erosion type chlorinator. Rainbow Lifeguard Products, El Monte, Ca.

were to place it between the filter and heater, the concentrated chemical would corrode the internal parts of the heater.

Pump systems

There are basically two types of pump chlorinators: peristaltic and standard pumps. Peristaltic pumps are generally mounted in the equipment area and pump a specific volume of liquid sanitizer into the plumbing lines during normal circulation. Figure 8-5 shows the working of a typical peristaltic pump. The rollers press against the flexible tube to create suction from a nearby bottle of liquid sanitizer. As with erosion systems, these pumps should be plumbed after any other equipment in the system.

Standard piston pumps are also used for chemical delivery, but they tend to corrode and require frequent repair or replacement. In the peristaltic pump, the liquid chlorine only touches the plastic tube, while the standard pump draws chemical into it and pumps it out into the circulation plumbing. The corrosive action of the concentrated chemical will, over time, corrode the pump, especially if even a tiny leak develops (and these systems are prone to leaks). A better pump uses a rubber diaphragm instead of a piston to create pressure and suction.

CHEMICAL SAFETY

Throughout this chapter, I have recommended that you take care when handling or dispensing chemicals or even small amounts of reagents. Here are a few additional safety tips:

1. Study the material safety data sheet (MSDS) found with each hazardous chemical sold. Familiarize yourself with the potential hazards and the remedies for exposure.
2. Sodium bicarb can be broadcast over the water, but soda ash should be dissolved in a bucket of water first. Dry acid (sodium sulfate) should also be dissolved in water. Cyanuric acid should be dissolved in a bucket of hot water, with ⅓ pound of soda ash for every pound of cyanuric.
3. When dealing with new chemical treatments, read all labels carefully. Apply in small amounts first, check for any unexpected results, and then continue.
4. Always take your time. Caution is the way.

1 Pumphead cover retaining knob
2 Pumphead cover
3 Pump tube assembly
4 Pressure roller assembly
5 Pumphead mounting screw
6 Pumphead spacer
7 Motor shaft seal
8 Pumphead
9 Motor mount screw
10 Timer control knob
11 Power on/off switch
12 Panel front with gasket
13 Indicator light
14 Cord stress reliever
15 Gearbox assembly
16 Electronic timer assembly
17 Motor assembly
18 Motor fan blade
19 Pump housing
20 Housing screw
21 Cord retainer screw
22 Rubber foot
23 Power cord
24 Junction box
25 Tube nut
26 Tubing
27 Injector assembly
28 Hose clamp
29 Ceramic footvalve weight
30 Footvalve/strainer assembly

8-5 Typical peristaltic chlorinator. Blue White Industries, Westminster, Ca.

9
Cleaning and servicing

Now that I have reviewed what makes a pool, spa, or water feature tick, including the components of water chemistry, I now turn to the most basic aspect of water maintenance, routine cleaning and servicing. No two technicians will approach a job in the same manner, so there is no right or wrong. At your regular weekly service accounts, you will develop a feel for what is needed week to week, season to season. At such calls, you will arrive at poolside with the right tools to do the job. At calls to new customers, you might study the pool before bringing in tools and chemicals, or you might try to anticipate what you will need and bring it with you. I fall in between, always bringing a few basics, doing the service work possible with those items while making a mental note of what else I need, then making one trip back to the truck for the extras.

Certainly everyone would agree with the idea that you save time if you bring all of your tools and chemicals with you, use what you need and then return to the truck when finished. Some technicians try to do this, using wheeled carts that are specially designed to hold hoses, brushes, chemicals, and tools. Unfortunately, my accounts are in locations with heavily landscaped, steeply graded hillsides where you only arrive at the pool by foot. A wheeled device is impractical. Therefore, I developed the style of going in with the basics and doing as much as possible before going back to the truck, usually a long way, rather than making a trip as soon as I discover what else is needed at the job site. As you start to work, invariably you discover other things you need that weren't obvious in the first place, so making that mental list while I work at what I can, then making one trip to the truck, works for me.

In any case, only experience and your personal style will determine how you approach a routine cleaning, a major cleanup, or other special service work. What follows is a description of what jobs you will face and what tools you will need to do the job. The rest is up to you. First I will review the tools of the trade, then describe how to use them in a typical service call and in special situations.

TOOLS OF THE TRADE

Figure 9-1 shows the basic service equipment carried by a professional water technician.

Telepoles

The heart of the cleaning system is the telepole (telescoping pole). The one you will use most on pools is 8 feet long, telescoping to 16 feet by pulling the inner pole out of the outer one. The end of the pole has a handgrip or a rounded tip to prevent your hand from slipping off the pole. The tip might also include a magnet for picking up hairpins or nails from the pool bottom. There are several sizes, from a 4-foot pole that telescopes to 8 feet, all the way up to a 12-foot pole that telescopes to 24 feet. To lock the two poles together, there is a cam lock or compression nut ring.

The cam lock (Fig. 9-2) is a simple device. The hub is mounted to the end of the inside pole by drilling a hole in the center of the hub, passing a screw through the hole, and screwing this assembly into the tip of the pole. The hub is serrated, like a cog wheel, to grip the interior surfaces of the outer pole. The screw is set just off center so that it can create a circumference larger than the interior diameter of the pole. When you twist the two poles of the telepole in opposite directions, the cam swings to one side or the other in this large circumference, locking the two halves together. By twisting in the opposite direction, you unlock the poles.

When you purchase your first telepole, take it apart and observe how this cam system works. Sooner or later, scale, corrosion, or wear and tear will clog or jam the cam. Rather than buy an entirely new telepole, you can take it apart, clean it up, replace the cam if necessary, and get on with the job.

The other locking device for telepoles is a compression nut ring, like the plumbing compression rings described previously. By twisting the ring at the joint of the two poles, pressure is applied to the inner pole, locking the two together.

I have used both types of locking devices and sooner or later they both fail because of normal wear and tear. My telepole is a hybrid. I use a good quality compression ring pole, but I have disassembled it and added a cam to the inner pole. This belt and suspenders technique gives me a pole that lasts forever (seemingly) and holds up no matter what pressures I apply, particularly important when doing a cleanup of an extremely dirty, debris and leaf-filled pool.

At the end of the outer pole you will notice two small holes drilled through each side, about 2 inches from the end and again about 6 inches higher. The various tools you will use are designed to fit the diameter of the pole. You attach them to the pole by sliding the end of the tool into the end of the pole. Small clips inside the tool have nipples that snap into place in one of these sets of holes, locking the tool in place. Other tools are designed to slip over the circumference of the pole, but they also use a clip device to secure the tool to the holes at the end of the telepole.

Telepoles are made of aluminum or fiberglass. The latter is more expensive but well worth the money. They are not only impervious to corrosion and virtually unbreakable, but they won't kill you if you inadvertently touch them to an exposed overhead wire (not that rare) or insert them in a pool with an electrical short (rare). I carry the 8-by-16-foot and 4-by-8-foot versions, because the long one is necessary for pools and the small one is more convenient for spas.

9-1 Typical cleaning and service tools. A: top row (left to right)—vacuum hose, tile soap, 8'×16' telepole, tile brush on wooden handle, vacuum head; middle row—test kit, leaf rake; bottom row—leaf vacuum, stainless steel brush, wall brush. B: top row (left to right)—submersible pump, skimmer suction diverter, vacuum hose barbed connector, spa vacuum socks, spa vacuum; middle row—drain flush (blow bag), pumice stone; bottom row—backwash hose, tile brush on pole, skimmer strainer, tennis ball, Leatherman Tool™, thermometer.

9-2
Twist-grip compression nut lock, cam lock, and tool clip.

For a time, my truck was at the repair shop and I worked pools from a passenger car, keeping my equipment in the trunk. Not too many cars can accommodate an 8-foot telepole, yet I needed the long pole for cleaning my pools. My solution was a rig I still carry. I took two 4-by-8-foot poles and removed the handle grip from one. I drilled holes in each side of the handle area, actually the top of the inner pole, and inserted a tool clip. I was now able to attach this telepole to the other one, like attaching a tool. Thus, I combined two 4-by-8-foot poles to make one 8-by-16-foot pole that was composed of four 4-foot sections with two cam locks and one clip—sort of like those breakdown billiard cues. This two-part telepole was easy to carry and store in the trunk of my car.

Leaf rake

Figure 9-1 shows a professional, deep-net leaf rake. The net itself is made from stainless steel mesh and the frame is aluminum with a generous 16-inch wide opening. There are numerous leaf rakes (deep net) and skimmer nets (shallow net) you can buy, but only the one pictured will last. The cheap ones are made from plastic net material and frames. Although the original price is about twice that of the cheap ones, metal ones last a long time and resist tearing when you are scooping out huge volumes of wet leaves after a windy autumn day. They also stand up to rubbing them along rough plaster surfaces, thanks to a rubber-plastic gasket that fits around the edge, unlike the plastic rakes that break or wear down when you apply such pressures.

The leaf rake shank fits into the telepole and clips in place as described previously. Be careful not to spill acid or other caustic chemicals on your leaf rake; both the metal and plastic mesh will deteriorate and holes will develop. Some leaf rakes are designed so you can disassemble them and replace the netting, which is fine if you have the time and patience to do it.

Wall brush

The wall brush is designed to brush pool and spa interior surfaces. Made of an aluminum frame with a shank that fits the telepole, the nylon bristles are built on the brush either straight across or curved slightly at each end. The curved unit is useful for getting into pool corners and tight step areas. Wall brushes come in various sizes, the most common for pool use being 18 inches wide. I carry an 18-inch brush and a 6-inch brush that I use for spas and tight places in the pool. I also carry a 6-inch wall brush with stainless steel bristles for heavy stains or algae problems. Don't ever use a wire brush that is not stainless steel in a pool or spa. Steel bristles can snap off during brushing and leave stains on the plaster when they rust. Also, if they are a bit rusty already, when you brush the plaster you will transfer the rust to the plaster, causing a stain.

Vacuum head and hose

There are two ways to vacuum the bottom of a pool or spa. One actually sucks dirt from the water and sends it to the filter. The other uses water pressure from a garden hose to force debris into a bag that you then remove and clean (leaf vacuum).

The vacuum head and hose are designed to operate with the pool or spa circulation equipment. The hose is attached at one end to the bottom of the skimmer opening and at the other end to the vacuum head. The vacuum head is also attached to the telepole. With the pump running, you glide the vacuum head over the underwater surfaces, vacuuming up the dirt directly to the filter.

Vacuum heads are made of flexible plastic, with plastic wheels that keep the head just above the pool surface. The flexibility of the head allows it to contour to the curvature of pool corners and bottoms. Adjustable-height wheels allow you to set the vacuum head to the best clearance for each pool's conditions. The closer to the surface, the better the removal of dirt. But if the suction is too great, it might suck the vacuum head right onto the surface, rendering it immobile. In this case, adjust the head height upward.

Wheels for vacuum heads are made of plastic or high-tech composite resins. Their bearing systems can be as simple as a hole in the wheel through which the axle is inserted or wheels with ball bearings to distribute the load and help the vacuum glide smoothly. I highly recommend you invest the few extra dollars for the resin, ball-bearing wheels for both your vacuum head and leaf vacuum. When you have tried each for a week, you'll agree.

Some commercial vacuum heads are made several feet wide and are built of stainless steel. Another type is a plastic helmet style, with a ridge of bristles instead of wheels. This vacuum head is used for vinyl pools, fiberglass spas, and when breaking in new plaster. In each of these cases, standard wheels can tear or score the surface. The brush vacuum is not only less harsh, but it brushes dirt loose from the surface being vacuumed for easier removal.

Hoses are available in economy models (thin plastic material) through "Cadillacs" (heavy rubber-plastic material with ribs to protect against wear), and in various lengths (10 to 50 feet). The hose cuff is made 1¼- or 1½-inch diameter to be used with similar vacuum head dimensions. Cuffs are female threaded at the end that attaches to the hose so you can screw replacement cuffs onto a hose. The best cuffs swivel on the end of the hose, so when you are vacuuming there is less tendency for the hose to coil and kink. Another valuable hose fitting is the connector. It is designed with female threads on both ends to allow joining of two hose lengths—a useful feature when you encounter a large or extremely deep pool. The connector in Fig. 9-1B has barbs on each and that insert into hose cuffs to join two hoses.

I carry a 50-foot hose for pool work and a 10-foot hose for spas. Joining the two gives me 60 feet, enough for even the largest residential pool.

Leaf vacuum and garden hose

Some products become so successful in setting the standard for the industry that the brand name becomes interchangeable with the product's descriptive name. Kleenex and Jell-O are examples of facial tissue and gelatin, but we all use the brand name regardless of the actual maker of the item. So it is with Leaf Master, the original water-powered leaf vacuum (also called a leaf bagger). The leafmaster (Fig. 9-3), which is attached to the telepole and a garden hose, operates by forcing water from the hose into the unit where it is diverted into dozens of tiny jets that are directed upward toward a fabric bag on top of the unit. The upwelling water creates a vacuum at the base of the plastic helmet, sucking leaves and debris into the unit and up into the bag. Water passes through the mesh of the bag but the debris is trapped.

Catch bag

Telepole

9-3
Leaf vacuum.

Garden hose

Leaf vacuum

Fine dirt passes through the filter bag, but a fine-mesh bag is sold for these units that will capture more dirt. When the bag has a few leaves in it, they will also trap much of the sand and other fine particulate matter that would otherwise pass through.

The only other drawback to the leafmaster is if you are in a location where water pressure from the garden hose is weak. The result is weak jet action and weak suction. The other result is that as debris fills the bag, the weight of it (especially wet leaves) tips the bag over, scraping the pool floor, stirring up debris, or tangling with the hose. The latter problem is easily solved by putting a tennis ball in the bag before placing it in the pool. The tennis ball floats, keeping the bag upright.

The water pressure problem can be solved if the pool is equipped with an automatic cleaning system that uses booster pump technology. The fitting for the automatic cleaner at the pool wall is usually a threaded two-part adapter. By removing the automatic cleaner's hose and unscrewing the adapter fitting, you are left with a ¾-inch female threaded opening. I carry a short plumbing nipple, available at the supply house, that is ¾-inch NPT threads on one end and ¾-inch garden hose threads on the other. By screwing this adapter into the automatic cleaner opening, I can attach the garden hose directly to the automatic cleaner booster pump line. I then turn on the booster pump and my leafmaster is turbocharged. Not all automatic cleaner fittings are the same, but a little disassembly and inspection will tell you what fittings you need to create an adapter that will work for that particular customer's pool.

The alternative is to create an adapter that fits the garden hose and terminates in a quick-disconnect, similar to the one on the automatic cleaner hose, for attachment to the automatic cleaner wall fitting without disassembly of that fitting. Again, a little experimentation will create the fitting you need for each style of automatic cleaner, but it will be well worth the trouble if the home has weak water pressure.

Leafmasters are made in rigid plastic or aluminum and for my money, the original still works the best. One word of caution, however. You will frequently see leafmasters by the side of the road or freeway, blown out of the back of open pick-up trucks. The design is a natural flying saucer, so make sure yours is secure when underway.

Your leafmaster will serve you better if you customize it slightly. Add resin, ball-bearing wheels, a shutoff valve at the garden hose fitting, and a double bag for catching all the dirt.

You would think that most homeowners would own and keep handy a simple garden hose. Well most don't, or it is too short, too kinked, or too leaky. Bring your own. Pool supply houses sell garden hoses that are resistant to kinks and float. A hose that scrapes the pool bottom will stir up dirt and debris before you can vacuum it.

Tile brush and tile soap

Tile brushes are made to snap into your telepole so you can scrub the tile without too much bending. Mounted to a simple L-shaped, two-part aluminum tube, the brush itself is about 3-by-5 inches with a fairly abrasive foam pad for effective scrubbing. I have found that these brushes are valuable for wiping algae off of ladders or other tricky spots in spas, but the elbow grease required to remove body oil, suntan oil, and scale from tiles is much more than you can get at the end of the telepole. Therefore, I also carry a barbecue grill cleaning pad. It has a convenient grip handle and an abrasive Brillo-type pad, which is much more effective at cleaning tiles. Because this also requires that you get on your hands and knees around the entire circumference of the pool or spa, carry a foam knee pad as well.

Tile soap is sold in standard preparation at the supply house, but I recommend mixing it into another container with one part muriatic acid for every five parts soap. This will help cut the stubborn stains and oils, but it will also eat into the plastic on the tile brush pads and plastic barbecue grill brush handle, so keep rinsing them in pool water after each application and scrubbing. Don't use other types of soap in place of tile formulations, because they might foam and suds up when they enter the circulation system (especially in spas).

Keep your tile soap in a squirter bottle, such as a kitchen dishwashing detergent bottle. You can control the amount of soap that flows onto the brush pad and this bottle serves another valuable function. When the wind is blowing ripples on the surface of the water, you won't be able to see the bottom to know what you are vacuuming. The squirter bottle allows you to send a stream of tile soap across the pool and, as the soap spreads out over the water, it calms the ripples. You don't need much, though you might have to repeat the process two or three times in the course of a service call to finish your work.

Test kit and thermometer

As discussed in the previous chapter, buy the best test kit you can afford and keep it in good working order. Your chemical testing is by far the most important aspect of your work. No one I know was ever injured or killed by a dirty pool, but many have been from improper handling and application of pool chemicals.

Because test kits and methods were discussed previously, I will not repeat the information here. One additional word, however, is that you should keep abreast of new testing materials and kits, especially electronic testers. Most advances in this field save time, money, or both and will certainly make your work more accurate and professional. In a business where competitors are differentiated by their service and knowledge, that can mean the difference between getting and keeping more clients than the next guy.

A thermometer is also an important item in your test kit. I keep a floating and a standard thermometer (with string). As noted in the chemistry chapter, certain tests are best performed at certain temperatures. You will also need a thermometer to check heater performance, spa temperatures, and to answer customer's questions about their water (especially useful at apartment building pools and spas when one resident wants the water warmer and another cooler, but both think 72°F is what they mean).

Spa vacuum

The spa vacuum is a miniature version of a leafmaster. It works on the same principle using a garden hose for water pressure to create suction. The dirt and debris are forced into a small sock and, like the leafmaster bag, fine dirt passes through the bag. I use a fine-mesh sock or ladies hosiery inside the bag provided when I want to pick up fine dirt.

The spa vacuum attaches to the telepole and is provided with various attachments, much like a household vacuum cleaner, for getting into crevices or brushing while you vacuum. I don't find any of these too useful and generally use the vacuum by itself. The spa vacuum is also a useful tool for sucking up small hairpins, nails, coins, or other hard to grab items from the bottom of pools.

Pumice stones

The soft pumice stone, made from volcanic ash, is abrasive enough to remove scale from tiles and other deposits or stains from plaster surfaces without scratching them excessively. Pumice stones are sold as blocks, about the size of a brick, and as small bladed stones that attach to your telepole for reaching tight spaces and underwater depths. Carry both. Since pumice stones disintegrate easily, it is wise to use them before you vacuum a pool or spa. Alternately, you can brush the residue to the main drain where it is carried to the filter. A good alternative to pumice, especially on fiberglass, which scratches easily, is a block of styrofoam or similar plastic foam.

Acid spotter

The acid spotter allows you to deliver full-strength acid to a stain at the bottom of a body of water. The disc portion attaches to the telepole for placement over a stained area. A small plastic hose runs from the disc to a bottle of muriatic acid on the deck. You start a siphon and drain acid into the disc, where it is kept in direct contact with the stain.

   The acid spotter is a useful tool, but time-consuming and not always necessary. For example, many leaves will stain plaster, but they bleach out with normal chlorination over a few days. Spending time removing every stain in a pool can waste a lot of time. Also, some stains simply cannot be removed, such as when rebar or a rebar tie has started to corrode from beneath the pool floor. The moral of the story is to have an idea of what the stain is before taking your spotter out of the truck.

Handy tools

The other basic necessities of the well-outfitted technician (as described in previous chapters) are
   - Drain flush bag (blow bag),
   - 100 feet of flexible backwash hose,
   - Skimmer diverters and adapters,
   - Spare garden hose ends (they often break and leak),
   - Basic hand tools,
   - Tennis balls (for floating the leafmaster bag as needed and for blocking off a skimmer when the pool has more than one),
   - Plastic trash bags, and
   - Business cards, notepads, and pens (you will need to leave customers notes more often than you can imagine and you will want to make notes of things that need maintenance).

POOL CLEANING PROCEDURES

Every technician, and every homeowner for that matter, approaches pool cleaning differently. After many years and hundreds of thousands of service calls I have discovered that there are a few basic procedures that are efficient and save time which any technician would do well to follow. Always start by determining the surface composition. You will work very differently on plaster than on a painted surface, for example.

Deck and cover cleaning

Most technicians overlook the fact that if they spend 30 or more minutes cleaning a pool, it will quickly appear as if they were never there when the leaves and dirt on the adjoining deck blow in on the first breeze. A quick sweep or hosing of any debris near the pool, at least 10 feet back from the edge, will keep your service work looking good after you have left. Similarly, remove as much debris as possible from the pool or spa cover before removing it. If the cover is a floating type without a roller system, be sure to fold or place it on a clean surface. Otherwise, when you put it back in place it will drag leaves, grass, or dirt into the pool. If it is a mechanized cover system, any small amount of standing water on top of the cover will slide off as you roll it up. If the motor is laboring you will need to use the water removal pump, the sump-type pump provided with these systems. Also be careful to avoid abrasive or sharp surfaces as you drag the cover off of the pool or you might be responsible for an expensive replacement. Finally, hose off the cover before returning it to the pool.

Water level

If you add an inch or so of water to the pool each time you service it, you will probably keep up with normal evaporation. If you wait a few weeks until the level is several inches low, it will take hours to fill, requiring you to turn on the water supply and go back later in the day to shut it off. You can leave a note for your customer to fill the pool, but most homeowners will forget or resent doing part of your job. Never leave the water on and leave a note for the customer to turn it off when they get home. The note might get lost or their plans might change and they could arrive home to a flooded backyard.

If the pool has a dedicated fill line, turn it on and leave your truck keys on the valve. This is one way to make sure you don't forget to shut it off. If the fill line is above the water surface, it will splash into the pool, obscuring the bottom and making it difficult to see your vacuuming work. In this case, use a garden hose instead of the fill line and put the tip of the hose below the water surface. You can also put it in the skimmer to diffuse the flow.

After rains you might need to lower the pool level. In this case, use your submersible pump and a backwash hose or spare vacuum hose for the discharge. Alternatively, you can run the pool circulation system and turn the valves to waste. If you use this method, don't forget to return the valves to normal circulation before you leave or you might get a late night call from an irate customer with an empty pool and burned out pump.

Surface skimming

It is much easier to remove dirt floating on the surface of the water than to remove it by any means from the bottom. Using your leaf rake and telepole, work your way around the pool raking any floating debris off the surface. As the net fills, empty it into a trash can or plastic garbage bag (which you will notice is on your equipment list). Never empty your skimming debris into the garden or on the lawn. Even if dumping a very small amount, a customer who sees you shaking out your leaf rake over his prize roses will quickly terminate your services. Also, the debris is likely to blow right back into the pool as soon as it dries out.

There is no right or wrong way to skim, but as you do, scrape the tile line, which acts as a magnet for small bits of leaves and dirt. The rubber-plastic edge gasket on the professional leaf rake will prevent scratching the tile. This action is often overlooked and is a rich source of small debris that will soon end up on the bottom.

One last tip. If there is scum or general dirt on the water surface, squirt a quick shot of tile soap over the length of the pool. The soap will spread the scum toward the edges of the pool, making it more concentrated and easier to skim off.

Tiles

Many technicians leave the tiles for last, but if they're fairly dirty you will remove the material from the tile and it will settle on the bottom you have already cleaned. Also, if you need to remove stubborn stains with a pumice stone, the pumice itself breaks down as you scrub, depositing debris on the bottom. Therefore, do the tiles first.

Tile cleaning is a real pain in the knees. Although body and suntan oils and other light dirt will come loose with a standard tile brush, most of the time you will have to get down on your hands and knees and do some real scrubbing. Generally even if the tiles appear clean, I scrub them once a month to knock loose any scale or other deposits before they become noticeable.

Whichever brush you use, apply a squirt of tile soap directly to the brush and start scrubbing. Remember, don't use soaps not designed for tile work because they will foam up. I carry a foam knee pad, available at any gardening store, and work my way around the pool. To break up this tedious job a bit, you might do half the pool, then do your chemical testing or equipment check before finishing. It gives your knees a welcome pause.

When cleaning tile, scrub below the waterline as well as above. Evaporation and refilling means the waterline is rarely at the same level, so clean the entire tile line. It takes very little extra effort while you are there. Refresh the soap on your brush as needed.

Finally, you need to scrub to remove oil, dirt, and scale, but you want to avoid scratching the tile. Small scratches might not be visible but they do remove the glazing that makes the tile impervious to staining. Once the tiles are scratched, oils and scale are absorbed by the porous pottery material below and the tile will never come clean. Some minor scratching can't be avoided and won't penetrate the entire glazing layer anyway, but use only as much force as needed to clean the tile. Never use really abrasive brushes or scouring pads to clean tiles.

Equipment check

At this point, many service technicians will start vacuuming the pool. Unfortunately they often get started only to discover that the suction is inadequate. It pays to check your equipment before vacuuming.

My technique is to review the circulation system by following the path of the water. Since you have already skimmed and cleaned the tile, you can now clean out the pool's skimmer basket without concern that it will fill up again while you work. Empty the contents of the skimmer basket into your trash can or garbage bag.

Next, open the pump strainer basket and clean it. If the pump has a clear lid, you might be able to observe if this step is necessary. Reprime the pump. Check the pressure of the filter. There is no point in checking it before cleaning out the skimmer and strainer baskets, because if they are full the filter pressure will be low and will come back up after cleaning the baskets. If the pressure is high, the filter might need cleaning. As mentioned in the chapter on filters, I don't believe in backwashing except as a very temporary measure. For example, if the filter is dirty and you need to vacuum the pool, you would not want to tear down and clean the filter, then vacuum dirt to it. In such cases you might want to backwash, if the filter uses that technology, vacuum, and then do the teardown.

Check the heater for leaves or debris. Look inside to make sure rodents haven't nested and to verify that the pilot light is still operating (if the unit is a standing pilot type). Turn the heater on and off a few times to make sure it is operating properly. While the heater is running, turn the pump off. The heater should shut off by itself when the pressure from the pump drops. This is an important safety check. Of course, some customers shut down the heater in winter and don't want the pilot running, so be familiar with your customer's needs before taking action.

Check the time clock. Is the time of day correct? Is the setting for the daily filter run long enough for prevailing conditions? If the pool includes an automatic cleaning system with a booster pump, is the cleaner's time clock set to come on at least one hour after the circulation pump comes on and set to go off at least one hour before the circulation pump goes off? Always check the clocks because trippers come loose and power fluctuations or outages can play havoc with them. Service work on household items unrelated to the pool can also affect the clocks, when other technicians turn off the breakers for a period of time to do some work. Also, electro-mechanical time clocks are not exactly precision instruments. One might run slightly faster than another, so over a few weeks one might show a difference of an hour or more, upsetting your planned timing schedule.

At each step of the equipment check, look for leaks or other early signs of equipment failure. Clean up the equipment area itself. Remove leaves from around the motor vents and heater to prevent fires, and clear deck drains of debris that could prevent water from draining away from the equipment during rain. This is your area, because when the customer looks at the equipment, he will judge you by the appearance of it.

Vacuuming

If the pool is not dirty or has only a light dusting of dirt, you might be able to brush the walls and bottom, skipping the vacuuming completely. If the pool or spa is dirty, however, you have two ways to clean it: vacuuming to the filter or vacuuming with the leafmaster.

Vacuum to filter

Vacuuming to the filter means the dirt is collected from the pool or spa and sent to the circulation system's filter.

1. Make sure the circulation system is running correctly and that all suction is concentrated at the skimmer port. Use your bronze skimmer diverter for this

process (as described in the chapter on basic plumbing) if dealing with a single port skimmer. If the system includes valves for diversion of suction between the main drain and the skimmer, close the main drain valve completely and turn the skimmer valve completely open. If there are two skimmers in the pool, close off one by covering the skimmer suction port with a tennis ball, concentrating the suction in the other one. On large pools, you might have to vacuum each half separately.

2. Attach your vacuum head to the telepole and attach the vacuum hose to the vacuum head. Working near the skimmer, feed the head straight down into the pool with the hose following. By slowly feeding the hose straight down, water will fill the hose and displace the air. When you have fed all the hose into the pool, you should see water at the other end (which should now be in your hand).

3. Keeping the hose at or near water level to avoid draining the water from it, slide the hose through the skimmer opening and into the skimmer. Attach the hose to the diverter (or with two-port skimmers, insert the hose cuff into the skimmer's suction port). The hose and vacuum head now have suction.

   On older pools, the suction port might be in the side of the pool below the skimmer. In this case you might need to put a tennis ball over the skimmer suction port to concentrate the suction at this wall port.

   In any case, make sure the hose does not contain a significant amount of air. When the air reaches the pump, you will lose prime. If this occurs, remove the vacuum hose, reprime the pump, then try again.

4. Vacuuming a pool or spa is no different than vacuuming your carpet. Work your way around the bottom and sides of the pool. If the pool is dirty, it will be very obvious where you have vacuumed and where you haven't. The only word of caution is to avoid moving the vacuum head too quickly. You will stir up the dirt rather than suck it into the vacuum. If the suction is so strong that it sucks the vacuum head to the pool surfaces, adjust the skimmer diverter or valves to reduce the flow. You might also need to lower the wheels on the vacuum head, raising the vacuum head itself.

5. Conversely, if the suction is weak, you might want to lower the vacuum head or you might need to move the head more slowly around the pool to vacuum it thoroughly.

   If the pool is very dirty, the sediment might fill the strainer basket or exceed the capacity of the filter. You need to stop when suction becomes weak to empty the strainer basket or, in extreme cases, clean the filter.

   If the pool contains both fine dirt and leaves, the leaves will quickly clog the strainer basket. You can now purchase a leaf canister, which is an in-line strainer that collects the leaves and allows fine dirt to pass on to the filter. You can empty the canister without losing prime in the pump, making your cleaning job faster.

   If the pool has an adjacent spa or wading pool operating from the same circulation system, you can vacuum this at the same time by quickly removing the vacuum head from the pool and putting it immediately into the spa or wading pool. If you are quick enough, only a small amount of air will enter the hose and you might lose prime for a minute, but there will be enough water in the line for it to reprime itself.

If you do vacuum both units at once, make sure the water you have vacuumed from the spa or wading pool is replaced. Normally, adjacent spas or wading pools have pool return lines in them so that water falls from one to the other. This will refill the spa or wading pool. If the systems operate from the same circulation system by changing valve positions, but are not circulating together, you will need to vacuum each separately so that water removed during vacuuming returns to the place of origin.

6. When you are finished, position yourself near the skimmer and remove the vacuum head from the water. The suction will rapidly pull the water from the hose. Be prepared to pull the hose off of the skimmer suction port before the air reaches there, otherwise you will lose prime. You can also let the vacuum run for a minute after you finish to make sure the hose contains clean water, then remove the hose from the skimmer suction port and drain the water from the hose back into the pool. Unfortunately, if any sand or small stones are hung up in the coils of the hose, they will wash back into the pool when you empty it that way, so I suggest the pump evacuation method just described. You can also pull the vacuum head from the pool and the suction end of the hose from the skimmer simultaneously, remove the hose from the water, and drain it on the deck.

7. Remove your equipment from the pool. Check the pump strainer basket and filter to see if they have become clogged with debris. Clean as needed. Replace the skimmer basket.

When a pool is exceptionally dirty you have another alternative to vacuuming and cleaning the filter several times. I carry a portable pump and motor, which is the same 1½-hp unit used in the circulation system, that simply plugs into a household electrical socket. Using this pump, I vacuum the pool and run the discharge through a second vacuum hose to a waste drain or the street. This method wastes a lot of water, but for extreme cases it might be the only alternative to draining the pool, which wastes even more.

Twice in recent years, Malibu endured devastating brush fires and all my pools were blanketed in soot and ash, as much as a foot thick on the bottom of some pools. I devised this portable pump after the first pool took six hours to vacuum to the system's filter. Since then, PoolTool and other companies have developed portable systems, but you can put one together yourself with a standard 120-volt pump and motor, a length of extension cord, and some hose cuff fittings on the suction and discharge sides of the pump. If the circulation system is equipped with a waste line, you can work this method using the pool's own pump, diverting the water directly to waste instead of the filter. Remember to refill the lost water so the circulation system does not run dry.

Vacuum to leafmaster

If the pool is littered with leaves or other heavy debris, you might need to use the leafmaster instead of the vacuum, or use the leafmaster first, then allow the fine dirt to settle and vacuum it to the filter. Most technicians would rather leafmaster than vacuum to the filter, because the leafmaster is faster and easier to use.

1. Attach a garden hose to a water supply then to the leafmaster. Clip the leafmaster onto the telepole.

2. Place the leafmaster in the pool. Turn on the water supply and vacuum. As with vacuuming to the filter, leafmaster vacuuming is just a matter of covering the pool floor and walls. Because the leafmaster is large, you can move it quickly and vacuum the pool in about half the time required to filter vacuum. Be careful not to move the leafmaster so fast that you stir up the debris. If your garden hose is not the floating variety, work in a pattern that keeps the hose behind your work so you will not stir up debris before you can vacuum it.

   Empty the collection bag periodically if it becomes too full (see step 3 for directions), but remember, some leaves in the bag will help trap fine dirt or sand in the bag, so leaving some debris in the bag is helpful.

3. When you are finished, remove the leafmaster by turning it slightly to one side and slowly lifting it through the water to the surface. If you pull it straight up, debris will be forced out of the bag and back into the pool. Never turn the water supply off before removing the leafmaster from the pool, for the same reason. The loss of vacuum action will dump the collected debris right back into the pool. When the leafmaster is on the deck, turn off the water supply and clean out the collection bag.

   Always be sure your collection bag is securely tied to the leafmaster. I can't tell you how frustrating it is to have the bag come loose as you lift the leafmaster from the pool, spilling the debris back into the pool. It can also come loose during vacuuming, especially if there is a lot of debris and strong water pressure.

   Remember, if the water pressure is poor, use the automatic cleaner booster technique described earlier or keep a tennis ball in the collection bag to keep it upright.

Chemical testing and application

You rarely know what chemicals the pool will need and unless you have a wheeled equipment cart you probably haven't brought any with you to the pool. By testing the water now, you can decide what is needed and make one trip back to your vehicle, carrying the cleaning tools out and the chemicals in.

1. Follow the general testing guidelines described previously, testing for chlorine residual, pH, total alkalinity, and acid (or base) demand. Unless you detect a problem, these are the basic tests you will conduct at each service call. Once a month, or if you find unusual readings in the basic four tests, you will test for calcium hardness and stabilizer.

2. Apply the chemicals as described previously. I use liquid chemicals for ease of distribution and price, but you will make your own judgments as you proceed. In any case, it is purely a personal preference. Apply sanitizer, following the guidelines described previously, then move ahead to the brushing step to give it time to circulate and distribute. Brushing accelerates that distribution. Apply acid or alkaline materials next. You don't want to apply them together because combinations of pool chemicals can be deadly.

   Be careful with chemical bottles on pool decks. Like the ubiquitous coffee cup ring, pool chemical bottles will leave stains on decks that are difficult to remove and even more difficult to explain. If you must set them on the deck, immerse the bottom of the bottle in the pool for a moment to rinse off any chemical. Splash some water on the deck where you set the bottle down so any residue will be neutralized and not leave stains.

Brushing

Having now spent 30 to 45 minutes on the service call, many technicians skip the brushing of the pool or spa. This is a big mistake. Dirt and minerals will build up over time, making the pool appear dingy and dirty even when you have cleaned it thoroughly. Also, algae starts out invisible to the naked eye, and brushing removes it from surfaces, suspending it in the water where chemicals will kill it before it can really take hold.

On pools or spas that are not very dirty, you can skip vacuuming and brush from the walls to the bottom and from the shallow to the deep end, directing the dirt toward the main drain where it is sucked to the filter. If you plan to use this technique instead of vacuuming, divert all suction to the main drain as described previously.

Cleanup and closeup

After you have collected all your belongings, take one last look at the pool before leaving. Did you add enough water and turn off the water supply? Did you pick up everything you brought with you? Did you leave the customer any notes as needed?

I can't tell you how many times, being in a hurry, I have skipped this simple 30-second step and regretted it. My pools in Malibu are often miles apart, up winding canyon roads, so nothing is more irritating than to arrive at the next pool and discover you left your vacuum head or test kit at the last pool.

Moreover, you might have done a fabulous job on the pool, but the customer will come home at night and only notice the debris you left on the deck, the cover you forgot to put back on, the equipment area gate you left open, or the hose you left on the deck. Look around and leave things better than you found them.

By now you see there is a reason for the order of the steps given in my service call procedure—saving time and extra work. You will develop your own practices, but I highly recommend you follow this basic order until you develop routines that you feel are as good or better. Time is the only product you have to sell in a service business. Especially in fall and winter when daylight is shorter, you can only perform so many service calls each day. If you make them complete, your work will satisfy customers who will retain your service and recommend it to others. If you make efficient service calls, you will be able to do more of them and make more money. It's as simple as that.

SPA AND WATER FEATURE CLEANING

The steps outlined for pool cleaning will work just as well for spas and water features. Here are a few special tips about what you might encounter that is unique to these bodies of water.

- Many spas are made of fiberglass, so take care when vacuuming to avoid scratching the surfaces.
- Vacuum the corners of water features and small spas with the spa vacuum described earlier.

- Evaluate the spa or water feature when you arrive, before you invest a great deal of time in cleaning. Test the chemistry first, so you can determine if you are better advised to drain the unit rather than clean and treat it. If the water or surfaces are very dirty, if the water is extremely hard or cloudy, if the dirt is in among rocks and gravel where it might be very hard to reach, pump out the water and clean the unit that way.
- If you do drain a spa or water feature, be sure the equipment is turned off at the breaker so the time clock won't turn it on before you are ready. You might need to set up your submersible pump and go to another job while the unit is draining. Your submersible has a small hole in the bottom to recirculate the last inch of water to avoid burning out it's seal, so you can let it run without worrying that it will run dry.
- Before you clean the spa or water feature, clean the filter and run some fresh water (from the garden hose) through the circulation system to purge any dirty water from the lines. Nothing is worse than draining, cleaning, and refilling a spa only to turn the circulation back on and watch dirty water contaminate your work.
- Be extra careful with chemical testing and application. Most spas and water features contain a tiny fraction of the volume of water in a pool, so they can't absorb a mistake the way a pool might. It is better to add chemicals more slowly and in less quantity than you think necessary. You can always add more, but it is a real problem to remove any excess.

10
Special procedures

The successful water technician should be able to perform the basic cleaning, chemical application, and repair services already outlined in this book. There are several additional procedures that many technicians leave to others, but which are easy to learn and can provide significant sources of income. Understanding how to maintain and repair the plaster of a pool or spa, for example, might make the difference between keeping a satisfied customer and losing the account to another technician.

DRAINING A POOL

Many of the procedures outlined in this chapter require draining the pool or spa. It might seem obvious that you use your submersible pump to do the job, but this simple task can create many problems if you don't take all factors into consideration. Therefore, before presenting information requiring this procedure, here are the steps to draining a pool.

1. Turn off all circulation equipment at the circuit breakers, so there is no chance it will start up from a time clock. Be sure that underwater lights are also switched off and not connected to a time clock. Often there are switches inside the home: tape over these with a note to make sure all family members know there is a reason to keep that switch off.

2. At a commercial pool, run yellow caution tape around the pool deck to keep unwary visitors from falling into an empty pool. Turn deck furniture on its side and use it as a physical barrier as well. Post signs around the pool deck about what is going on and letting people know the pool will be closed for several days. With today's busy families, it is a good idea to do the same for residential pools.

3. Drain the pool. But make sure you won't be the victim of hydrostatic pressure. Concrete does float; in fact boats are made from concrete. If it has rained recently, is likely to rain during the course of the job, or the area is known to have a high water table (groundwater), don't drain the pool. In any case, it is wise to have your customer sign a simple hold-harmless statement stating that they accept responsibility for any damage to the pool caused by hydrostatic pressure which you cannot predict.

When draining a pool with your submersible pump and hoses, direct the flow to a deck drain if possible. This will send the waste water through an intended channel, rather than over a backyard garden or down a hill where erosion damage can occur from such a large volume of fast-moving water. Once you have started pumping, watch the flow into the drain for several minutes. Sometimes debris will back up in the line and it will overflow, but not until it has filled several hundred yards of pipe. It might take time, but the clogged drain can flood backyards, living rooms, and might actually flush the water back into the pool.

If deck drains cannot accommodate the flow, connect several vacuum or backwash hoses together and run the waste water into the street where it is carried to storm drains. Again, watch the flow and make sure that it does not back up the storm drain and flood the street.

In areas where water conservation is a concern, you should wait to drain the pool until the chlorine residual has gone below 1 ppm, then let the flow irrigate lawns and gardens. Obviously a 20,000-gallon pool will flood the typical backyard garden if it is pumped all at once, but in some areas, particularly in the arid south-western United States, you can pump for a few hours a day and drain the pool over a week or two to make use of the waste water. Another technique along these lines is to punch holes along the length of a backwash hose (or old vacuum hose) and seal up the discharge end (tie a knot in a backwash hose or use plumbing fittings to close a vacuum hose). This then acts as a huge sprinkler, evenly distributing the water all along the length in various gardens, perhaps even across the lawns of several neighbors (with their permission, of course). In other words, be creative about your method of pumping and the potential uses of the waste water.

One other word of warning about deck drains. They are usually made of PVC, but because they don't carry water under pressure they are not usually pressure tested. If ground movement or other erosion has destabilized them, the pipe might have separated and much of your waste water will end up eroding the backfill around the pool. If you have any doubts about the integrity of the deck drain or its ability to handle the water, run the hoses into the street.

In some jurisdictions there might be restrictions on pumping out pools relative to the permissible volume and even the permissible chemical makeup. Extremely low-pH water might have to be neutralized before pumping it into municipal stormwater or sewer lines. Check your local codes before turning on the pump.

When lowering a submersible pump into a pool, never do so by the cord. Attach a nylon rope to the bracket or handle and lower it that way to avoid pulling electrical wires loose. Not only will the pump fail if the wires come loose, but when you plug in the cord it might electrify the water.

One last tip about submersible pumps and pool draining. Buy a ground fault interrupter (GFI) that can be plugged into the wall socket before plugging in the cord of the pump. You can also rig a remote on/off switch that plugs into the socket as well, allowing you to operate the pump while you are working in the pool. Check your local electronics store for a combination remote receiver/GFI or simply buy one of each and plug them in together before plugging your pump into the outlet. It will be the best $30 to $50 you ever spent. I have been in water with a pump that had a slight short and felt the tingling of the electricity conducted through water and have known technicians who have been killed by not taking this aspect of operational safety seriously.

BREAKING IN NEW PLASTER

An excellent example of a simple special procedure that can bring you new customers is the break-in of a new plaster job. Whether it is a pool and spa that is newly constructed or an older vessel that is being replastered, the technician who gets the break-in work usually gets the service contract.

I try to maintain good relations with the pool builders and plastering companies so they will recommend my company to perform the break-in of new plaster. Most builders subcontract the plastering and most plasterers don't have the time to perform the break-in service. Therefore, they subcontract it to a local service company or recommend the technician directly to the pool owner. About 20 percent of my new customers over the years have come from these referrals because I offer the pool owner a monthly service contract after the break-in.

With this in mind, it is obvious that new plaster break-in is one of the most important examples of your professionalism and knowledge of pool and spa maintenance. By following the procedures outlined here, without taking shortcuts, you will find new plaster break-in to be one of your most effective calling cards.

Why break in plaster?

There are two basic reasons to break in plaster rather than just turn on the pump and start swimming. First, you need to remove the plaster dust from the water, which will otherwise settle and build up as hard, rough scale. The pool is filled before the plaster dries, so the actual drying (curing) takes place underwater. Plaster shrinks somewhat as it dries, so if you allow the plaster to dry before filling the pool, the weight of the water against the brittle, dry plaster will cause it to crack. Adding the water before the plaster dries allows it to be pushed into place against the pool shell, pressed evenly by the weight of the water. It doesn't shrink, become brittle, or crack.

By allowing this curing process to occur underwater, you are left with the by-products of the process to remove from the pool or spa. Plaster dust—calcium carbonate—will adhere to the surfaces of the pool as scale, especially noticeable in dark-colored pools. The curing process, called *hydration*, can take up to four weeks, but most of the excess plaster dust is removed in the first week if the break-in process is properly executed.

The second reason to break in the plaster is to balance the water chemistry so that the water itself does not destroy a good plaster job. Tap water is never exactly balanced to reflect the ideal chemistry components for pools and spas (as detailed in the chapter on chemistry). Therefore, the break-in process is designed to create plaster-friendly water that is neither etching nor scaling.

Procedure approval

The break-in procedure is the same for a new pool or a replastering of an older pool. In either case, the most important first step is often overlooked. When approaching the break-in, I always confer with the pool builder and/or plasterer to outline my break-in procedure and ask if they have any suggestions or preferences. This collaboration is essential in keeping the builder and plasterer responsible for their work. If there are discolorations, delaminations, roughness, or other problems with the plaster, the builder or plasterer will almost invariably blame the break-in procedure. If they reviewed and approved your procedure, they will be more likely to accept responsibility for their construction and plaster work.

To accomplish this simple task, I keep a step-by-step list of my break-in procedures, detailing the brand names of chemicals used and the daily break-in schedule. If you can arrange to have the builder and plasterer sign a copy of your procedure list, you will ensure their future participation in the warranty. At the very least, be sure a copy is delivered to them and keep notes of who you reviewed the procedure with, when you did so, and any comments they had.

Further, it is important to document that you followed the break-in procedure that was agreed upon. Keep a clipboard in the equipment area to list each action taken each day as the break-in proceeds. The notes should include the date, time, weather conditions, test readings, and chemicals applied or actions taken. Again, this is insurance against anyone blaming your break-in for later plaster problems. When you can produce detailed evidence of the original game plan and the execution thereof, anyone seeking reasons for plaster problems will readily accept that it was not the break-in process and will look elsewhere. This is important because plaster discoloration or roughness often are the result of a faulty break-in and it is much easier to blame that process than to investigate the plaster job or other hidden mistakes.

Break-in step by step

There are many variations to the steps needed for a successful break-in of new plaster, but there are not many disagreements about what those steps should be. What follows are my recommendations after breaking in literally hundreds of plaster jobs over the years.

1. My first step always is to review the process with the builder and plasterer, making my recommendations but following any reasonable variations they might suggest. By the way, if you encounter severe disagreements at this point on the right way to break in the plaster, don't take the job. It isn't worth the value of the job to do a break-in that you know will give unsatisfactory results. Even though you can say you followed the advice of the builder or plasterer, you are still the one being paid to perform the break-in and, therefore, you are the one who will be blamed for poor results. If you can't agree on the procedure and you feel the requested variations will damage the plaster, walk away from the job.

2. As a part of setting up the job, inspect the plaster work. Look for trowel marks or footprints, and look for signs of sloppy work, such as plaster on the tiles or deck. Examine the steps and anywhere that ladders, rails, or other hardware meet plaster surfaces. They should be seamless and smooth, without trowel marks or high spots. Bring any imperfections in the plaster job to the attention of the builder or plasterer for correction before filling. Make the homeowner aware so you won't be held responsible for imperfections in the plaster work. I have seen footprints in the plaster on

numerous occasions after the break-in, and regardless of the fact that I took charge of the pool after it was full of water, believe me, it's the technician that is always expected to remove those marks and stains.

3. Test the source of the fill water. In the chemistry chapter, I discussed the tests for heavy metals such as copper, iron, and manganese. In normal water maintenance you won't be testing for such metals, but now is the time for these tests. If you detect high levels of metals in the source water, you should be advised that additional chelating agents will be required and you can adjust your chemical balancing in general to avoid precipitating these metals onto the new plaster.

Similarly, if the testing of the source water reveals it to be of exceptionally high or low pH and/or total alkalinity, you should be prepared to adjust it. If, for example, you are facing one extreme or the other, you might want to employ the barrel-fill technique (outlined later). If the water appears to require little adjustment, you should proceed to fill the pool and balance it gradually.

When testing fill water, test all of the sources. In some areas, water is trucked into the pool, with each tanker coming from a different location. Test each truckload. If the water pressure at a particular location is weak and you will be filling from more than one hose bib, or from the house and the neighbor's, or from a hose bib and pool fill line combination, then you should test each source. There can be significant differences between sources that you would expect to be fairly similar, so test them all.

One last word about testing the source water. Let the water run a few minutes before testing from any source to evacuate any concentration of minerals or water that might have an altered chemical makeup from sitting in a garden hose or copper pipe for a long period. This is particularly true of a pool that has been empty for an extended period, where the fill water hasn't been used in that time and the first gallons have been sitting in the pipes. Also, for this reason, it is wise to evacuate the lines into a bucket so as not to bombard the fresh plaster with a dose of water that is high in mineral or metal content or that has unusual pH.

4. Fill the pool or spa. This step seems the easiest part of the break-in process but it can be tricky. Usually the plasterer will fill the pool and call you in for the break-in, but it is important to know the procedure in case it becomes your job.

First, never just turn on the fill line if the pool has one. The water cascading several feet down will damage the soft plaster below. You must provide a cushion for the water, which can be a few feet of water provided by a hose before you fill with the fill line. You can also place a bucket under the flow of water to provide a cushion, but be sure to place rags under the bucket so that the weight of the water in the bucket doesn't leave circular scratches in the plaster.

When filling with a hose, wrap the end with a rag so the hard fitting on the hose end does not scratch the plaster. Better yet, remove the fitting if you can, or cut it off and replace it later. Don't lay lengths of hose in the bottom of the pool either. The coils will leave marks, and if the pressure is enough to make the hose fishtail, the sweeping action will leave marks as well. Never walk on the fresh plaster; this will leave footprints that stand out like the proverbial sore thumb when the pool is full, and they are very difficult to remove.

The objective is to fill the pool quickly and without stopping. If you stop at any point and then continue, you can expect to see a ring at the level you stopped. If you fill slowly, the plaster that is allowed to air dry will be a very different shade than the plaster that cured underwater. In extreme cases, the air-dried plaster might also crack.

For the same reason, don't add chemicals when filling the pool. The concentrations of chemicals while the pool was at one level versus another will tend to leave stains on the wall at the places where the chemistry changes took place. The exceptions to this are the addition of chemicals by the barrel method (described later), because you are evenly changing the chemistry of all of the fill water before it enters the pool, and chelating or antistaining agents.

To accelerate the filling process, add water from as many sources at once as you can. Use more than one hose, the fill line, and even a hose from a neighbor's yard (with permission, of course). Calculate the time required for filling so you can be present when the pool is full. Don't rely on the homeowner to turn off hoses, however; they always agree to do so, but rarely follow through.

One last point about the fill water source. If the water tests out especially corrosive and low in calcium hardness, check with the homeowner to see if a water softener is a part of the household plumbing. If so, there are bypass valves on top of the softener to close it off from the system while you are filling the pool. Most softeners are added in a utility closet at a point in the plumbing where they will only impact the water inside the home, but sometimes they are inserted in a system at a point prior to this.

5. Check the equipment. It might take a day for the pool to fill, so this is a good time to make sure that you are prepared to circulate the water when that happens. Make sure that you clean the filter and that the pump works. Be careful not to force any dirty water into the pool that might have been standing in the filter or plumbing. At this point, turn the heater off and leave a note on it that it must remain off for at least three weeks after the pool is full.

6. Add a chelating agent to keep any metals in the water from precipitating out of solution and depositing on the new plaster surfaces. There are many products for this purpose with names like Metalgone or Stainout. Consult your supply house and follow the instructions on the label regarding application techniques and volume. Generally you will add a quart of liquid product as soon as you start the filling process. Repeat as directed by the product label.

7. When the pool or spa is full, turn on the circulation equipment. If the filter is a DE type, add the DE immediately so that the filter grids do not clog with plaster dust. Note the pressure on the filter so that when the pressure is 10 psi higher you know to clean the filter. You might need to break down and clean the filter several times during the break-in process. Remove the trippers from the time clock and leave any necessary notes on it; the circulation equipment runs 24 hours a day for the first week.

Some technicians argue that 12 hours of circulation per day is adequate. I disagree and can point to specific cases where staining and other problems occurred when the pump was shut off for several hours during the first few days. Since it costs so little to leave the pump running, I would rather be safe than sorry. There is no disadvantage to running the pump 24 hours a day, while there might be several disadvantages to running it less. Believe me, this is important. Run that pump 24 hours a day for the first week.

8. Purge all air from the system, especially from the filter. Filtration at this point is critical. You will recall from the filter chapter that air in the filter represents media area that is not filtering because it is not in contact with circulating water. Bleed the air from the filter regularly.

As noted earlier, do not allow the pool to be heated for at least three weeks after start-up. Similarly, don't turn on any waterfalls or fountains during this period because the force of the water can damage the fresh plaster.

9. Test the water for pH, total alkalinity, and calcium hardness (at this point, there should be no chlorine residual to test). Plaster is made from calcium compounds, and if the water is low in calcium hardness specifically and total alkalinity in general, it will leach calcium out of the plaster, effectively dissolving it. When it is fresh and soft, plaster is especially unable to resist such corrosive water.

One dramatic example that illustrates the importance of this step is a technician who tested the fill water and found it had a pH of over 8, a calcium hardness of 38 ppm, and a total alkalinity of 52 ppm. After a day and a half of filling and not treating the water during that period, he tested again and found the pH over 11 and the hardness and alkalinity up to 362 and 172 ppm, respectively. Because the only source in the untreated 20,000 gallons of pool water was the calcium carbonate in the fresh plaster, the technician was able to calculate that this rise in test readings was the result of 54 pounds of dissolved plaster.

The first component to balance is total alkalinity. Since this will also raise the pH, test and adjust that after you have adjusted the total alkalinity. Use sodium bicarbonate as directed on the product label to raise total alkalinity to about 120 ppm. *Bicarb*, as it is called, is extremely soluble and there is little risk of the product saturating the water and precipitating out of solution.

Bring the calcium level up next (unless it is already in the 200- to 400-ppm range) by adding calcium chloride to the water (follow the instructions on the product). Make sure it is completely dissolved by adding it to a bucket of water, dissolving it, then adding it to the pool water. Add chemicals near return lines for even distribution, then brush the pool for more thorough circulation.

Finally, adjust the pH. Generally you will need to lower the pH, but this must be done gradually. As noted in the chemistry chapter, pH readings take awhile to stabilize, particularly during the break-in process when no conditioner has yet been added to the water. Test the pH every four hours and, using the acid demand test, determine the amount of acid needed. Regardless of the result of that test, however, add no more than 1 pint of acid per 10,000 gallons of water at any one time. In this way you will avoid spiking the pH. Remember, you can easily correct water that needs more acid, but your plaster will immediately suffer if you add too much.

The total alkalinity of the pool or spa represents the amount of alkaline material dissolved in the water. Because you are not finished dissolving alkaline material in the water, there is not much point in further adjusting this until the break-in process is complete. The calcium carbonate of the plaster dust, the calcium chloride you might have added, and other factors will affect the final total alkalinity measurement. Therefore, don't attempt to adjust it daily.

Also, do not add sanitizer or stabilizer at this point. Algae doesn't prosper on smooth surfaces (like new plaster), in alkaline conditions (like water curing new plaster), or on surfaces that are constantly being brushed (as you are about to do). Therefore, the sanitizer only has the potential to harm the plaster during the curing process not help it.

Finally, continue to test the water for all parameters at least once per day. Better yet, test with each brushing (three times per day) and adjust the chemistry as needed in gradual increments.

As soon as the circulation begins, so must the brushing. Brush every inch of the plaster surface thoroughly and vigorously at least three times each day. Try to divide the day into equal parts, rather than brushing three times in just one part of the day.

Brushing knocks loose the plaster dust, avoiding scale build-up. It also keeps any precipitate or algae from staining the new plaster. Brush from the top of the walls down and from the shallow to the deep end, with the suction concentrated at the main drain to pull the dust and minerals into the filter. When you have finished brushing, return the suction to half main drain and half skimmer. As always, it is better to skim any dirt or leaves into the skimmer than to find them at the bottom of the pool.

10. Break down and clean the filter each time filtration slows and/or the filter pressure exceeds 10 psi over its clean operating pressure. Plaster dust will quickly clog the filter, so be prepared to clean it as soon as the first day after starting the circulation.

11. After 48 hours, vacuum the pool at least once each day before brushing. Use a brush vacuum rather than one with wheels that might leave marks on the plaster surfaces. Vacuuming will quickly fill the filter with plaster dust, so be prepared to clean the filter as needed.

12. After 72 hours, begin to add sanitizer, gradually raising the residual to normal levels. As in all chemical applications, add gradually. Because the plaster is still soft and receptive to stains, you must avoid precipitating minerals and creating stains, which can happen even with the correct amounts of chemicals. As always, the key is to add chemicals gradually and make sure the distribution is thorough by adding near a return outlet, by walking the chemical around the pool as you add it, and by brushing immediately after you add chemicals to guarantee there are no harmful concentrations of chemical in any one area. After your desired sanitizer residual has been reached, add stabilizer.

Continue these procedures for at least one week, preferably two. If you see no plaster dust coming off when you brush, the surface is curing but not yet finished. Continue the brushing routine for at least three days after you no longer observe any plaster dust.

Keep swimmers out of the pool during the break-in period. Body oil in the water can stain the plaster and add organic material, while the varying levels of chemicals during break-in can irritate skin and eyes.

One last aspect of the break-in process is worth noting. When the fill water requires extreme adjustments, some technicians try to add chemicals as the pool is filling with the *barrel technique*. In this method, the fill water is passed through 55-gallon drums with the appropriate chemicals added before it is passed into the pool. The chemicals can be slowly added in the barrel by hand or with chemical feeders. The problem with this system is that it is labor and equipment intensive, subject to failure (leaks and breaks), and requires detailed monitoring and tweaking. The result can still be adding too many chemicals at once. If the pool will fill in less than 24 hours, I don't recommend this method, unless the fill water is extremely unbalanced. Even then, you can add small amounts of balancing chemicals directly to the pool and brush them for even distribution as the pool fills. As noted previously, any dry chemical can be dissolved in a bucket before adding it to the pool to ensure dissolution and distribution.

Acid break-in

In recent years, the United Chemical Company has developed a break-in procedure that requires only a day or two and uses no circulation, no brushing, and no filter cleaning. The process remains controversial, but it has been used successfully in enough break-ins to warrant explanation here.

1. Fill the pool or spa. There is no need to test the fill water because this procedure uses acid to bring water chemistry down to dramatically low levels regardless of where you start out. The objective is still to fill the vessel as quickly as possible to avoid the problems described previously. **Do not** turn on the circulating equipment.
2. Add the chelating agent to prevent stains as described previously, but make sure it is a chemical that can operate effectively in extremely acidic water (which this pool will be shortly).
3. Add 4 gallons of muriatic acid per 10,000 gallons of water by walking it around the pool and brushing to ensure even distribution. The acid will dissolve the plaster dust without brushing or vacuuming.
4. Brush daily, but do not vacuum or add chlorine until the plaster dust has disappeared, usually in a few days. For white pools, brush the entire pool. For dark-colored plasters, brush only the areas that appear white (scale deposits).
5. When there is no more plaster dust, neutralize the acid by using the base demand test to determine the amount of soda ash needed, applying it as directed (follow product label instructions). Wait 24 hours and retest, repeating this step as needed.

6. When the pH reaches 7.2 or more, start the circulation and clean the filter when the pressure increases 10 psi over normal operating level.

7. Add small amounts of sanitizer until the residual is at appropriate levels.

Comparing the description of the first break-in method with this acid method, you can see the appeal to the technician. It takes less work and time. It does work when administered perfectly, but the reason for the controversy about this method is that few technicians can balance water chemistry so perfectly without going over or under.

For example, if the pH goes down too far, you will damage the plaster. Certainly the calcium carbonate dust is dissolving, actually escaping the pool water as a gas, but so is the plaster itself. Conversely, if the pH remains too high, the dust will not dissolve completely and it will be left as scale deposits. In this method there is no vacuuming, so there is no way to remove such deposits. Also, because vacuuming is not part of this method, you have no way to remove dirt and debris from wind or construction. These can stain the plaster.

Finally, when you raise the pH after the process is complete you use soda ash, which is a carbonate material, thus returning to the pool what you have just evaporated out. If you do this too fast or in excess, you will scale the pool anyway.

Having listed these caveats, there is agreement from no less than the National Spa and Pool Institute that the method works . . . if administered properly. I think the potential risks outweigh any benefits (less work and time), and it's not a method for an inexperienced technician. Moreover, as noted, you cannot use it in a wind-prone or construction area where debris needs to be vacuumed out of the pool.

LEAK REPAIR

As previous chapters have detailed, there are many ways that a pool or spa can leak through the equipment and plumbing. There are an equal number of ways that the vessels themselves can leak.

When a customer reports an unexplained loss of water, it makes sense to look for the obvious first. Signs of water in the equipment area and exposed plumbing will usually yield results; however, buried plumbing (especially copper) commonly leaks where it can't be seen. While the worst leaks require excavation of plumbing or rebuilding of the pool itself, the most common leaks can be repaired effectively. In this section, I will examine how to find those more common leaks and discuss their repair techniques.

Finding leaks

As noted previously, the first place to look for leaks is in the exposed equipment and plumbing, but there are often visible signs of leaks that are otherwise hidden from view. Cracks in the pool or spa interior might be a sign of leaks. Tiles falling off the walls or loose coping stones often suggest structural leaks from shifting ground (which might have been caused by a water leak eroding the soil). Even cracked or lifting segments of deck might indicate the source of a pool leak. Tree roots might lift not only the deck, but the plumbing or pool wall as well. If however, there are no such visible signs, you need to look elsewhere. Some leak detection methods will help to both verify a suspected, visible leak or locate a hidden leak.

Evaporation test

Before elaborate leak testing, you will want to verify the customer's complaint of water loss. The most simple method is to fill a bucket and place it on the deck next to the pool or spa. Mark the level in the bucket with an indelible felt-tip marker and do likewise for the water level in the pool or spa. Turn off the circulation to eliminate any variables in evaporation. After several days, mark the new level of water in the bucket and pool or spa. They should evaporate an equal number of inches. If the pool or spa level lowers significantly more, a leak is likely. If both vessels lower a similar number of inches, then there is no leak.

Dye testing

Perhaps the simplest method of detecting a leak in the vessel itself is to shoot a dye test. As the name implies, a colored dye is disbursed in suspected areas and, as the dye disappears, the leak is found.

1. Clean and brush the pool or spa thoroughly. Cracks can sometimes be hidden by dirt or other material settled in the crack. Pay careful attention to steps, corners, and around fittings.

   Turn the circulation off and begin the examination on a calm day. Wind rippling the surface makes it difficult to see small cracks. You might also want to squirt a little tile soap across the surface to sharpen visibility further (and you might need to repeat the process from time to time during the exam).

2. Examine the pool for cracks, beginning with the tile line. You might not need to go much further if you see gaps or missing tiles. Tap the tiles with the handle of a screwdriver (or gently with a hammer) to see if any fall off, are loose, or sound hollow. Note any positive locations.

   If you find probable leak locations along the tile line, don't just stop there. If tree roots, shifting ground, earthquakes, erosion, or other problems have caused leaks at the tiles, they might also exist elsewhere. It is important to find and repair all leaks, because you might work long and hard on the ones you find only to discover the pool still leaks and the customer won't pay for a job he perceives isn't done (or done right). This is also important because you might discover that leaks lower in the pool require major dismantling and rebuilding of the pool, in which case your work will have to be redone and, again, the customer will likely not pay you.

   Therefore, continue to visually inspect the interior surfaces, looking for cracks or discolored patches of plaster (or plastic in the case of vinyl-lined pools or acrylic spas), noting anything you suspect.

3. The dye test can easily be conducted using an old test kit reagent bottle or similar squeeze bottle filled with food dye (available at any grocery store). Some technicians use phenol red, but you will need to check many locations and it is unwise to inject that much acidic chemical into the water.

   Work around the pool, particularly in the locations of suspected leaks. You will need to get into the pool or spa, including diving to the main drain, to do a thorough job. At each crack or suspected area, aim the nozzle of the bottle at the crack. Squeeze a bit of dye into the area and watch it. If the dye simply swirls around the crack without being sucked in, then there is no leak in that area. If the dye is sucked into the crack, it is riding on a flow of water leaking from the pool. The speed with which the dye disappears will help you estimate the size of the leak.

4. As with the visual inspection, continue around the entire pool looking for leaks to exhaust all possibilities. Be especially careful around skimmers, steps, rails, ladders, or other fittings. Light niches are often the source of unexplained leaks, so pay close attention to the entire light fixture area. If you are having an especially difficult time finding a leak, you might even wish to remove the light fixture and dye test directly into the niche itself, concentrating on the area where the cable passes through. Don't forget the interior of the skimmer and the main drain, as well as the return outlets.

After you have thoroughly examined the pool with the dye test, you will know what repair problems you are faced with.

Drain-down test

If you have tried the evaporation test and dye testing, or if you are not a good diver and wish to skip the dye test, try the drain-down method.

1. Turn off the equipment and mark the level of the water in the pool or spa.
2. Mark the level again at the same time each day to establish a rate of leak. Because of normal evaporation, the level will continue to decrease indefinitely; however, the objective is to determine when the level stops lowering as a result of the leak. If, for example, you record a loss of 2 inches per day for four days, then the rate slows to 1 inch every five days, you will know the level at which the rate slowed was the level of the leak. Mark that level of rate transition.
3. Examine all possible leak areas along the transition level. The leak must be along this line. For example, if the water loss slowed when it reached the level of a particular return outlet, you might reasonably suspect the leak in that plumbing line. If the water slows when it lowers to the area of the light niche, the leak will likely be in there.

   The only fault with this method is that it is an indicator, not a precise tool. Since water seeks the same level in all plumbing and parts of the vessel, the water might stop at the level of a certain plumbing fixture, but actually be in an entirely different location that is coincidentally at the same level (Fig. 10-1). In any case, the new level will tell you where to look further and where you need not look.

10-1 Water seeks its own level.

Leak detectors and pressure testing

When the previous methods fail to help you locate the leak or you wish to further verify your assumptions, there are two other methods of leak location. There are electronic listening devices called *geophones* that can actually hear water dripping or flowing. By applying such devices around the pool and related plumbing, an operator can identify where water is moving out of the system. Because these devices are expensive and their operation requires a great deal of experience and skill, most service technicians don't buy or use them. There are numerous professionals who do, however, and they are easily found in the phone book or through referrals at your supply house.

The second method used to find leaks is pressure testing equipment. It is not difficult to pressure test a plumbing system with the knowledge already presented in this book, but the amount of time and additional equipment (plugs, adapter fittings, compressed air, and related fittings) makes this type of testing impractical for most technicians. It is wiser to contract out this work and make your money on the repairs. Many pool builders and plumbing contractors are equipped to pressure test pool or spa systems. Companies that conduct leak testing might also conduct pressure testing.

Skimmer repair and replacement

One frequent location for leaks is the pool or spa's skimmer. Because these are essentially separate devices hanging on the side of the pool or spa wall, they are especially prone to leaks when earth shifting takes place. When pools were plumbed with copper, skimmers were a particular sore spot for leaks. Chemicals collecting in the elbow of the copper plumbing (exiting the skimmer and turning toward the equipment area) and constant erosion from swiftly moving water make small leaks in this spot inevitable. Leaks in this location are accelerated by the practice of adding chemicals directly to the skimmer (or leaving chlorine tablets in the skimmer). The high acidity of the chemicals eats away at the copper in these fittings.

Some pools and spas built in the 1970s began using PVC plumbing but still used copper fittings in the skimmer and main drain. If you suspect a leak at one of these locations but have dismissed it because the plumbing is PVC, be aware that it might be in the copper portions of such combined plumbing systems.

Plastic skimmers are prone to separation at the seams. To determine if that is the cause of a leak, use the drain-down method in just the skimmer. Turn off the equipment and plug the skimmer suction ports. If the water level drops at a rate faster than evaporation, the skimmer is leaking. PVC plastic skimmers can often be repaired in place, without excavating the skimmer. Follow the acrylic repair procedures described later.

Repair or replacement of a skimmer is a simple matter of some excavation, masonry, and basic plumbing. Because it seems like major surgery, many technicians miss an opportunity for a very profitable job by being intimidated by the prospect of these repairs. Using the following simple steps (Fig. 10-2) will enable you to take advantage of such opportunities.

1. Turn off the circulation equipment at the circuit breaker. Drain the pool or spa to a level well below the lowest point of the skimmer plumbing, usually 2 to 3 feet below the normal water level.
2. Plug the suction ports with rags so that debris from the job will not contaminate the plumbing lines. Similarly, plug the throat of the skimmer—the weir area—to keep debris from the pool.

3. Using a concrete saw (available at tool rental stores) with a diamond blade, cut a 24-inch square into the deck around the skimmer. Make sure the cut goes through the entire deck, at least 4 to 6 inches deep. This will make excavation easier and prevent radiation (spreading) of the cuts, which can crack or damage the rest of the deck.

4. Using various width masonry chisels, chip out the concrete deck within the 24-inch square. Slowly chip and remove each piece, taking care not to chip other plumbing that might be in the area, making the original problem worse. Take your time. When you have completely penetrated the deck, the skimmer will be surrounded with dirt below. This can be carefully scooped out to totally expose the skimmer and its plumbing. Chisel the old skimmer into pieces for removal.

5. Pressure test the plumbing. You might assume the leak was in the old skimmer or the joint of the plumbing and skimmer, however there might be additional leaks in the related plumbing.

6. Clean the hole of loose debris and set the new skimmer in place. When you are sure you have adequate room for the new unit, plumb it to the existing suction lines using basic plumbing techniques.

7. Fill dirt about two-thirds of the way up the excavation to stabilize the skimmer in place, covering any pipes and plumbing connections with dirt. Anchor the throat of the skimmer to the throat in the concrete of the pool wall with anchor bolts for added strength. Seal the skimmer-to-pool wall joint with concrete (use standard mix concrete or quick-set available at hardware stores, and follow the package instructions). In most skimmer replacements, you will not need additional reinforcement. If you had to excavate more than the 24-inch square, you might want to cut short lengths of ¼-inch rebar and lay them horizontally in the hole before pouring in the concrete to provide more lateral strength. Lay them 4 inches apart, so there is enough room for the concrete, running in the same direction. Unless the excavation is extremely large, you need not run additional rebar at right angles as you would in the original pour.

8. Make a form in the pool deck around the skimmer, using cardboard or bender board that is no more than ¼ inch thick. Fill the hole around the skimmer with concrete up to the deck surface. Trowel smooth and finish in a similar fashion as the rest of the deck.

9. Remove the form materials when the concrete sets up. Fill the expansion joint created by the removal of the form material with silicone deck sealer, available at your supply house or hardware store.

If the size of the new skimmer doesn't exactly correspond to the opening in the pool wall or if you chipped any of the pool wall or plaster, repair it as outlined later. Don't be afraid to adapt—cut the pool wall opening larger as needed or use short lengths of flexible PVC to adapt to odd plumbing angles.

Similarly, if the new skimmer is not the same depth as the old one, each manufacturer provides depth-extender collars that can be glued to the top of the skimmer and mounted in the concrete of the deck to bring the level up to the correct height and provide a receptacle for the skimmer lid, which sits at the level of the deck.

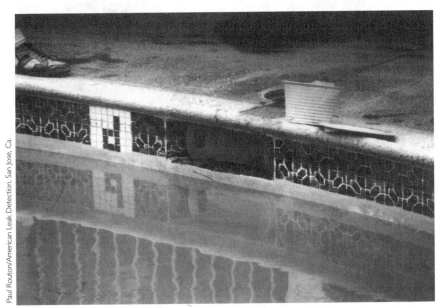

10-2 Replacement of a skimmer.

10-2 Continued.

10-2 Continued.

10-2 Continued.

When making these repairs, use plastic sheeting to cover the pool walls or adjacent deck areas to avoid dripping wet concrete on clean areas. Concrete is very hard to clean off.

On older pools that include an equalizer line through the pool wall at the base of the skimmer, patch the wall and install a two-port skimmer instead, plumbing one port to the suction line and the other to the main drain (or permanently closing it off if there is no pipe to the main drain from the skimmer location).

On fiberglass or vinyl-lined pools, the replacement will be more a process of unbolting and disassembly, replacing the skimmer and gaskets that separate it from the pool wall. Each assembly is different, but by carefully noting the steps of the disassembly you will make later reassembly easier.

Patching and repairing

As noted in the section on skimmer repair, you might need to be familiar with simple patching techniques applicable to pool walls and plaster.

Plaster

When the pool is drained for any reason, you have an opportunity to look for plaster blisters or pop-offs, areas where the plaster has come way (delaminated) from the underlying surface. Such areas can be anything from the size of a quarter to dinner plates to several square feet. Beyond that and you might need to recommend replastering.

The causes of delamination, also called *calcium bleed*, are numerous. The original plaster job might have had poorly prepared surfaces, poorly mixed materials, or too much drying before the water was added. Most likely, however, improper water chemistry has created aggressive water, stripping calcium from the plaster, weakening the plaster layer and allowing it to separate from the underlying surface.

Some blemishes will be obvious, the plaster having cracked and fallen away. Others will appear as discolored spots from the water that has entered between the plaster and the subsurface. You can also look for plaster faults by tapping suspected areas with the handle of a screwdriver or with a hammer to listen for hollow sounds (assuming the pool is empty). If you are committed to making repairs, poke around the entire pool with a chisel or screwdriver to pop the blisters. They have to come off anyway, so you might as well discover them all. Once you've been around the entire pool you can then evaluate if patching is the answer or if replastering is more logical.

When you have identified blisters, chip the loose plaster away all around the area until you reach solid, dry plaster. Use a hammer, chisel, or screwdriver depending on the size of the blister. Once completely exposed, clean the area of all water and loose debris. To make sure the final patch blends in and appears even, clean up the jagged edges of the blister area by sanding the perimeter. The most effective way to smooth out these edges, especially if there are many blisters to repair, is to use a power grinder with a small (6-inch diameter) diamond grinding wheel. The objective is to create a clean, smooth edge all around. You are now ready to patch.

The other major source of plaster problems stems from rusting rebar. If the gunite was not applied evenly over the rebar, water will rust the steel, creating red rust spots on the bottom of the pool or spa. These are very obvious against white plaster. Steel bleeds are more commonly caused by the steel wire used to tie the rebar together before the gunite is applied. If the tie wire is not closely clipped, the tails can stick up through the gunite and act as a wick to draw water and rust deeper into the rebar.

In either case, the spot of rust you see might only be the tip of the proverbial iceberg. Once the rust starts it spreads like a cancer and can affect vast areas. When you chip off the plaster, keep excavating to expose all rusted rebar. If you leave any rusting steel in the subsurface, it will leech up to the plaster and restain it.

When you have exposed all the rusted steel, cut it out with a hacksaw. Unless the area is large, you won't need to replace the rebar. If it is a large area, try to replace the same amount, size, and distribution pattern that you have removed (basically, crisscross the bar 4 inches apart). If you have excavated more than an inch or two, fill the hole with quick-set cement and allow it to dry before going on with the surface plaster patching. The objective is to repair the hole to the level of the surrounding subsurface so that the plaster patch will be no thicker than the surrounding plaster.

The best way to learn how to work with any repair material is to practice in the workshop before attempting the repair in the field. A little wasted material is less costly (in time and money) than making mistakes on a customer's pool. All these materials are available at your supply house.

Mixing a batch of plaster for patching (or practice) and the actual application are simple.

1. Start with a clean 5-gallon bucket, adding 4 cups of water.
2. Slowly stir in calcium chloride (the powder used to make the mixture set up) until the liquid becomes warm to the touch. The reaction of the calcium chloride will actually make the water heat up. You will use less than a cup, but keep monitoring the temperature (the chemical will not harm your fingers). Too much calcium chloride can actually melt plastic buckets, so don't be too heavy-handed. The general rule of thumb is one part calcium chloride for every ten parts cement used. You can also make patch material

without it for a slower drying mixture. This might be helpful if the area to be patched is not underwater and it is a hot, dry day (or a windy day when the material will dry quickly anyway). Faster drying means more likelihood of cracking. Again, practice with several mixtures and compare results so you become familiar with both the art and science of plaster patching.

3. Add 8 cups of white cement and 4 cups of sand (the supply house will have a specially mixed grain size of sand called *aggregate*). Mix the materials thoroughly with your hands, adding water if needed to achieve the texture desired for the particular patch job.

The mixture created is called a 50/50 *admix*, even though there is twice as much cement as sand. That is because the weight of the sand is half that of the cement, meaning the final mix is actually equal weights of each.

If the plaster in the pool or spa is colored, add the color powders slowly and keep mixing until you reach the approximate shade required.

An alternate method of mixing the patch material, which is especially useful if you are preparing a very dry mix for an underwater patch, is to combine the dry materials in a bucket first, then slowly add the water until the desired consistency is achieved. If you dry-mix, put the sand in the bucket first, then stir in the calcium chloride and cement (because these are lighter than the sand, they will combine more readily if they are added to the sand, rather than the other way around).

For patching vertical surfaces, pool walls for example, the mixture should be thicker than if patching a horizontal surface. The texture of bread dough will make the patch material adhere and stay in place better as it dries. The looser texture for horizontal surfaces gives more flexibility and drying time when making the repair.

When patching underwater (which is most common), the texture should be as dry as you can make it and still work the material. Obviously, as you fill the patch area underwater, the material will pick up additional moisture and the texture will loosen.

The most important consideration in creating patch mixture is to consider the drying time and conditions. If the plaster dries too fast, it will shrink and crack. If the material is too thick, it will not smooth out when you apply it and the result will be rough and unsightly. As noted previously, practice before going out to the job site.

The above recipe will make enough material to fill an 18-inch diameter patch about ½-inch deep. Remember, once you start the job, it might not be practical to mix more material and, if you do, it might not appear the same on the repaired area as the first batch. Therefore, always mix more than you need.

4. Apply the patch with a trowel using broad strokes, trying to fill the entire area on the first application. Try to fill so that the patch material is higher than the surrounding plaster level, then remove the excess by scraping the trowel edge across the surface, making the patch clean, smooth, and level. Actually the best finishing tool for this smoothing is a straight piece of cardboard or plastic.

The other application method is underwater, where you work by hand with a fist-size ball of material and push it into the patch area with your fingers. Again, use more than you need so you can scrape off the excess, feathering the edges into the existing plaster for a smooth finish. Don't be intimidated by working underwater, even if it means donning a face mask and fins. It saves time, money, and water over draining the pool. Unless you are a certified SCUBA diver however, don't try to use air tanks for deeper work. In that case, drain the pool.

One last plaster patching technique is used to fill small surface cracks. Not all visible cracks allow water to penetrate the subsurface or cause delamination. Some are from too rapid drying or minor settling of the surrounding land or the pool itself. If there are no blisters, patch only the cracks.

The objective of patching small cracks is to slightly widen the crack so that it will accept patch material. You can use a small chisel to create a slight V shape along the crack. Follow the patching directions as outlined previously, either troweling on the patch material or rubbing it into the crack with your fingers and smoothing over the resulting repair with a straightedge.

Perhaps the most important aspect of any cosmetic patchwork is to give the customer realistic expectations. Even the finest patchwork will be visible for a few weeks, simply because the new material is clean and the existing plaster has darkened with age. Colored plaster takes time to mottle and blend. If the customer understands that the reason for patching is primarily to avoid further delamination (or spreading of the rust in the case of steel bleeds) and secondarily for cosmetics, he will have more realistic expectations of the result.

Coping and tile

When the ground around a pool shifts, which might be the result of settling or water leaking from the pool and causing erosion, the clues that this is happening might include popped tiles and coping stones. Coping stones are made of cast concrete, set edge to edge, so there is little room for expansion. A loose coping stone or one that has risen from the edge of the pool means there is some more extensive problem underneath. Rarely has a loose coping stone simply come free from the mortar in which it was originally set.

The objective then, is to remove the loose coping, excavate the underlying deck, determine the cause, and relieve the pressures that created the problem so the same coping, or more, won't come loose again. Loose tiles are also an early sign of pressures behind the location that need greater attention. The steps to diagnosing and repairing loose tile or coping stones are as follows (Fig. 10-3).

1. Lift off any loose coping stones and look for others by tapping each one with a hammer or the handle of a screwdriver. A hollow sound will quickly reveal other stones that might still be firmly in place, but that are not actually supported by anything.

   In many cases, you will need to remove the tile beneath the affected coping. It might have come loose already. Drain the pool to a level a few inches lower than the bottom of the tile line. On larger jobs, it is easier to work inside the pool, so you will want to drain it completely.

2. Remove the tile beneath the affected area by cutting through the grout and mortar that separates the bad section from the good. Also cut into the pool wall beneath the affected tiles. The cut should be about ½ inch, made with the diamond blade on a hand-held electric grinder (described earlier). This cut prevents spreading cracks or chipping of the adjacent undamaged sections of plaster and tile.

3. On small repair jobs where you have not completely drained the pool, you might want to float a piece of plywood under the work area to catch as much debris as possible. Using a broad, flat chisel, chip the tile away to expose the mortar bed and/or pool wall beneath. If you are careful and have made adequate cuts, the tile might remain intact, but don't count on it. Make sure the homeowner has replacement tiles of the same design or that you can purchase an acceptable substitute. To make reassembly easier, mark the tile or lay it on the deck behind its location so you can return it to the same spot. When you remove each tile, place scratch marks (or use an indelible felt-tip marker) to note the exact positioning of each tile to make replacement easier.

4. Remove the stones. If the coping stones were not loose already, cut the grout joint between each one to make removal easier. Use a concrete saw, available at any tool rental store, with a 12-inch diamond blade. Cut along the mortar at least 4 inches deep to free each end of the stone. The back side of the stone should not be connected with mortar, but rather with a flexible expansion joint mastic or silicone that will not hinder removal.

   Standing on the deck, you should be able to grab the stones by the nose (the side facing the pool) and pull them free. If not, you might need to drive your chisel underneath the stone, which is now possible because you have removed the tile. As with the tile, mark the stones and lay them out on the deck so they can go back where you found them.

5. Chisel and clean the underlying area as needed. Remove old expansion joint material and examine the area between the pool deck and the bond beam. If coping and tiles have come loose, you will probably discover that there is no space between the two to allow for shifting and expansion. This is the cause of the problem, and it will cause more loose stones and tiles in the future unless the pressure is relieved and an expansion space provided.

6. The objective is to create an expansion joint area about ½-inch wide and deep, enough to totally separate the deck from the pool wall. Chisel or cut away any material that is pressing against the pool wall. Never cut the pool wall or bond beam to create the expansion joint.

7. When the expansion joint is complete, fill in any dirt that might have eroded away to complete the backfill area. If severe erosion has occurred, you might need to demolish a larger portion of the deck to expose the amount of backfill lost, replacing it before continuing. In such cases, work with a general construction contractor to rebuild the deck.

8. Prepare the plaster patch material as outlined previously. Some technicians add a latex additive to give the patching compound resiliency, but the best advice is to match the surrounding material. If the existing mortar is somewhat flexible, the new material should be as well. You can also use a premixed waterproof product like Thoroseal, applying two thin coats before resetting the stones. The advantage of using a waterproof mortar is that you prevent water from weeping or leeching into the backfill again.

9. Clean the stones of dirt and old, loose mortar. Apply a light coat of patch material to the underside of the stone, then sufficient patch material to the mortar bed to bring the stone up to its original level. You want to raise it slightly higher than the adjacent stones so that when it is pushed into place, it will settle down to the correct level. Tap the stones in place with a rubber mallet.

Be careful not to allow the patch material to fall into the pool. If it does, be prepared to clean it up quickly so you don't create unsightly plaster chips in the pool.

10. Prepare a brown coat of mortar to reset the tiles. The preparation of the bed is the most important step, because there might be high spots of old grout or mortar left after the original removal of the tile. Grind these down with a hand-held electric grinder. It is better to grind too much, which can be filled with new mortar, than to leave high spots that prevent the tiles from reseating. Follow the marks you made to replace the tiles in the same locations.

    Thoroseal-type products work well and prevent water intrusion when the job is done. You might need to mix it with less water so it will hold the tiles on the vertical. Apply the tiles in the same manner as the coping stones.

11. Regrout the stones and tiles. Grout can be premixed material purchased at the supply store or hardware store, or it can be mixed by combining one part white cement with two parts sand. Use #60 silica sand unless the joints are over ½-inch wide (then use #30). Mix the grout to a loose enough texture to ensure it will completely fill the voids between the stones. Overfill, allow the mixture to set up slightly, then wipe away the excess to a smooth, level surface. Do the same with the tiles, carefully wiping all excess off the surface of the tile. A wet sponge wiped over the finished tile and coping will remove any grout film, which will otherwise leave a discoloration looking like paste wax.

    Some pools use a colored grout on the coping stones or tile. Powdered dye is added to the grout and mixed until the color matches the existing installation. Remember, the finished job will not match until the new work has had a chance to weather and acquire the same shade as the surrounding work.

12. Complete the expansion joint. Fill the joint with sand up to the last ½ inch. Fill the last ½ inch with flexible mastic or silicone joint sealer, which can be poured as a liquid or injected like caulk. Follow the product label directions for application, especially concerning temperature and humidity ranges.

If you understand the basic underlying construction of the coping and tile area of the pool or spa, you will be able to make these basic masonry repairs. Many will not be as complicated as described, requiring only rehanging a few tiles or resetting a single loose stone. If that single tile or stone keeps coming loose however, or if more than one are loose, follow the procedures described earlier to determine the cause and effect a long-term repair.

Vinyl liners

There are numerous manufacturers of vinyl liners for pools, using several different compositions of vinyl, PVC, or rubber-based materials. Because of the different qualities of each make of liner, there are no standard procedures for repair or replacement. Having said that however, it is much easier to repair or replace a vinyl liner than to repair a plaster pool because the liners are sold with detailed instructions and repair kits.

Locating leaks in a vinyl liner is the same as in any other vessel. Start with an evaporation test to verify that there is a leak, then use the dye test or drain-down

10-3 The coping stones and tile line.

method to locate the exact leak location. Since vinyl-lined pools are often assembled aboveground, it might be a simple matter of crawling around the pool and under any adjacent decking with a flashlight looking for wet areas.

Most of us are more familiar with glue and vinyl patch material than with plaster and cement. Most repair kits detail the process of drying and cleaning the area to be patched with a solvent, roughing up the area to be glued with sandpaper for a better bond, then gluing a patch of the same material over the tear or puncture. These are simple procedures that you have used since childhood on various repairs around the home, and they should present no unusual problem on the vinyl liner.

The only caveat is to be sure to use the right kit for your liner. Because the composition of vinyl materials differs widely, glue and patch material designed for one liner might not adhere to another. The only other problem with vinyl-liner repair is that you have to drain the pool. Some glues say they are formulated to work underwater, but I have always found them to fail in a short time.

Replacing a liner can be a difficult task, unless you enlist the help of several people to assist in stretching out the liner evenly and hanging it uniformly. Again, follow the directions provided with the replacement liner and make sure enough help is available to execute the procedure without wrinkling or tearing the new liner. The secret to successful liner replacement is to avoid sharp edges and prepare the base with adequate sand to cushion the final result.

Acrylic spas

Cracks in molded fiberglass or acrylic spas can be easily patched. To detect the leak, follow the previously described methods using evaporation, dye, or drain-down. Leaks in spas often occur where plumbing meets the spa shell. Shells often move away from their deck supports when the ground shifts and are especially susceptible to soil erosion problems. Since bather loads and displaced water are high in relation to the total size of a spa, water frequently washes away fill or base materials allowing the weight of bathers to shift the spa and separate plumbing.

Cracks or breaks in the acrylic can be repaired with materials available at the supply house. The technique is similar to patching small cracks in plaster, where you widen the crack to accommodate a fill material. The only difference with acrylic spa material is that the filler is not the final step; there is also a topcoat.

1. After identifying the hole or crack, drain the spa to a level below the work area and dry it thoroughly.

2. Drill a small hole at each end of the crack to prevent it from spreading. The hole should be slightly wider than the crack, but not more than ¼ inch.

3. Using a small blade on a hand-held electric grinder, open the crack to a V shape wide enough to accommodate the fill material. Clean the crack with acetone to remove any dust or loose material.

4. Patch the crack with acrylic filler, which is available in kits from the supply house. Modern filler is designed to expand and contract at the same rate as the spa shell, which is subject to extremes of temperature, chemicals, and bather load weight stresses. Follow the package instructions for mixing, curing, and drying times. Generally, acrylic patch sets up rapidly and dries in a few minutes, so be prepared to work quickly. As with plaster patching, make more material than you think you need so you don't run out in the middle of a job. Slightly overfill the crack, then wipe off the excess. Slight shrinking takes place during drying.

5. When it is completely dry, sand off the excess down to the level of the surrounding surface. Avoid sanding too much of the surrounding surface. Because the acrylic finish is usually very high-gloss, imperfections in the sanding will stand out like the proverbial sore thumb. Take great care to smoothly sand the patchwork.

6. The patch must be painted to match the color of the spa. Color powders or liquids are provided in the kit to mix with the base enamel paint, also provided. Mix very small amounts and paint a test area, inside the skimmer for example, until you achieve a good match. Remember that when the paint is wet, it appears somewhat darker than the final color.

   Paint the patch, smoothing out the color and feathering it into the surrounding area. Spraying instead of brushing will achieve smoother results. In either case, advise the customer that the match will not be exact.

7. A clear topcoat seals the paint job and adds luster to the finish. The topcoat is also provided in the repair kit. Adding a little of the final color mix from the base coat to the topcoat will also help the finished repair blend better with the background material.

Do not use automotive body resins and bondo-type products to repair spas. That was the only repair material available for many years and experience taught that these products quickly shrink and leak when exposed to the harsh spa environment of heating, cooling, chemicals, and weight pressures.

Larger fiberglass repairs are detailed in the chapter on spas.

REMODELING TECHNIQUES

There is a trend today to rebuild the large pools of the 1940s through 1970s by actually creating a smaller new pool inside the old one. Large pools require more expensive cleaning, heating, chemical maintenance, and repairs. Unless the homeowner has a large family, most builders today offer shallower, smaller pools that provide adequate room for swimming laps and recreation in a small backyard, but require less attention and upkeep. Builders today will actually build a new, smaller pool inside the shell of a large older pool, using the original plumbing and equipment, and filling in the difference with backfill and landscaping.

There are several other ways to remodel a pool or spa. New decks and landscaping will improve the look of an older pool (best left to a general contractor and landscaper). As for the vessel itself, the water technician can offer resurfacing of the interior, new tile, and recaulking of the expansion joints to help the look of an existing pool or spa.

Plastering and replastering

In most areas there are numerous subcontractors who will plaster a new pool or replaster an old one. It is usually best to refer the customer to one of these professionals and earn a referral fee and the fee for the break-in service as described previously.

Because nothing is more critical to the successful maintenance of a pool or spa, it is valuable to understand the plastering process. It also allows you to evaluate a good subcontractor and a good (or poor) plaster job. As noted previously, the water technician will always be held responsible for discolored plaster, cracks, or blisters, so it pays to know the process to be able to accurately analyze the plaster work. Your customers will also come to you for advice on when to replaster and how to analyze the quality of the work provided. The plaster also significantly impacts the chemistry of the pool, so it is important to understand the plastering process to know how chemicals will affect the plaster surfaces and vice versa. Because there are several approaches to replastering a pool or spa, it is important to understand the methods so you can make recommendations to your customers and evaluate the needs of their pool.

New plaster

Let's begin with a discussion of the plastering process, then review the variations for replastering.

1. The gunite or concrete pool shell must be clean of loose dirt, algae, and water before plaster is applied. To accomplish a contaminant-free environment, wash and scrub the shell with chlorine and water to remove organic waste. Next, wash and scrub with an acid solution of one part muriatic acid to four parts water to eliminate minerals and concrete dust (called *latiance*).
2. Plaster will not adhere to an acidic surface. Every bit of acid must be removed from the concrete shell. Neutralize the acid with trisodium phosphate (TSP) or baking soda. Use 1 pound in 5 gallons of water of either product, creating a wash. Thoroughly scrub and pump out the shell. The concrete of the shell must be moist (hydrated) but not have standing water. The shell is now ready for plastering.
3. Remove all plumbing outlet nozzles and any removable hardware such as light fixtures or ladders. Stuff rags in the plumbing to keep plaster out of it and tape off the threads of any fittings. Finally, tape the bottom of the tile line to keep it clean.

4. The plaster material is a mixture of one part white cement for every two parts aggregate (sand and marble dust, also called *marcite*; or sand and limestone). Calcium chloride is added to help the mixture set up more quickly, but never more than 2 percent of the weight of the cement (not the total mix; only 2 percent of the cement itself). Plaster is mixed in a cement mixing drum using a paddle mixer. Be sure that any gray cement residue from previous work is not still on the inside of the mixer because it will contaminate the color of the new plaster. Color powders are added to the mix if the final plaster is to be colored. Water is added, roughly 2 gallons for every 100 pounds of dry mix, until a heavy oatmeal consistency is achieved. The more water, the weaker the mix, so caution in mixing is advised. The entire mix must be completed in 10 to 12 minutes.

5. The plaster is applied by trowel. The first coat should be scratched on thin to fill and smooth out the roughness of the concrete. Then, two more thicker coats are applied while the underlying coats are still wet. It takes several plasterers to apply the material fast enough to finish in a few hours so that the pool can be filled before the plaster dries out and cracks. Care must be taken to avoid leaving plaster on the tile or fittings. When cleaning these, avoid running water over the fresh plaster; this can leave furrows or stains.

6. Break in the new plaster as described previously.

When plaster cures, the topcoat becomes almost transparent, allowing any imperfections in the shell to show through. The first coat of scratch plaster is important in leveling out such imperfections, although it cannot completely hide them. If gray lines or slightly darker spots appear at various places after the pool is full, this is usually the reason. Therefore, the finish job of the shell is as important as the scratch coat.

In locations where gunite rigs are unavailable and the concrete shell is poured by hand, such imperfections are common. In these cases, it is valuable to grind or chisel down any high spots prior to plastering.

Replastering

A good plaster job over a well-made gunite shell will last up to 20 years. When plastering the shell for the first time, the plaster permeates the pores and roughness of the gunite surface, creating a firm mechanical bond. When replastering, the new plaster must adhere to the old smooth surface with either a new mechanical bond, a chemical bond, or a combination of both.

There are several ways to prepare the old surface to enhance the bonding of the new plaster. Depending on the condition of the old plaster, one or more of these preparations might be used. The surface preparation is the most important step of the replastering process, because fine workmanship will be of little value if the plaster doesn't stick to the old surface.

The first preparation method is to etch the old surface with acid. As discussed in the chemistry chapter, muriatic acid will dissolve plaster, leaving it rough and pitted and thereby promoting a better mechanical bond for the new plaster. The old plaster is washed with raw muriatic acid to deeply etch the surface.

The limit of this method however, is that acid will not dissolve organic waste embedded in the old plaster. It does not remove loose dirt, although scrubbing and washing after application of the acid might handle that. Finally, if any acid residue is left in the porous old plaster, it can create gas when activated by the moisture in the new plaster and lift the new material from the old surface. Acid etching is usually a good step, as long as it is thoroughly cleaned away.

The next method is to chip the old surface with hammers and/or axes. Workers literally chip away at the old plaster, working around the pool creating a jagged, rough surface for the new plaster. Again, this enhances the mechanical bond for the new material.

Sandblasting literally sands away the old plaster down to the concrete. It does not, however, make the surface rough to aid mechanical bonding. Sandblasting is usually used to remove old paint from a plaster job before etching or chipping.

Finally, the removal of the old plaster can be accomplished, although expensively, by jackhammering the old material off of the pool shell. The danger of jackhammering is that it can crack the pool shell, causing structural failure when the weight of the water is added.

Having prepared the surface with one or more of these methods, a chemical bonding agent is added to the scratch coat (the first coat of plaster as described previously). The bonding agent helps the scratch coat adhere to the shell on one side, then helps the plaster adhere to the scratch coat on the other side. Called a *brown coat* because some bonding agents slightly darken the scratch coat, the chemical bond rarely lasts as long as a good mechanical bond. Any replastering job should last at least ten years, however.

Other than these preparation differences, the original plaster job and the re-plaster job are alike. When advising your customer on the various methods of surface preparation and their relative costs, remember the value of mechanical bonding over simple chemical bonding. Chipping or sandblasting followed by etching and cleaning can be costly, but a few hundred dollars now might double the life of a $2000 plaster job, making the extra preparation very cost-effective.

One last way to evaluate the quality of a plaster contractor is to look at a recent job site. A clean site with no plaster on the tile or white staining on the deck or walkways leading to the pool are a reflection of a good contractor. The sloppy ones are also sloppy in their plastering work, taking shortcuts that shorten the life of their work. Also, since many problems can develop with plaster that might be related to the contractor, choose one that has been in business awhile and will likely be there to back up a warranty tomorrow (or at least one that will be around to offer technical advice for any problems that you or the homeowner might have caused).

Fiberglass coatings

Few gunite pools have an original interior surface of fiberglass because such applications are relatively new. The technology for surfacing gunite pools with fiberglass was imperfect ten years ago, often leaving these pools with surfaces that were discolored, cracking, or delaminating.

In the past five years, however, numerous companies have improved the technology to the point where a fiberglass surface will give excellent results for many years. There is insufficient data available to evaluate if such coatings exceed the performance of plaster, but the obvious benefit of a fiberglass coating is that water and maintenance chemicals cannot corrode and destroy it like plaster.

Fiberglass coatings (Fig. 10-4) require the same surface preparation as replastering. The best mechanical bond is achieved by roughing up the surface prior to application. Fiberglass sheeting is laid on the pool interior and painted in place with chemical resins and fixatives.

10-4 Fiberglass coatings.

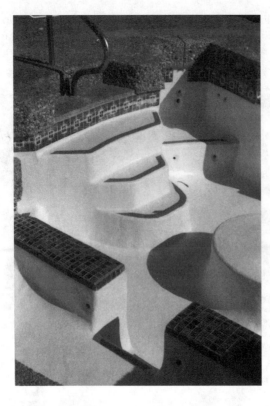

10-4 Continued.

Because fiberglass coatings and their applications are a unique technology, recommend your customers to a reliable subcontractor that specializes in such work. There is no break-in for fiberglass coatings, but most companies will pay the technician a referral fee.

Paint

The purpose of coatings of any kind over the gunite material of the shell is to prevent leaking, because concrete itself is porous. Modern pool paints are an attractive, inexpensive way to coat gunite, fiberglass, plaster, or any other interior surface with a smooth, colorful, waterproof coating.

Like fiberglass coatings, older formulations of paint were unreliable, often peeling and fading in two years or less. Paints were easily damaged by heat and harsh chemicals and were considered a temporary fix at best. Today, many paints will last up to five years, providing an inexpensive alternative to other coatings. Surface preparation and painting can be done by any careful technician, providing a good source of work and profits.

As with other coatings, the success of painting depends largely on the preparation and qualities of the subsurface being covered. If the paint is to cover old plaster, it won't keep delaminating material from peeling from the shell surface. In other words, before painting, evaluate the surface being covered. If it won't last for three to five years, then there is no value in painting over it. It is time to replaster or remove the old plaster before painting.

It is also important to give the customer realistic expectations about paint as a surface covering. Irregularities in the subsurface will show through the paint and will still be felt underfoot in shallow areas. Colors might be vivid initially, but will fade throughout the life of the paint—especially the brighter the original color. Finally, paint will not last much longer than three to five years and might begin to peel or dissolve prior to that.

Pool paint is manufactured in three types. Chlorinated rubber is designed to be flexible when applied, therefore it is durable and appropriate for rough, previously unfinished surfaces, acting like a stretch-rubber cover. Solvent-based paints are also good on rough surfaces because they are designed for thinning with additional solvent, making them easy to spray on. Water-based epoxies are used for painting fiberglass pools or spas and are more resistant to chemical variations and temperature extremes.

Generally, cover old paint with new paint of the same type. You can use water-based epoxies to cover chlorinated rubber, but not the other way around. If you're unsure, sand off the old paint or consult the manufacturer (both good ideas anyway). Another simple test method is to rub a little of the new paint over the old painted surface. If the old material dissolves or bleeds, don't use the new product without removing the old.

Also, dark colors will cover more evenly and hold up longer than lighter colors, the worst being blue. If the customer wants a blue pool, remind him that plaster is white but the pool looks blue when the water is added and chlorinated. Paint white or a dark color.

If you are prepared to proceed, follow the paint manufacturer's label directions and these guidelines.

1. Prepare the surface. Empty, clean, and remove any loose materials from the pool. Sand any loose plaster or high spots. If there are blisters, follow the plaster repair procedures to prepare a level surface. Remember, paint can't replaster or level out a pool surface. If the surface is fiberglass, rough sand the entire area to be painted for a good mechanical bond.

   Because sanding, especially if it requires removing old paint, requires enormous amounts of time and sanding discs (for your electric sander), test a few square feet before estimating the cost of the job for your customer. I have spent more on the preparation, in time and materials, than I quoted to do the entire job, simply because I thought it would take less to get it done.

2. Clean the finished preparation job. If you have spent many hours creating a level, paintable surface don't waste that effort by leaving dust and dirt on it that will prevent the paint from adhering. Wash down and scrub the surface with TSP (1 cup per 1 gallon of water and 1 ounce of tile soap to cut any oils on the surface). Make sure that you are using pure TSP, not products labeled as such but containing additives that might leave residues.

3. Acid wash the entire area to be painted, following the acid wash directions detailed later in this chapter. When acid washing for a paint job preparation, the idea is to etch the surface to create a roughness that will help the paint adhere. In regular acid washing, the idea is to clean the plaster but not to etch. In other words, for a paint job use a stronger acid mixture and leave it on the surface longer.

4. Scrub and rinse the entire surface with TSP again to remove any trace of acid, this time without any soap. Rinse and scrub again with clean water. This is not the time to skimp on labor, because any remaining acid or dirt will prevent the paint from adhering. The entire pool might be flawlessly painted, but the customer will only notice the 3-inch blister or the small crack in the final paint job.

5. Allow the surface to dry thoroughly and don't plan to paint on foggy or humid days when the moisture content in the air is high. Similarly, avoid working during hot afternoons when the surface to be painted is hot. Either condition will keep the paint from bonding to the surface or cause blisters. As with all painting, follow the product directions. Mix the paint, enough to do the entire first coat. Mixing might mean simply stirring or it might mean adding catalysts according to the manufacturer's directions.

6. Use good-quality tools to apply the paint. Thin, cheap rollers will buckle under the pressure needed to force the paint into the crevices of the roughed-up surface, especially on the first coat. A 9-inch wide, ½-inch nap, lambskin roller on a sturdy frame works well. Cut in (detail paint that can't be done with a roller without overpainting) the tile, drain, light, and other detail areas with a good bristle brush first, then roll on the paint. Apply a thin but even coat. The first coat will use two to three times as much paint as subsequent coats, except when applying over fiberglass.

7. After waiting the manufacturer's recommended time between coats, apply a second and, if necessary, a third coat. On all coats, but especially these finish coats, paint the entire surface at one time. If you stop and restart, no matter how careful you are, the dividing line will still be visible with slightly different tones of color. On the last coat, sprinkle some clean sand over the wet paint on the steps and in the shallow areas to aid traction for swimmers.

8. Let the paint cure as recommended by the manufacturer, then refill the pool or spa. Avoid using the heater for a few more days, just to be sure the paint has thoroughly cured.

As with any repair procedure, but especially with paints, work carefully and clean, taking care to spread drop cloths on decks and over deck furniture. Clean up any spills promptly and clean tools according to the manufacturer's suggestions, preferably back at your workshop. Remember, the customer will only see the blemishes of the job, no matter how few and no matter how good the rest of the job looks.

Expansion joints

The most overlooked aspect of remodeling a pool are the expansion joints (Fig. 10-3). Take a look around the pool deck. The sealant in the expansion joints has probably dried, shrunk, and cracked. Water has probably washed into the joint for years, eroding the underlying backfill that supports the pool and deck.

Repairing expansion joints is an excellent profit center for your company, and it can be done at almost any time. You can plan the work for times when you might not otherwise be very busy. The work is time-consuming, but easy, and might be the single most important elective surgery you can provide a pool.

1. Remove the old sealant. This is the time-consuming part. The old material can only be gouged out of the joint with chisels, screwdrivers, or other scraping implements. Try to select a tool that is the approximate width of the joint, usually ½ inch or less. There is no shortcut here—work your way around the pool, scraping and lifting all of the old material from the joint.

2. Inspect the joint. If there are areas where the deck is pressing against the pool wall or coping stones, chisel out the joint as described previously. Since you have exposed the joint, this is the time to repair any stresses that might create greater problems to the pool wall, coping, or tile later.

3. Clean the joint. New sealant will not adhere well to surfaces inside the joint that are covered with loose dirt, grease, old sealant residue, or concrete dust. Using a wire brush and plain water, thoroughly clean the joint. Allow it to dry.

4. There are several types of mastics and silicone caulk sealants that can be used to close expansion joints. The idea is to keep water out of the joint which can erode the subsurface materials. Review the options with your supply house to determine the material that is best suited to the deck, coping, and weather conditions at the pool and in your area.

   Mastics are available as liquids that you pour into the joint and allow to harden. Silicones are available as caulk, shot from a standard caulking gun. As always, follow the directions on the package.

   Because the application of either product will likely spill over to the edges of the joint surfaces, it is helpful to apply masking tape along each edge of the work before applying the sealant. Any excess can then be removed when you remove the tape. Be careful not to let the tape sit in the sun for more than a few hours or it will become permanently stuck to the surface and require scraping to remove it.

Acid washing

Perhaps the simplest remodeling technique is an acid wash to clean the plaster surfaces of the pool or spa. As discussed earlier, it is necessary to drain the pool from time to time, to remove scale causing minerals and chemicals. This is a good time to acid wash the pool, to brighten the appearance and remove any deposits that might have built up.

As with other cosmetic procedures, the key to success is realistic expectations. If there is severe etching and staining, an acid wash will not smooth out rough plaster or remove every discoloration. It will make the pool look better, but it might still feel rough underfoot and show signs of metal deposits and/or scale from years of abuse. Especially with colored plaster, you must be careful not to remove so much scale and/or stain that the acid also removes substantial amounts of plaster, leaving pitting or gunite showing through.

Before acid washing, it is important to know how old the plaster is and if it has been acid washed before. If the plaster is nearing the end of its useful life (10 to 15 years), there might not be enough material to wash without stripping the surface down to the gunite. Similarly, if the pool has been acid washed two or three times already, it might be time to consider replastering, painting, or some other recoating.

In each case, regardless of what the pool owner has told you, once you drain the pool, test a few areas before continuing the job. If your tests reveal the plaster is thin or already badly etched, discuss expectations and alternatives with the owner.

There are two accepted methods for acid washing a pool or spa. The traditional method requires draining as described. The newer method, requiring somewhat more careful application, uses lowering the pH without actually draining the pool or spa.

Standard drain method of acid washing

For many years the only way to acid wash a pool or spa was to drain the water and work directly on the plaster. Many technicians today still perform only this procedure, as opposed to the drainless method discussed later. If you have chosen the drain method, the following steps work best.

1. Turn off the circulation equipment and in-water lighting at the circuit breakers.
2. Drain the pool, following the techniques described previously. The best technique to keep downtime to a minimum is to start pumping out the pool in the late afternoon. Watch your disposal technique for a long enough period to know it will evacuate all the water safely, then allow the pump to run overnight. In the morning, most residential pools will be empty and ready to acid wash. On larger pools, use multiple pumps to achieve the same results.

    Position your pump near the main drain, unless you know the lowest point of the pool to be elsewhere. I have often found that the lowest point is several feet away from the main drain, but that can be difficult to determine until the pool is virtually empty. In any case, arrive at the pool with enough time to finish pumping the last of the water. If you're down to the last 200 to 300 gallons, you might want to leave it in there as a dilution pond for the acid you will be using. If the pool also includes a spa, you need to drain that as well before beginning work. If you use the technique of directing the waste water to a lawn or garden for irrigation, remember to redirect the discharge hoses before you actually begin the acid wash. Direct the hoses into a deck drain or run them out to the street.
3. When the pool is drained and you arrive to perform the acid wash, bring your acid and tools into the empty pool. Set up your equipment on the stairs. Hose off the deck as soon as you arrive and continue to keep it wet during your work time to prevent inadvertent staining or spills. While we're on the subject, once you begin, wash your boots off each time you exit the pool. They carry acid as well.
4. Remove the light fixture and any other loose or removable hardware such as ladders, rails, or metallic return outlet nozzles. Remove the main drain cover. Clean out the main drain and light niche. If the pool is plumbed with copper, stuff the outlet pipes, skimmer, and main drain with rubber plugs or rags soaked in water and soda ash to neutralize and/or keep out as much acid as possible. Rinse (and scrub if needed) any other organic material (oils, leaves, and dirt) from the plaster, because the acid will not dissolve these and therefore will not clean the plaster beneath.

5. Inspect the pool for plaster blisters as described previously. Make any necessary repairs or patches before acid washing, so the subsequent acid wash will clean all surfaces to approximately the same color, helping any patches to blend.

6. Before starting the acid wash, put on old clothes (a few small acid splashes will put holes in cloth), rubber boots, gloves, and a respirator (rated at least R25). The boots provide skin protection and insulation against shock if the pump is shorted out in the waste water around the main drain. I have done acid washes in bare feet with no protection and have known technicians who have not used protection during 30 years of performing acid washes, but the cumulative effect of inhaling acid fumes, exposure to raw acid on the skin, and the effect of fumes on the eyes can't be measured. Moreover, I have experienced some really scary incidents when acid washing in varying wind conditions, where a cloud of acid fumes suddenly changed direction and enveloped me, choking and burning with sobering effect. Remember too that if any chlorine is still present on the plaster, when the acid hits it, it will create lethal chlorine gas. In short, believe me when I say **always** use a respirator, gloves, and boots.

7. Finally, prepare a garden hose that can reach all parts of the pool and keep the water running. Although leaving the hose on will waste a bit of water, it will also keep the floor of the pool wet in the area you are working, neutralizing the acid as it runs toward the main drain and the pump. In this way you can control the area the acid covers and the length of acid contact.

8. Take a garden-type sprinkling can, add a squirt of tile soap, fill it two thirds with water, one-third with acid (always adding the acid to the water, not the other way around). The soap will help the mixture cling to the walls of the pool and will cut the fumes somewhat. If you add the soap first, you won't need to stir or mix the solution. The effectiveness of acid washing is a function of the strength of the mixture and the length of time it contacts the plaster. To determine what the particular job needs, test the solution on a small area by pouring a stream from the sprinkling can. Allow it to sit on the plaster for about 30 seconds while scrubbing the area with a stiff-bristle broom or brush, then wash it off with clean water from the hose.

   If the stains haven't disappeared, you might need to leave the mixture in contact longer, strengthen the mixture, or go over the area more than once. Trial and error is the only way to become familiar with these variables, especially since you might encounter several different problems within each job. The one-third acid mix might work fine for most of the pool, but a few areas might need a 50 percent solution or even straight acid to budge tough stains. I have had to apply pure acid and leave it for up to 45 seconds on some stains to get them clean, so be prepared to spend some time experimenting on each pool to determine the needs of the job.

   One word of caution however: you don't want to cure the illness, but kill the patient. It's not worth removing every last bit of discoloration or scale, just to end up with extremely rough surfaces or plaster so thin that you can see the gunite showing through. Therefore, the other reason for conducting some trial spots and solutions is to determine the effect on the plaster as well as the stains.

If you find stronger solutions and longer contact times having little effect you might have an older plaster mix composed of a silica sand base, rather than the more erosive modern mixtures of crushed marble (marcite) and calcium chloride. Don't be afraid to try stronger mixes or pure acid if that's what it takes to achieve good results.

Another reason for the test is to determine the amount of acid you will need for the job. Whatever you think you need, bring at least 4 gallons more (an extra case). Nothing is as aggravating as running out of acid with just a few more feet of plaster to wash. By taking note of how much you use in your test area, extrapolating that over the pool, and adding a little extra, you will generally be prepared. As a rule of thumb, a typical 20,000-gallon residential pool usually requires about 20 gallons of acid. You might not use it all, but it is better to be prepared.

9. Once you have determined the acid strength and contact time for the job, continue around the pool with the sprinkler can of acid solution, the scrub broom, and hose for rinsing. Keep rinse water flowing on the pool bottom when you are not actually rinsing an area that has been acid washed to neutralize acid on the pool bottom. Keep the pump operating to remove the waste from the hopper (the low point in the pool around the main drain). Even though the acid solution is somewhat neutralized by the plaster it is dissolving, it is still very potent. If you allow it to run undiluted along the bottom as you work the entire pool, when you finish you will have acid washed the bottom and the hopper several times.

If your area establishes a minimum allowable pH for waste water, you might have to toss a cup of soda ash into the hopper and mix it with the broom from time to time to bring up the pH of the discharge you are pumping. Some jurisdictions prohibit a discharge with a pH less than 5.0 and it can take up to 8 pounds of soda ash to neutralize 1 gallon of acid, so make sure you have adequate soda ash on hand when you do the work (soda ash is very inexpensive).

The purpose of scrubbing with a broom is to force the solution into the pores of the plaster for thorough contact. All the scrubbing in the world won't actually remove stains, so let the acid do the work by making sure it is in complete contact with every part of the plaster. The most effective way to work is to have one person apply the acid and have another scrub. Either can rinse after they finish. If you must work alone, use a more diluted concentration, allowing you to leave it in contact with the plaster longer and giving you ample time to apply, scrub, and rinse.

10. For safety reasons it is better to work with a helper. Acid washing not only cleans the plaster, but leaves it smoother than before as it removes scale and rough spots in the plaster. Since the smooth surfaces are also wet, it is easy to slip and fall, especially on the steeper slopes of the deep end. For this reason, I try to work around the sides of the pool, then the bottom, always standing so one foot is on the lower part of the slope and one on the higher. If you stand with both feet facing the shallow or deep end, it is easier for your feet to slip out from under you. Also if you face uphill, you are less likely to fall than if you face

downhill. No matter how careful you are or how slowly you work, you will experience what I'm talking about first hand sooner or later, so work with a helper.

11. Take advantage of the pool being empty and clean the tiles. Under normal conditions, with water sloshing along the tile line, it is often difficult to remove all of the oils and scale on the tile. With the pool empty, scrub the tiles thoroughly, using tile soap, pumice, or the barbecue brush and, while you have it, some acid solution. If the tiles are particularly dirty, you might want to clean them before the acid wash so that any runoff washes over the plaster before it is cleaned. Use a pumice stone or block to scrub out any especially stubborn stains in the plaster.

12. Finally, thoroughly rinse the pool and lightly scrub any remaining acid or soap residue, so that when you refill the pool it will not have negative affects on the pH. You might want to go over the steps and shallow area, where feet contact the surface, with an electric sander and some fine to medium wet/dry sandpaper (80 to 100 grit) to smooth out any rough spots.

13. Give the light fixtures and other hardware a good cleaning before reinstallation. Metal fixtures can be lightly acid washed and the plastic ones can be scrubbed with cleanser and a wire brush. Use a cup to remove any acidic water from the main drain before reinstalling the cover. Reassemble what you took apart and remove your equipment from the pool. To avoid trailing acidic mixtures across pool decks and through access areas. I rinse and then put the pump, hoses, sprinkling can, and empty acid bottles in big plastic garbage bags before removing them from the pool. Another option is to put everything in a large plastic garbage can as a way to collect and carry everything back to the truck without spilling. Even a few drops spilled on walkways and steps will be noticeable, and you might find yourself acid washing all of the concrete around the house for free to erase the evidence.

14. Fill the pool. Unlike fresh plaster, there is no concern after an acid wash of bruising the plaster, so turn on fill lines and garden hoses to refill the pool as soon as possible. Be careful to gauge the refilling time so that you or the homeowner are around to turn off the hoses before overflow and flooding occurs. Before restarting the circulation, make sure the filter and strainer basket are clean so that dirt doesn't flush into the pool and onto the clean plaster.

15. Brush the pool while circulating the water for a few minutes, then check the chemicals and balance (as described in the section on restarting after a new plaster job). If the pH or total alkalinity are extremely low, turn off the circulation, balance the chemistry, and brush to circulate your chemicals before restarting. If you didn't do a good job removing excess acid, resulting in corrosive water, you might strip copper or other metals from the pool equipment, staining the plaster all over again.

Drainless method of acid washing

The drainless method of acid washing requires patience and skill and should not be attempted by a technician that has little experience in balancing pool chemicals. It is a process that is effective, although not as labor saving as its supporters would tell you.

The problems with this method are

- Because you don't drain the pool, you don't get the opportunity to inspect the plaster in detail and evaluate the severity and extent of staining or blistering. This is especially important if the history of the pool is unknown, such as how old the plaster is, how many times it has been acid washed, or whether the pool has ever been painted.
- If you lower the pH too far or allow an excessive length of acidic water contact with the plaster surfaces, you can permanently damage the plaster, creating extreme roughness or dissolving it down to the gunite. When restoring the pH, it is easy to exceed the necessary chemical additive requirements, spiking the parameters and making draining necessary anyway.
- This method requires up to five days, meaning the pool is out of service far longer than with the traditional acid wash method. It also requires three daily brushings during that period, meaning it is labor-intensive.
- To avoid corrosion of the light fittings, ladders, rails, and any other metallic fittings, you must dive in the pool to remove them before beginning the procedure and dive again to replace them.

The only real advantages of this method are

- Conservation of water and the opportunity to avoid draining a pool if hydrostatic pressure might crack it or pop it out of the ground.
- Where there is no safe place to discharge the waste water, this might be your only choice.
- The cost of materials is about the same for either method. Water in most areas is not free however, so the customer might save several hundred dollars by not draining. But I charge more for the drainless acid wash because of the extra labor of three daily trips for up to five days plus the time to start the process and several days for restoring the chemical balance afterward.

Being aware of the concerns, if you need to use the drainless acid wash, follow these steps.

1. Turn off the circulation equipment at the circuit breakers.
2. Remove all metal fixtures from the pool, such as lights, ladders, rails, etc. You can smear them with petroleum jelly for protection, but unless you are thorough and cover every inch of metal, they will corrode. It is probably more work to try to cover every part of a ladder with jelly than it is to remove it, but if it is anchored into the plaster, you might have no choice. Some proponents of this method will tell you that stainless steel fixtures are immune to the acid and you need only be concerned with chrome-plated fixtures. But my experience is that imperfections in the stainless steel and/or corrosion from age means that the acid will attack this metal as well.
3. Pour 2 gallons of muriatic acid for every 5000 gallons of water into the pool. Add a metal chelating agent according to the instructions on the label of the product you choose.
4. Brush the pool immediately to distribute the acid. After a few hours, test the water. The total alkalinity should read 0. If not, use your test kit to calculate the amount of acid required to lower the total alkalinity to 0 and add acid accordingly, brushing each time you add more to distribute it evenly.

5. Continue three times daily brushing. You will see plaster dust clouds of dissolving material when you brush. Feel the surface with your hand to monitor any developing roughness in the plaster. Stop the process if the surface becomes extremely rough or pitting begins. Otherwise, continue the routine until the staining disappears. On white pools this is easily detectable, whereas with dark pools the color will be an intense, dark shade of the plaster's color. In that case, when the scale and mottling disappear, the process is complete.

6. Continue to monitor the total alkalinity of the water during the process. It must be 0 during the entire process. If a reading appears, add 1 gallon of acid per 10 ppm of total alkalinity per 15,000 gallons of water. As always, brush the chemical around after adding it.

7. When the stains are gone or you feel the plaster is becoming rough, stop the procedure by testing the water for base demand and add the required soda ash to raise the pH to 7.2 or higher. Brush the soda ash for even distribution and retest as needed, adding further soda ash until you have achieved the desired pH.

8. Conduct a total alkalinity test, adding the necessary chemicals to raise it to normal levels. Twenty-four hours after raising both the pH and total alkalinity, test again and adjust as needed before turning the equipment back on. If the calcium hardness exceeds 500 ppm, you will have to replace the water anyway. If draining is a problem, remove as much as possible for several days until you have changed sufficient water to lower the hardness levels.

9. Reassemble the fixtures removed and vacuum the pool. Clean the filter and chlorinate normally.

As you can see, this process requires patience, skill, and work. One variation of the two methods, which I have heard about but have no personal experience with, is a combination method. You use a drainless acid wash for a 24-hour period to loosen the stains, then drain and finish with a traditional acid wash. Adherents of this technique say the drainless portion removes much of the scale and staining and leaves the surface extremely smooth. Draining then gives you a chance to inspect the overall condition of the plaster and be guided accordingly, changing out water that might be excessively hard anyway and cleaning up any remaining stains with a much weaker solution than would have been used in the drain method alone.

11
Spas, hot tubs, and water features

The fastest growing segment of the water industry is the spa, hot tub, and water feature market. Spas and hot tubs are popular for their therapeutic value and low initial and upkeep cost, in addition to the fact that most busy people today get more use out of a spa or hot tub than a pool. With the creation of portable spas, even renters and apartment dwellers can enjoy getting themselves into hot water. Water features, meaning decorative fish ponds, waterfalls, and fountains, have grown in popularity as a living design enhancements in homes, businesses, and public buildings.

Everything presented to this point has been information that is applicable to any water vessel from the smallest portable spa to the largest swimming pool. There are a few repair and maintenance techniques that are unique to spas, hot tubs, and water features that are worth learning for two reasons. First, to be a well-trained technician, you should understand the details of maintaining water in a smaller vessel, especially because many of your pool customers will also have a spa or water feature to be maintained. More importantly however, as a businessperson you will discover that many of your potential customers have only a spa or water feature, therefore a significant source of income can be derived from this area of the market. Further, you will find that the work on these smaller vessels requires less time and material, meaning you can make more profit from spa and water feature work than much of your pool maintenance work.

SPAS

Spas today are made from many different materials and are designed to be integral components of a pool system, stand-alone units, or completely portable units. As noted earlier, the techniques learned previously for plumbing, cleaning, and repairing, and a general understanding of the operation of a pool are largely identical to that for a spa. This section presents those concepts that utilize these skills and details any important variations.

Spa materials

Virtually any container that holds water and is large enough to accommodate bathers can be made into a spa. In various parts of the world I have seen spas made from large metal or plastic barrels, old water tanks, and bathtubs, as well as the more traditional fiberglass, acrylic, and gunite construction. An exciting new variation is the soft-sided spa, allowing even the smallest home and budget to enjoy hot water. Wood is also used to build spas and is detailed in the next section, Hot tubs. It is important for the technician to understand the basic construction alternatives, to be aware of what type of vessel is being serviced, and to adjust procedures accordingly (Fig. 11-1).

Gunite

Some spas are built just like a pool, using an excavation, reinforced concrete, and plaster. These are stand-alone units with their own equipment and plumbing or units built adjacent to a pool, using the same equipment and plumbing as the pool (as illustrated in Fig. 1-1). In areas where a gunite rig is unavailable, the spa might be constructed with forms and poured concrete. The gunite or concrete spa can be viewed as a small pool and serviced accordingly.

Fiberglass

Spas are made of fiberglass in molds or hand-laid to conform to a specific design requirement. Sheets of fiberglass material provide the strength, while chemical resins coat the material and provide stiffness and strength. The final shell is covered with a glossy coating called *gel coat* to provide a surface that is virtually impervious to water and maintenance chemicals. Fiberglass spas can be serviced in much the same way as gunite spas, with the exception that there is no plaster to dissolve and raise the pH. Because of this, maintaining a careful watch on chemistry is even more important and is detailed later. Various manufacturers provide the fiberglass shell by itself, units preplumbed for easy installation, or units built into a self-standing cabinet with a package of all the necessary equipment. The self-contained unit requires only the addition of water and a 30-amp circuit to plug in the equipment.

Acrylic

Acrylic spas are essentially big plastic bowls molded over a fiberglass shell (Fig. 11-2) and that can be treated much the same as fiberglass spas. Both fiberglass and acrylic spas are built as shells that can be plumbed and set into the ground, above ground, or built into a self-contained cabinet for portability. Although completely self-contained, the weight of the shell, cabinet, and equipment package make these units at least as difficult to move as a grand piano.

Soft-sided

In recent years, several manufacturers have developed a lightweight, very portable, self-contained spa made from reinforced foam. As the name implies, the soft-sided spa is made of stiff (but flexible) foam rubber, reinforced with aluminum framing, and covered with a waterproof fabric to hold water. These spas are built with a self-contained equipment package, usually covered in the same materials to protect the equipment and reduce noise. Many apartment dwellers are buying these spas because of their true portability and low cost.

A

B

11-1 Various types of spas. A: redwood hot tubs; B: all-tile gunite spa; C: Hot Shell™ in redwood tub; D: jetted bath tub.

C

D

II-I Continued.

Acrylic

Bonding coat

Fiberglass

Bond
beam

Insulating foam

11-2 Acrylic/fiberglass spa construction. Caldera Spas and Baths, El Cajon, Ca.

Jetted bathtubs

Another choice of spa, especially for homes or apartments where space is an issue, is the jetted bathtub. As the name implies, a standard bathtub is fitted with a booster pump and jets to create a spa. Some units are built to accommodate two people and some are round instead of the traditional rectangle. Jetted tubs are filled with hot water like any other bathtub from the normal home water heater supply, so no additional heater is used. Because they are indoors and are drained after each use, jetted tubs do not have skimmers, strainer baskets, or filters. Jetted tubs are popular as personal spas because they are inexpensive, requiring only one pump and some plumbing, they heat up quickly because they are filled with water that is already hot, and they require no water maintenance or special chemicals. Obviously the jetted tub is maintained like any other bathtub, drained and cleaned after each use.

Installation

In this section I will deal with the installation of a shell-type spa—fiberglass or acrylic—because gunite construction is the same as pool construction and the soft-sided spa is a manufactured item.

Jetted bathtubs are installed like any other home bathtub, with the addition of the jet pump, plumbing, and wiring. Since most of these are designed into the construction of the home, most technicians will not install jetted tubs. I have been asked to remove an old bathtub and replace it with a larger jetted tub. In these few cases, I bring in a plumbing subcontractor to provide the necessary in-home expertise of installing the tub, while I concentrate on the pump and related specialty plumbing. It usually pays to let others do what they know best and make your profits doing what you know best.

This section on installation assumes that the customer has chosen a style of spa and a location. If that is not the case, you can evaluate their uses and needs, show them catalogs of available styles from the supply house, then let them make their choice. Some guidelines that might help in this process include the following.

The shell

Ask the customer the maximum number of people that will use the spa at any time. Spas are sold as two-person, four-person, etc., indicating the number of bodies the shell can comfortably accommodate. The molded-in seating or lounges will make that determination fairly obvious. It is important to consider the customers priorities as well. For example, the customer might have one or two parties each summer when eight or ten people will use the spa, but the remainder of the year it will be used by only two people. The amount of space the ten-person spa requires on the deck or in the yard and the annual costs of maintenance, heating, and electricity might not make sense if it is used by only two people.

Consider the manner in which the spa will be used. The customer will want straight bench seating if it is used mostly for socializing with several people, but will prefer lounges if it is used for therapeutic bathing.

Note also the size of the most frequent users. I was once asked to install a spa for a very famous actor/musician in Southern California who is very short. His builder had already purchased the shell, preplumbed with skimmer, drains, and jets. When the project was complete, this person sat in his new spa and found the water level up to his eyeballs. All of the spa seating had been designed for a taller person whose upper body length would leave their shoulders at the waterline. Most spas have seats of varying depth to accommodate a variety of human proportions, but this one hadn't counted on its only occupant being so short. The solution was to lower the skimmer so the waterline would be lower; something we should have thought about first.

The plumbing

Similar considerations need to be made when plumbing the shell. Depending on the intended users, you might be able to install a shell that has been preplumbed to accommodate the average bather. Typically two jets per person are included in a spa, set in the walls at varying heights. An air bubble ring is also provided to create general turbulence in the water for a massaging effect.

If the users have specific preferences about number, type, and location of jets, it makes more sense to buy the shell and plumb it to their specifications. By inquiring at your supply house, you can make a list of options available to the customer, such as small pinpoint jets, larger standard massage jets, rotating jets, and those that can be turned on and off to put more or less pressure in one specific area of the spa. Some customers only care about the general turbulence of the hot water and require one or more air bubble rings, while some customers are only interested in the massage jets and don't want to invest the extra money in the air bubble ring and electric blower required to power it. A blower can also be plumbed to turbocharge the jets. In short, familiarize yourself with the customer's preferences and design the spa accordingly.

If they don't seem to have strong preferences, I recommend a preplumbed, one-size-fits-all spa. In either case it is valuable to force the customer to make the decisions. You can provide advice, but if you make the decisions for them, you will be blamed if the results are not what they expected. If they make the choices, they live with the results.

When selecting a preplumbed spa, consider the cost and feasibility of transportation. Figure 1-11 shows a spa shell that is preplumbed. As you can see, the exposed fittings and plumbing can easily be cracked or snapped off during transport and handling to the job site. I learned my lesson early when I failed to personally supervise the delivery of a preplumbed spa and simply installed it for a customer. After many hours of excavation to expose the plumbing all around I found four leaks, cracks that were probably caused by careless handling during transportation. There is an irresistible urge to use the plumbing fittings as handles, which invariably weakens or cracks them. The moral of the story is to have a truck of adequate size and enough handlers (remember, these things are large, bulky, and heavy) to get the spa into the yard without damage. If the pathways are tight or the location is far from the truck access area, recommend plumbing the spa at the installation site.

Finally, price makes a difference to almost every customer. The self-contained spa, built with the same shell and plumbing you might provide for a custom design, will cost anywhere from $2000 to $5000. By the time you factor in the installation of the shell, plumbing, equipment, and subcontractors, you can easily spend up to ten times that for a built-in spa. Before investing a great deal of time talking about custom options, ask your customer about their budget for the job.

The equipment

At the same time as you are choosing the number and styles of plumbing options, you need to review the equipment requirements of each choice. The pump needs to be of sufficient strength to provide adequate power to the jets. The general rule of thumb is that each jet needs ¼ hp, so a 2-hp pump could theoretically give adequate performance to eight jets. Having said that however, you must also consider the total dynamic head of the system, as described in the chapter on pumps, especially if the equipment is located a great distance from the spa. Long distances are often required to satisfy customer demands that the equipment noise be kept away from the spa environment. Remember also that the turnover rate must be satisfied and that most jets, to operate as designed, require about 15 gpm of flow each.

Using those general guidelines, decide on the size of the pump needed and if more than one pump is required. In larger installations, you might want to divide the spa into two halves and operate each side from a separate pump. When making pump selections, consult your electrical subcontractor to determine that adequate electricity is available for your choices and that the customer understands the costs of any upgrading that might be required of the electrical breaker panel.

Cartridge filters are the preferred choice for spas, following the selection and sizing criteria outlined previously. Consider the volume of use when choosing a filter for a spa. If the customers will use it frequently (many people jump in the spa when they get home from work before greeting the rest of the family), choose a filter size larger than actually needed to reduce the frequency of required tear-down and cleaning. If the customer is using a regular cleaning service, you might be less concerned about the frequency of teardown than if the customer plans to perform their own routine maintenance. Customers rarely clean the filter as often as needed.

The most important consideration in choosing a heater is the customer's intended use. If the customer's use is unpredictable, they will need a larger heater to raise the temperature quickly. If you know the spa needs to be heated daily at 6 P.M., you can choose a smaller heater and set the system to turn on with a time clock early enough to reach the desired temperature by the desired time.

Consult with your plumbing contractor about the availability of gas supply and advise the customer of the cost of upgrading lines and meters and adding the supply line to the equipment location. If there is no gas available in your area, consult with your electrician about adequate electrical supply for an electric heater. Since even the largest electric heaters take a long time to heat a spa, choose the largest one your customer and the electrical supply can afford.

Decide if the customer wants one or more blower rings, which force bubbles into the water for a general massage. Rings are often located in the floor and/or seats of the spa. You might add a second blower to turbocharge the jets, or install a three-port valve to divert the air flow of a single blower to the air ring and jets. Again, consult with an electrician about available electrical supply and the cost of wiring to the equipment location.

To provide a true overall estimate of the job, provide your customer with a list of additional equipment they might want to consider.
- Cover
- Chlorinator
- Lighting
- Mist spray
- Fill line (automatic or manual)
- Control system (electronic, manual, or air buttons)
- Time clock (might be included in the control system)
- Handrails, built-in drink holders, and padded neck rests

The supply house in your area can provide catalogs of available products to help your customer make these choices.

Step by step
Having made the decisions about what to install, the following list outlines the steps required.

1. Planning. Obtain all necessary local building permits prior to starting the job. The complete installation might include building a deck, extension, patio, or enclosure on the customer's house, so if you do not hold the appropriate contractor's license in your area for the work required, collaborate with a general contractor for the planning and permitting of the job. As in other jobs requiring construction methods beyond the normal work of the water technician, I prefer to subcontract the work to someone else and let them do what they are trained to do, while I install the shell, plumbing, and equipment. In addition to hiring a general contractor to work on the construction and permit portions of the job, you might want to subcontract moving the electrical, gas, or water supply connections to the equipment area. You might also help your customer by recommending landscaping and masonry subcontractors as well.

2. Foundation. Working with your general contractor, carpenter, mason, or other subcontractors, plan the site preparation and coordinate the spa installation with any deck or other foundation work.

   Whether you will be preparing the deck and foundation or the subcontractor is doing it, a few concerns are often overlooked. First, be sure the deck is level. This might seem obvious, but if there is the slightest tilt, it will become very apparent when the spa is filled with water. A spa that is sitting on a deck that is not level will have a waterline that is high on one side and low on the other. This could be important if, for example, the high end is the side of the spa with the skimmer. To keep enough water in the skimmer to keep it from running dry, it might be necessary to fill the spa to the point of overflowing at the low end. Remember also that people in the spa will displace a great deal of water, so even if the variations are minor, they will become significant when the spa is in use.

   Of course you might start out with level surfaces, but if the foundation for the spa and deck are not stable, the spa might not remain level. A spa that is full of water and three or four people will weigh over 2 tons. In a relatively small area, that weight creates a great deal of force for movement if it is not absolutely secure.

   You will need to follow the building code in your area, but the objective is to secure the spa from shifting by supporting it in as many locations as possible, evenly set on the deck. Figure 11-3 shows how the spa lip rests on the deck. It also shows how the spa rests in the excavation, surrounded by sand for general lateral support. The level is maintained by the deck. Obviously, if the spa shifts the sand will compress and fail to support the spa. The advantage of that is that the sand also gives enough to prevent damage to the plumbing.

   Once the location of the spa is determined, a simple hole is dug in the ground, large enough to accommodate both the spa and plumbing. Figure 11-3 shows how the excavation is filled with sand, but remember, you might need to excavate the sand at some later time to repair leaks, so dig a hole that is also large enough to accommodate a technician making repairs or inspection.

   For this reason, I prefer an installation that supports the spa without filling the excavation with anything. As long as the base of the spa is supported

11-3
Spa foundation.

(Fig. 11-3, items A, B, and C) and the lip is secure on the deck, there is no need to cover the plumbing with sand. Believe me, at some point during the life of the spa you will need to access the plumbing, so preparation for that now will pay dividends later. Your subcontractors can help you install simple post and pier supports (Fig. 11-4) to create a framework to stabilize the spa, leaving the plumbing exposed. If the excavation is large enough and a deck access (or opening in a slope) is created, you will be able to easily diagnose and repair leaks.

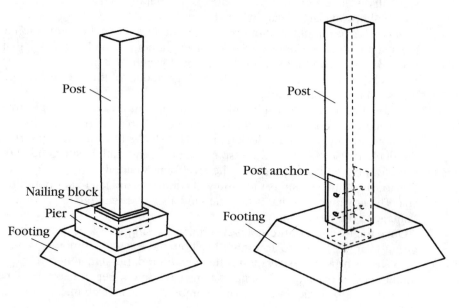

11-4 Post and pier supports.

You can also prevent leaks in this way. If the spa shifts for any reason, the pressure of the plumbing against the sand can crack joints and fittings. Erosion, earthquakes, and even frequent passing heavy trucks can cause earth movement and shifting of the spa. My experience is that no matter how well installed it is, every spa leaks sooner or later. Site preparation makes the difference between a simple repair and tearing out the entire installation and starting over.

3. Bring the spa to the site. Acrylic and fiberglass gel coats scratch easily, so take care to cover the top of the spa to prevent tools, dirt, or sand from damaging the surface. You will need to work inside the spa and painter's dropcloths make good working covers.

4. Plumbing. If the spa is preplumbed, skip ahead to step 7. If not, plumb the spa at the job site. Figure 11-5 shows the plumbing of a typical spa shell. Working with your customer, decide on the number and location of the jets, marking them with a pencil or felt-tipped pen. Plan on installing a skimmer even if the spa is indoors in a clean area, because the skimmer will remove body oil or suntan lotion from the surface of the water. Locate the skimmer on the side of the spa closest to the equipment, but consider the seating pattern of the spa and avoid placing it where it will prevent someone from using a lounge or molded seat. Purchase the jet fittings, drains, and skimmer before cutting any holes in the shell.

5. Install the skimmer and main drain. As shown in Fig. 11-5, suction plumbing runs from a skimmer and main drain, joined with a T fitting as shown or an optional three-port valve. Position the skimmer as discussed previously, then determine the appropriate height. The top of the skimmer opening will normally be only 2 or 3 inches below the lip of the spa so that the normal water level, in the middle of the skimmer opening, will be about 6 to 8 inches below the lip. Remember that bodies displace water, so if the skimmer is located too high and the water level is raised to fill the skimmer, when you add bathers the spa will overflow. In addition to the obvious problem of a wet deck and an unsatisfactory installation, the water loss means that when the bathers leave, the water level will drop below the skimmer opening and the pump will run dry. If the skimmer is located too low, the water level will be higher than the skimmer and the spa will have no skimming action. If you are in doubt about the positioning of the skimmer, take a look at a preplumbed spa or a unit already in service to get a feel for the best placement.

The actual installation of the skimmer is simple. The skimmer is supplied with a faceplate and gasket that fits over the opening on the inside of the spa. Using the faceplate as a template, determine the exact position of the skimmer opening by holding the faceplate against the inside of the spa and tracing the opening with a pencil. Mark the location of the screw holes. Remove the faceplate and drill out the screw hole locations to the size of the screws provided. I drill the screw holes before cutting out the skimmer opening because the spa wall is stronger before removing the material from the opening and it is less likely to chip or crack than if I were to drill the holes after that portion is removed. Next, drill a starter hole within the perimeter of the skimmer opening to create an access for your saw blade.

11-5 Typical spa plumbing. Sta-Rite Industries, Delevan, Wisc.

Use an electric reciprocating saw, jig saw, or manual keyhole saw (if you have a lot of patience) to cut out the skimmer opening. Hold the skimmer against the back of the spa and apply the gasket and faceplate to the interior side of the spa. Screw them together with the hardware provided. Apply some silicone sealant to the bolts before slipping them through the holes. When you tighten the bolts, the sealant will fill any gaps.

Install the main drain following the procedures described in step 6 for jet fittings. Some building codes require two main drains attached to the same suction line, joined by a T fitting or one drain with two suction ports, as shown in (Fig. 11-5). Since spa suction is extremely strong, should one drain be covered by a hand or foot, the other one will then pull water rather than a single one that might injure the bather. In this case, the second drain is usually mounted in the side while the primary drain is located in the floor of the spa.

Install any light fittings following the instructions with each different model. Some will require creating a light niche in which the actual fixture is mounted. The niche is installed on the shell in the same manner as the skimmer. Others are designed more like a main drain, where the fixture is threaded to fit through a hole in the spa wall and screws into a body on the exterior like a jet fitting.

6. Plumbing the jets. A typical 1½-inch plumbed jet fitting requires a hole in the shell of 2¼ inches to accommodate the throat of the jet (Fig. 11-6). Using your electric drill and a 2¼-inch hole saw, cut holes for the jets in the locations marked. The jet body slides through the hole and a nut on the back secures it in place. Be sure to use the gasket provided on the inside (the water side) of the fitting to prevent leaks. Tighten the nut with channel-lock pliers, applying enough force to tighten but not crack the plastic nut.

The other type of jet uses a jet body that is held against the exterior (back side) of the spa, while a threaded throat passes through the hole in the spa wall and screws into the body. Jet wrenches are provided by the makers of these jets for tightening. On the face of the throat you will find two holes. The wrench fits into these holes to grip the throat fitting. While holding the jet body (usually requiring a helper), tighten the throat fitting into the body. Use enough force to tighten the throat against the gasket for a watertight seal, but not so much that you crack the fitting. Make sure to install all jet fittings with the water plumbing below and the air plumbing on top.

Some manufacturers make microjets, smaller versions of the 1½-inch standard jet, built to accept 1-inch PVC plumbing. These are most often used in jetted bathtubs, but some customers prefer the microjet in combination with the standard jet because you can use more of the microjets with the same equipment than the standard units.

11-6 1½-inch plumbed jet fitting and hole cut.

Whichever design of jet you use, be sure to buy the ones designed for thin-wall spas. Jets are also made with identical components for hot tubs, except that the threaded throat is several inches longer to accommodate the greater thickness of the wood as opposed to the thinner fiberglass or acrylic walls.

Using flex PVC and the plumbing techniques described previously, connect the jets as illustrated in Fig. 11-5. If you are using one pump, the return line that sends water through the jets should be diverted with a T fitting, sending half of the flow to half of the jets and the other half to the remaining jets. If instead the flow runs the length of the jets, the pressure will be extremely high for the first few and extremely low on the last one or two. By equally dividing the flow you avoid that problem. You might also use a three-port valve instead of a T fitting, so that you can equally divide the flow or divert all of it to one side or the other.

If the spa is large and more than one pump is being used, divide the number of jets into two groups and plumb the flow from each pump accordingly. Since one pump will circulate water through the filter and heater, plumb it with fewer jets. For example, if the spa has ten jets, plumb the filter and heater pump with four jets and the unrestricted second pump with six jets.

In each case, cap the dead (terminal) ends of each jet line with 1½-inch PVC caps attached to a short length of flex PVC pipe. I prefer this method to inserting a 1½-inch PVC plug into the jet, because if the plumbing needs to be changed for any reason, the plug makes that side of the jet permanently inaccessible. If you do need to use the other end of the jet that is capped, you can cut off the cap and glue on a slip coupling, plumbing anything on from there.

Connect the air side of the jets together. If the jets are to be turbocharged from a blower, plumb them like the water side, with a T fitting or three-port valve between equal numbers of jets. Cap the dead ends as described previously.

If the jets will be aerated instead from the atmosphere, there are two choices. First, on the top of each jet is a ¾-inch opening to accommodate a length of pipe that extends up to the lip of the spa, aerating each jet individually. The other choice is to glue these openings shut with the plugs provided and stub one or two common air lines to the lip of the spa, as shown in Fig. 11-5. With this method, an air control fitting can be mounted in the lip of the spa to which the common air pipe is glued. This air control fitting allows the spa user to close off all air supply, resulting in a jet that moves only water, or the air supply can be opened for the desired amount of added aeration. If the spa has been constructed with one or more air blower rings, they will be provided with a 2-inch stub in which to glue the PVC pipe.

7. Before installing the spa, perform a leak test on the plumbing. If the spa is preplumbed it is important to perform this test to be sure the factory did a good job and that nothing was damaged in transit. It is much easier to repair or replace the plumbing above ground than to wait until the unit is

installed. Glue flex PVC into the T fittings or three-port valves and run enough length to bend the pipe over the top of the spa. You will have one line for the main drain, one for the skimmer, one for the air ring, and one for the return water to the jets (or two if you have installed a two-pump system). Cut the PVC and tie the lengths above the top of the spa, then fill the spa with water. There should be no leaks before you set the spa in the ground. Let the water sit for an hour or more to be sure there are no slow leaks. Take care when filling not to wet the outside of the spa or the ground, making it difficult to observe any leaks. If you have the equipment, pressure testing the spa is a good idea (and might be required by building codes in your area), although I have found that if the plumbing is done with care and the spa passes this simple fill-up test, it will not leak under pressure. When you are sure there are no leaks, drain the spa.

8. Set the spa in the ground, onto the deck and any other supports. Carefully lower the plumbed spa into the ground, using enough helpers to avoid dropping the unit or resting it on the fragile plumbing. If the spa is to be backfilled with sand, once the unit has been stabilized and secured, perform another leak test. This might seem redundant, but believe me it is easier to spend an extra hour at this point than to make repairs later that require excavating the sand to find the leaks. If you took my earlier advice and used supports, you won't need to leak test at this point, because without the sand, you will be able to see if leaks are present once the job is complete.

9. Set the equipment. The equipment should be located as close to the spa as possible to avoid excess head in the system and to avoid heat loss by running water through long pipe runs. The level of the equipment should be as near the water level of the spa as possible to avoid priming problems or excess head in the system. Working with your subcontractors, prepare the site with a concrete pad of adequate size to accommodate the equipment, leaving enough room between components to allow access for repairs or maintenance. If you need to provide your own equipment pad, follow the instructions in the section on foundations for hot tub installations. Complete the gas and electrical supply connections to the equipment and plumb the components using the methods described in previous chapters.

10. Trench as needed to run plumbing and electrical lines between the equipment area and the spa. Plumb the equipment and spa stub-outs together. If there are long runs of aboveground pipe under a deck or in other exposed areas, wrap them with insulation to prevent heat loss.

    Special attention should be given to the blower lines to ensure that water doesn't travel back up the line into the electric motor. See the section concerning blowers for details. Before closing the trenches, connect any control devices or switches planned for the installation, such as air switch buttons with air hoses running to the equipment area through a pipe in the trench.

As with any installation procedure, you will encounter variations in the spa installation process. Nothing in this procedure is set in stone, and each step requires planning and creativity to complete the job in a way that allows for correction of problems or later repairs. Remember, it is much easier and less expensive to change the design on paper than in the field. Planning and forethought will make a big difference in the satisfaction of your customer.

Special equipment

The spa is essentially a small pool, a vessel of water that is circulated, filtered, and heated. There are, however, a few pieces of equipment that are unique to spas and require further discussion.

Air blowers

An electric-powered mechanical device that forces air into the spa is called an *air blower*. Air blowers are rated by horsepower—1, 1.5, or 2—and are available as aboveground units or burial blowers, designed to be buried in the ground to reduce operating noise. Blower casings are generally made from plastic or metal.

Blowers are centrifugal air pumps, just as the circulating pump is a centrifugal water pump. The motor is the same design used on vacuum cleaners. Air is pulled in from the outside, drawn over the motor to keep it cool, then forced out to the spa. Air is pulled into the blower from impellers, again much like a water pump. Larger units use two impellers to draw more air. It is this air suction that creates the noise associated with blowers, not the motor itself.

The air blower adds general turbulence to spa waters for a general massaging effect or it is plumbed into the air line of spa jets to turbocharge the jet massage. In either case, selecting the correct size for the installation is important. If the blower is too large, it will try to push more air than the system can handle and the blower motor will overheat and burn out. If it is too small, the air delivered will be weak or unable to overcome the system's total dynamic head and will deliver no air to the spa.

When replacing an existing blower, you can generally replace it with one of the same size. This is, however, a good opportunity to make sure the original unit was sized correctly. Air blower sizing is calculated on the depth of water through which the air must be forced. Water has weight, and the greater depth, the greater the weight. With this increasing weight, the blower requires increasing force to push air through the water. For easier calculations, the resistance created by horizontal plumbing and angled fittings is expressed as depth so you can determine the total depth of the water that the blower must overcome. To calculate the size of the blower required, proceed as follows.

1. Measure the actual depth of the water above the air outlet. This is not the depth of the spa, unless the air outlet is a bubble ring in the floor. It might be the depth of the jets if the blower is turbocharging the jets; it might be the depth of the seat if the bubble ring is in the seat. Remember that water displacement by bathers will raise the water level several inches.
2. Blowers are plumbed with 2-inch PVC. Measure the total length of all pipe between the blower and the spa. For every 10 feet of pipe, add 1 inch to the total water depth.

3. For each 90-degree angle (one 90-degree fitting or two 45-degree fittings), add ½ inch to the total water depth.
4. Add the three components to arrive at a total water depth. If it is less than 36 inches, a 1-hp blower is required; from 36 to 48 inches, a 1.5-hp blower is required; for 48 to 55 inches, a 2-hp blower is required.

   Figure 11-7 depicts a typical aboveground blower installation.
   Water depth = 32.0 inches
   Total pipe length of 41 feet ÷ 10 = 4.1 inches
   Five 90s × 0.5 inch = 2.5 inches
   Total depth of water = 38.6 inches

Rounded to the nearest inch means this example is 39 inches of total water depth, requiring a 1.5-hp blower. As noted above, you don't want to oversize the blower, but when your calculation is right on the edge, move up to the next higher unit.

5. Make sure that your system can handle the volume of air. If the blower is serving an air ring, use the following calculations for the number of holes and their respective size, regardless of horsepower. Of course, if you want fewer holes, each of which is supplying more air, use a ¼-inch hole. If the spa is large and you want an even distribution of many smaller holes, use the ⅛-inch holes or something in between for your installation.

   If the air ring has
   ⅛-inch holes, drill 30 of them;
   $\frac{5}{32}$-inch holes, drill 73 of them;
   $\frac{3}{16}$-inch holes, drill 50 of them; or
   ¼-inch holes, drill 30 of them.

   If the ring has too many holes or they are too large, you can plug some with plaster (on plaster spas) or silicone sealant. If the holes are too small or there are not enough, select the proper drill bit size and enlarge the existing holes and/or drill new ones. As noted previously, if the blower is supplying standard 1½-inch spa jets, make sure there is one jet per ¼ hp of blower.

Now that you have selected the proper blower, the installation is critical to peak performance. To prevent water from entering the electric motor, install the blower higher than the water level and plumb in a Hartford loop (Fig. 11-7) and a check valve. The loop must be 12 to 18 inches above the water level to be effective. When reducing from the 2-inch pipe of the blower to the 1½-inch jet plumbing, make the reduction at the T or three-port valve, not near the blower. A check valve is also recommended in the system of not more than ½-inch pressure. If you use the loop and theantisurge valve provided with most blowers, the check valve is one restriction in the system too many. The other function of the check valve and/or loop is to keep the water as close to the spa as possible and fill the lines from the blower with air as much as possible. The result is that the air reaches the spa faster than if it must force itself through water in the pipes before forcing itself through the water in the spa. Therefore, the loop and check valve should be located as near the spa as possible. Antisurge valves and check valves built in at the blower are for added protection.

11-7 Plumbing installation of an air blower. Conger Manufacturing, Paso Robles, Ca.

When designing the plumbing, use two 45-degree angles to make pipe turns rather than one 90. As described in the total dynamic head section of the chapter on pumps, less resistance is created in this way. When gluing the plumbing together, allow the fumes to dissipate by waiting at least 24 hours before operating the blower. An electric spark can ignite the fumes. Most manufacturers recommend dry-fitting the blower to the stub-out pipe rather than gluing. Silicone sealant will keep moisture out of this connection and hold the blower in place. Over time, a dry-fitted blower without silicone or glue will vibrate loose and literally blow off the plumbing.

Blowers are provided as side discharge, bottom discharge, or convertible. Figure 11-7 depicts a bottom discharge, while a side discharge is the same product but with the air discharge on the lower portion of one side. The convertible type has discharge ports on both the bottom and side, and a plug is used to close off whichever discharge port is not needed. The plug is either screwed in place or attached with some silicone sealant. I have found that on plugs that are not designed to be screwed in place, you can touch a dab of PVC glue to one side of the plug and insert it. Again, give the fumes adequate time to dissipate before turning on the unit for the first time.

In addition to the check valve and Hartford loop, some manufacturers provide a small rubber flapper insert called an antisurge valve that is inserted in the discharge port before mounting it to the pipe. This check valve provides security that the air can exit the unit but nothing else will enter. By the way, because the antisurge valve is not listed by Underwriters Laboratory (UL), it cannot technically be called a check valve, thus the alternate term, antisurge valve. Some manufacturers build the check valve inside the discharge pipe of the unit, so look inside to determine if you need one. Always read the manufacturer's literature provided with the blower.

Burial blowers are installed in the same way, but the unit is buried for noise reduction, usually no more than 12 inches deep. An air vent stack runs above ground for ventilation. Otherwise the sizing and plumbing recommendations are identical to the aboveground unit.

Because the blower vibrates somewhat, be sure it is secured. In Fig. 11-7, I would drive a stake or length of rebar into the ground and use a plumber's strap to secure the pipe to it. If the deck is concrete, bring the blower pipe up adjacent to a wall so the blower can be strapped to the wall. If the blower is a side-discharge unit, it can be mounted to a shelf or other surface directly.

Finally, the blower must be connected to the electrical supply. Blowers are supplied with either 110-volt or 220-volt motors. Unlike pump motors, these are not convertible, so be sure to buy the unit that corresponds to the equipment electrical supply. Some blowers are supplied with a built-in switch; but whether they have one or not, a wiring J-box is mounted to the unit for connection of the conduit (use waterproof conduit for outdoor installations) and wiring from the electrical supply.

Each unit includes a ground wire for bonding to the electrical panel's ground bar, but if the unit is installed within 10 feet of the spa, additional grounding must be used. A bonding wire terminal is provided on each blower to directly run a #8 copper grounding wire into the ground for this purpose or for attachment to the bonding system of the other equipment. Consult with your electrical subcontractor and/or follow local building codes. Also follow the recommendations and wiring diagram provided by the manufacturer.

For the blower to operate correctly, it must be provided with adequate amperage. A 2-hp blower will draw up to 13 amps at 110 volts, meaning the electrical service required is substantial. If you are having performance problems with a correctly sized blower, check the amperage to be sure it is supplying that required by the blower. The amperage is listed on the rating plate of the blower, along with the manufacturer's name, voltage, and horsepower. Sometimes the horsepower rating is obscured (on some manufacturers it is not printed on the rating plate). Since more horsepower needs more amperage, you can use this general rule of thumb to estimate the horsepower of the existing blower:

- 1 hp   6.5 amps at 110 volts   3.25 amps at 220 volts
- 2 hp   13 amps at 110 volts   6.5 amps at 220 volts

The motor, brushes, and other components of a blower can be replaced, but an entire unit costs less than $100 for most sizes. Since the labor and parts for a repair can cost that much, I have never recommended fixing a blower to a customer unless the unit is reasonably new.

Most blower failure is caused by overheating, which in turn is caused by too much restriction on the system. In other words, before replacing a fairly new blower, check the sizing carefully. Blowers heat up even when used in a well-designed system, so failure is inevitable. A blower should last three to five years in outdoor installations with regular use, and twice that for indoor installations.

Some blowers are built with a thermal overload switch similar to that on pump motors. When the operating temperature exceeds the designed amount, the switch opens and the blower shuts off. When the air cools, the switch cools as well, closing it and restoring electric current to the motor. If your blower cuts on and off, it might be a sign of inadequate ventilation around the blower (often), system pressure creating an overheated motor (most often), or a faulty thermal switch (rarely).

Booster pumps

As already mentioned, some spas use the circulating pump to power the jets, while some installations add a second pump that is not restricted by the other components. There are no special instructions regarding a booster pump, which is identical to the circulating pump. The booster pump should be plumbed with its own suction drain so it is not competing with the circulation pump, one starving the other for water flow. Some main drains are designed with two ports under one cover so you can connect a suction line for each pump, and this might be preferable to adding a second drain fitting in a small spa.

Similarly, don't try to force the return lines together into a few jets. The system can only handle so much water, so plumb some jets to one pump and some to the other as described previously.

Finally, booster pumps require a strainer basket just like the circulation pump. Because the circulation pump is connected to the skimmer, its basket is more likely to contain leaves or other debris, while the booster pump strainer basket gets mostly hair and grit. Since spa pumps are powerful and close to the bather, booster pump baskets also often contain jewelry.

Small spas use two-speed motors as described previously in the chapter on pumps. The low speed is used for circulation and heating, the high speed for jet action.

Electric heaters

Larger spas are heated by gas-fueled heaters. Smaller spas or those in areas without a gas supply use electric-fueled heaters. Electric heaters are also used where there is no venting capability, which is required for a gas-fueled heater. The electric heater is described in the chapter on heaters. Electric heaters take a long time to heat even a small spa, so I recommend a cover for any spa that uses one to retain as much heat as possible.

Skid packs

A *skid pack* is a combination of spa equipment mounted on a metal frame called a skid. The pump and motor (usually two-speed), filter, heater, blower, and control devices are all combined on the skid pack for compact installations, mostly on portable spas. Most skid packs are designed to be plugged into a wall outlet in the home at 110 volts, but larger ones that require more amps are designed for 220 volts supplied by a dedicated circuit from the breaker panel. Some skid packs are convertible for either voltage.

Components in a skid pack are susceptible to the same maintenance and repair problems as individual components. The difference is that to repair a component on a skid pack you might have to disassemble several other components from the pack to access the one that needs repair. When estimating a repair on a portable spa that uses a skid pack, keep that in mind. On many such jobs, I have spent twice as much time extracting and reassembling than I spent on the actual repair of the damaged component.

The second major failure of skid packs is the control system. There are far too many designs to profile a common control system for skid packs, but they might include a time clock, air switches, various thermostat controls, light controls, and high/low pump speed controls. Some manufacturers provide a plug-in circuit board where each component plugs into an outlet designated for that item. The switching system controls the electrical flow to each outlet. When a blower fails to operate, for example, you can literally unplug the blower and test the outlet when the switch for the blower is on. If there is no current, you know the problem is in the switch, wiring, or another control circuit rather than the blower. If there is current, then the blower can be removed and replaced. Testing for current can be done with your multimeter or a special plug-in tester provided by the manufacturer.

Electronic sensors and controls are far more accurate (in temperature control and precise time settings, for example), but are far more susceptible to failure from moisture, corrosion, salt air, or delicate wiring and connections being broken. Electronic controls also require a greater degree of troubleshooting sophistication, specialized to each manufacturer's electronic engineering designs. When I determine that a skid pack's problems are the result of the electronic controls rather than a specific component, I either advise the customer that the required time (and therefore the labor charges) will be more, or I call a factory representative to perform the repair. Believe me, after spending hours on frustrating troubleshooting and repair of delicate electronic systems, I usually call the factory representative. But because many skid pack makers have come and gone from the business, you might have no choice but to fool around with the controls or tear them out and replace them with a more reliable air switch system. Finally, if the skid pack is fairly old, it might not make sense to replace expensive components when an entire skid pack can cost as little as $700 (wholesale).

Sometimes performance problems can be traced to inadequate amperage if the skid pack is operating on a 110-volt system. Check the electrical supply circuit if the blower appears to be running slowly or the motor overheats and shuts off frequently. Call in an electrical subcontractor if the circuit needs upgrading.

Finally, as with any pool or spa equipment repair, make careful note of how the skid pack comes apart to make reassembly easier. This point is extremely important with skid packs, because they often require a lot of disassembly to get at a particular component and because so many manufacturers are no longer in business to offer technical advice.

Sprays

The last item of specialized spa equipment worth mentioning is the cooling spray. While the bather sits in the hot water, some enjoy a spray of cool water occasionally. This is something that can be inexpensively added to any spa.

The spray is supplied with water from a garden hose. Run the hose to the edge of the spa and plumb it to a ¾-inch PVC pipe mounted about 3 feet above the spa. Secure the pipe to a wall or to a stake in the ground or deck. Plumb a suitable nozzle onto the end of the ¾-inch pipe to create a mist of cool water. Some shower heads have mist settings, but I use a bronze garden mist head, available at any garden supply store for a few dollars. To give the bather control over the mister, install a ball valve between the garden hose and the ¾-inch pipe at a convenient location on the edge of the spa.

Repairs

Spa repair is mostly a matter of adapting the pool repair techniques already discussed and applying them to a smaller arena. A few specific repairs come up more often with spas, however.

Leaks

Leak detection and repair were outlined in more detail in the previous chapter, but it should be noted that spas are far more prone to leak problems than any pool I have every seen. In addition to the likely leak locations already detailed, many fiberglass and acrylic spas leak at the light fixture, skimmer assembly, or jets. Spas are designed to vibrate with turbulent water and bend under heavy bather loads, so these add-ons are subject to loosening and leaks.

Conducting a dye test (described in chapter 10) is easier on a spa than a pool because you can test virtually every corner without getting very wet. Also, the dye remains close to your eye level for easier observation. So unless you have trouble finding a leak, I would not use the drain-down test but move directly to the dye test when troubleshooting spa leaks.

If the jets are found to be leaking around the faceplate, they probably need to be tightened. Use your jet wrench to check each fitting. Time, harsh chemicals, and heat can deteriorate the gasket between the jet body and throat, so tightening might not seal the leak. Unfortunately, you can't usually replace gaskets. If you remove the throat, the jet body behind the spa will invariably shift slightly, making realignment for reassembly impossible. If you can access the jet body, replacement of the gasket is possible.

If the gasket seal is found to be leaking, try draining the water level down below the jets and unscrew the throat slightly. Allow the area to dry completely. If the gasket is brittle and/or extremely deteriorated, remove it. Take a new gasket and cut it along one side so you can open it and slip it around the jet throat. Obviously, the place where you cut the gasket will not seal when you tighten the throat down, so fill the gap with silicone sealant, then tighten. Wipe the excess sealant that oozes out as you tighten away from the face of the jet. Wipe in an even, steady motion, forcing the material into the seal area and smoothing it over the joint of the jet face and spa wall. The objective is to form a watertight seal and thoroughly cover the work with the sealant. Allow the work to dry completely before refilling the spa and dye testing again.

Similarly, the integrity of other joints, such as the skimmer, drains, and light fixtures, is often compromised. In each case, tighten whatever you can and then seal the entire assembly with silicone sealant. These seals will often leak again as the silicone shrinks from constant temperature extremes and harsh chemicals, but this provides an inexpensive and quick way to extend the life of the spa. Of course, the best repair is to completely disassemble the leaking component and replace its gasket or the component itself. But if the spa installation does not allow that an option, silicone is your best bet.

If you have made these repairs in numerous places in a spa, recommend a complete overhaul to the customer. In this case, you remove enough of the deck and backfill to expose the plumbing, replace the fittings and fixtures in the spa, then rebuild the assembly. Estimate these jobs carefully because they can be very labor intensive. I usually advise the customer that the charges will be actual time and materials, at a preapproved labor rate and parts estimate, with a probable total but no guarantee. The advantage of the rebuild is that the customer doesn't keep throwing $100 repairs into the spa for prolonged periods. One other choice is to fiberglass the leaking fitting(s) to the shell as described later.

Cracks

The other type of leak that a spa is subject to is the crack. Earth settling or shifting, kids jumping in the spa, and age can create cracks in the spa. Acrylic spas can be repaired as described previously, and fiberglass spas can be repaired with the techniques described in the hot tub section.

Equipment

The last area of spa repair is in the equipment itself. There are no special repairs to spa equipment beyond those outlined in the chapters that dealt with each component and the special equipment repairs described previously in this chapter. The only comment to be made about spa equipment is that it might need more frequent repair than comparable pool equipment, because spas tend to get more use and abuse than pools. Similarly, because of the relatively small amount of water and the high number of bathers, spa equipment might need more frequent cleaning, especially filters.

Cleaning and maintenance

Spas require cleaning, chemical balance, and periodic maintenance like any other body of water. There are a few concerns that are unique to spas.

Draining and acid washing

Because of the concentration of use and the small amount of water, it is often more effective to drain a spa and clean it than to spend countless hours trying to rebalance the chemistry or clean the shell with the water in it. Of course, I don't advocate draining the spa with every service call, but you need to evaluate whether a standard cleaning makes more sense than draining.

When the spa is drained, pay special attention to the condition of your submersible pump. Since there is less water than in a pool and it is more contained, if the pump has a short, you can be seriously injured by electric shock. In a pool, the electricity dissipates throughout the large volume of water. I have worked in a spa with my hands or bare feet submerged and the pump operating and felt that unique tingling that says the pump cable is leaking electricity. Be careful.

Also, you can acid wash most fiberglass or acrylic surfaces, but remember that acid has no effect on organic material and oils. Still, most plastics will benefit from a wash of diluted acid (about 20 percent) and tile soap and a good scrubbing. More than with pools, you must take care to remove all traces of soap before refilling. Even small amounts of soap will foam up when heated and agitated by jets and blowers.

Draining is also appropriate when the water chemistry is so far out of balance that it will take relatively large amounts of chemicals and frequent return trips to restore the chemical balance. Because of heavy bather loads, particularly in commercial spas, total dissolved solids are frequently high and fresh water is the only answer to restore balance and clarity.

Chemistry

The only real difference between pool and spa chemistry is that it is much easier to make mistakes with the smaller volume of water in the spa. Moreover, plastic spas have no alkaline material to counteract acidic water, which is more often caused by body acids and urine than a technician adding too much muriatic acid. The results are more devastating than acidic water in a pool because the heat of the spa water accelerates the etching of metals from the interior surfaces of the equipment. If the spa is not plastered, there is no staining that can't be easily removed, but damage will be done to the equipment. In short, take extra care balancing spa chemistry, especially pH, and don't be afraid to drain the spa or shut it down from use until the water has been rebalanced.

Temperature

Most health departments recommend spa temperatures not exceed 104°F. Every commercial spa I have serviced has come with at least one resident who wanted the spa hotter and others who wanted it cooler. Whatever management decides in those instances, be sure to lock the equipment area or fit the heater controls with a locked cover.

If someone is injured in a spa, the technician will be held at least partially responsible. This is particularly true with temperature. You are responsible for keeping the temperature reasonable for the users, so pay attention to the temperature each time you service the spa, making a note somewhere for your own protection.

The other extreme is the spa that won't stay hot enough, especially in cool or windy weather. You can try the obvious raising of the thermostat and longer circulation time. Failing that, make sure the spa is covered and that all exposed pipes are wrapped.

HOT TUBS

Wooden hot tubs enjoyed great popularity in the 1970s and then largely disappeared during the 1980s. In recent years they have made a comeback as people seek simple, natural surroundings and as they discover the economy of wooden spas. Figure 11-8 shows a cross section of a typical wooden hot tub. Made from any dense hardwood that is resistant to moisture and decomposition (usually redwood), hot tubs are assembled from dozens of vertical boards called *staves*. The edges of each stave, which look like a standard 2-by-4-inch plank, are beveled (angled) slightly so that when they are assembled edge to edge, the result creates a circle. The bottom of each stave is carved with a groove, called the *croze*, to accommodate the width of the floorboards.

The floor is composed of similar horizontal boards, but with straight edges. One edge of each floorboard has two or more holes and the other edge corresponding pegs. This aids in assembly, keeping the floorboards properly aligned, although the pegs provide no actual structural support. The ends of each floorboard are curved slightly and the center floorboard is the longest, with each succeeding floorboard on either side being slightly shorter so the ultimate result is a circular floor. The floor is supported by 4-by-4-inch joists. The tub does not actually rest on the staves, but rather the floor of the tub rests on the joists.

The tub is assembled in the ancient tradition of cooperage (barrel making), with the staves being held together by curved steel bands that are tightened by passing each end through a hoop lug (a hub connector fitting) and secured by nuts screwed down on each threaded end. The tub is made watertight when water is added,

11-8 Cutaway view of a wood hot tub.

swelling the wood and sealing the seams between the staves. Wood seats are added by fastening bench supports with stainless steel screws.

Wood is specially selected and prepared for hot tub construction. Smooth, even, vertical grain is preferred, the best being taken from the heart of the tree where the grain is best and the most tannin is located. Tannic acid makes the wood more impervious to the decomposition effects of heat, water, and chemicals. The wood is then kiln dried so it is devoid of almost all moisture. Thus, when the finished tub is complete, the wood quickly absorbs moisture and swells for tighter seam sealing.

The woods used for hot tubs differ in various parts of the world depending on their availability and price. It is generally accepted that California redwood is the best hot tub material, but teak, cedar, mahogany, certain oaks, and several exotic rainforest hardwoods are also used.

Hot tubs are made in various sizes—the most common unit is 6 feet in diameter by 4 feet deep. I have assembled and installed tubs up to 10 feet in diameter by 4 feet deep with 20 separate jets.

One advantage of using wood for a spa is its natural insulation property. Simplicity and blending with the environment of modern gardens and homes is the other. The disadvantages are that it is not very comfortable (you can't mold lounges in a hot tub as you do in a plastic spa), the wood maintenance requirements, leaking problems, and bacterial questions. Although a properly maintained wooden hot tub is as clean and safe as a gunite or plastic spa, many health and safety codes prohibit wood for commercial or public use because bacteria can be held in organic materials like wood and transmitted to others.

Installation

Each manufacturer provides assembly instructions with their precut tub components. The manufacturer I use will assemble the tub for me for an extra $100; well worth it to avoid the job myself. Unfortunately most tubs are too big and heavy to carry through the narrow passageways alongside most homes, so assembly in the homeowner's backyard is usually required.

Installation of the hot tub proceeds along the same lines and requires similar equipment and plumbing as a plastic spa. The following guidelines detail the few procedures that are unique to hot tubs.

1. Foundation. After working with the homeowner and/or general contractor to determine the best location for the hot tub, prepare a stable foundation. Hot tub wood needs to breathe, to be surrounded by air. I have been called to installations that were backfilled with sand or dirt, usually for leak problems, and have always discovered extensive wood rot. The tub requires lateral support (side to side), and the most important aspect of the installation is the foundation. Because the weight of the tub, water, and bathers is enormous, do not mount a hot tub on a patio or wooden deck that has not been reinforced.

   If you are not using a general contractor, prepare your own foundation. Prepare a level, well-compacted area of ground. Create a form for the concrete using 2-by-4-inch lumber as shown in Fig. 11-9. Excavate the area inside the form to a depth of at least 2 inches. The completed foundation sits in the ground, rather than on top of it, and will be resistant to shifting or earth movement. Lay ½-inch rebar in a crossed pattern as shown, creating 1-foot squares. Tie the intersections with tie wire to secure the grid. Slide small blocks of wood or brick under the resulting grid to elevate it about 2 inches off of the ground. Prepare a concrete mix, available at the hardware store, and mix as directed with sand. Pour the form full of concrete, forcing the mixture under and around the rebar grid. Make a screed of a 2-by-4 and level the concrete as shown. Allow the foundation to cure for at least four days before assembling the tub on it.

   An alternate foundation can be made of precast concrete piers, available at masonry supply yards. Set the piers 18 to 24 inches apart in lines so that the tub's joists will rest on the piers. Pier foundations can be used on any solid ground that is not subject to erosion or further compaction.

2. Plumbing (Fig. 11-10). Because air must be left around the tub, you can usually set it in place before plumbing it and still have enough room to work. If working in tight quarters, preplumb the tub, as described in the section on spa plumbing. In either case, work with the homeowner to determine the location and height of the jets. Drill holes in the wood taking care not to cut the retainer bands on the outside of the tub. Jet fittings are installed as described previously, but be sure to use jet throats designed for the thicker walls of a wooden tub (the throat is longer; the jet body is the same). Plumb the spa as outlined previously.

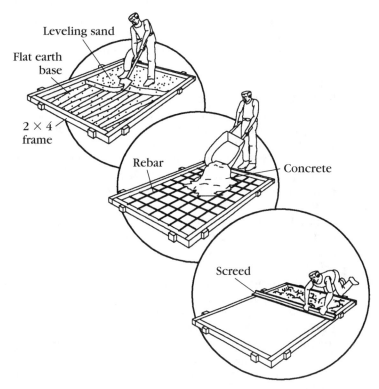

Leveling sand

Flat earth
base

2 × 4
frame

Rebar

Concrete

Screed

11-9 Building a concrete foundation.

Jet

Air control

Floor drain

Air blower
ring

From air blower

Return line

Suction line

11-10 Hot tub plumbing.

3. Air bubble rings. If an air bubble ring is desired, you must create it from flex PVC. Install an additional main drain-type fitting near the bottom of the tub. The inside of the fitting is threaded for 1½-inch plumbing. The supply house will have a threaded T fitting that screws into this extra drain fitting, stubbing out to the left and right with ribbed nipples. Lay a circle of 1½-inch flex PVC around the edges of the floor of the tub. The ends of the pipe will fit over the ribbed nipples, creating a ring. Using the guide in the blower section, drill the number and size of holes into the top of the ring. The pipe from the blower to the tub can be glued into the back side of the drain fitting. When air is forced into the ring, the fitting will hold down one side but the other will float. Secure the ring to the floor with a stainless steel hook and a strap made from plastic tubing or some other nonmetallic material. Some air rings in hot tubs are installed under the seats, toward the front edge (the edge at the middle of the tub). Again, the fitting will hold one side but the ring needs to be secured to the bottom of the seat in at least one other place.

4. Additional fittings. Most hot tubs do not use a skimmer, however if that is desired, install it as described previously. The other plumbing connections to the equipment will also be the same as with plastic spas. Since hot tubs require frequent draining, many technicians drill a hole in the floor of the tub and install a drain that leads through a 1-inch pipe to a hose bib outside the tub. This allows you to connect a garden hose so you can direct the discharge from the tub away from its base.

Figure 11-10 shows an air blower line that serves an air ring. As with plastic spas, it can be plumbed with a three-port valve to alternate between the ring and the jets or a second blower can be installed. Figure 11-10 shows jets that are aerated from the atmosphere, but as described previously, the hot tub can be fitted with jets that are charged from a blower.

Break-in

A plastic spa can be filled with water and used immediately, but a wooden hot tub requires break-in. When you have completed the installation, fill the tub and keep the water running. Water will run through the seams for as long as 24 hours while the wood absorbs water and swells, sealing the seams. If you only fill the tub partially, the seams will close only up to that level, so keeping the tub full is important. Allowing the water to overflow will also moisten the outside of the wood, speeding absorption and swelling. Some tubs will seal in as little as three or four hours. In any case, keep the water supply flowing until the tub is no longer leaking.

As soon as there is enough water holding in the tub to start circulation, begin to run the pump. Turn the heater on as well, because that will speed the swelling of the wood and the leaching portion of the break-in. For several days, tannic acid and natural wood oils will leach from the timbers, turning the water coffee-colored. The best wood is taken from the heart of the tree where these acids and oils are most abundant, so the time it takes to leach them from the wood is an indication of the quality of the timber in the tub. I have installed some hot tubs that barely discolored the water, indicating the wood contained little protective tannic acid or oils.

Continue to run the circulation 24 hours per day for two days. Brush the inside of the tub vigorously once or twice each day to scrub out as much of the acid and oil as possible. There is no way to filter or chemically remove this discoloration, so drain and refill the tub at the end of the second day. It might take two or three such refills before all the discoloration disappears from the tub. Some technicians will tell you various methods to speed this process, including bleaching and soda ash preparations, but believe me, I have tried them all and the only thing that works is patience.

Although customers are advised not to use the tub during the break-in, some have done so anyway without suffering any ill effects from the discolored water. It doesn't stain and apparently has no detrimental health effect on the skin, but I still ask the customer to stay out of the tub until the water is cleared. Each time you drain the tub, clean the filter and drain as much water from the circulation system as possible to remove all the affected water. When the water is clear, add sanitizer and operate the tub normally.

If leaks persist in a particular seam or part of a seam, take note of the rate of the leak. If it seems to be leaking a little less each day, it might be an imperfection in the wood and it will seal in a few days. If the rate of the leak continues unabated, follow the leak repair procedures outlined in the next section.

Repair

Hot tubs are more prone to leaks than other types of vessels, simply because there are so many more places for leaks to develop. In addition to plumbing and through-wall fittings, hot tubs have dozens of vertical seams and the entire length of the croze seam where leaks can develop. In extreme cases, wooden tubs can also rot through the entire thickness of the wood itself.

Leaks in stave seams

If leaks develop between the staves, it could mean that the bands need to be tightened. Rarely does a seam leak along its entire vertical length, rather a section of several inches to 1 foot will leak along one seam. Tighten the band closest to the level of the leak. If the wood has dried out, fill the tub and allow the wood to reexpand.

If tightening the bands is unsuccessful, the seams can be caulked. Some technicians dry out the wood and apply silicone sealant, others use sealants designed for use on wet surfaces. These measures will work temporarily, but the wood and sealant expand and contract at different rates. This means that in a short time, the repaired area will leak again.

For centuries, builders of wooden ships have caulked seams with lamb's wool. Because it is organic, it will expand and contract very similarly to wood. Wool contains natural resins and oils, and it will remain impervious to decay for several years. Some technicians recommend twine or other organic materials, but lamb's wool will last the longest.

The best source of lamb's wool is in the local grocery store. Mop heads are often made from lamb's wool (be sure you are not buying a synthetic imitation) and provide strands of wool about the right thickness when unbraided. Knitting yarn of pure wool will also work. Repair is a simple matter of pulling out strands of wool from your source material that are thicker than the seam being sealed and forcing them into the seam with a putty knife. Work from the inside of the tub with the water drained below the work area to make access easier. Don't allow the wood to dry out however, because that might create even more leaks when the wood reswells.

When you have completed the caulking, refill the tub and heat the water. It might take a day for the wool to swell and seal the seam. During this process, water leaks out, pulling the wool deeper into the seam. If after two or three days the seam is still leaking, try the repair again.

Another method of sealing seam leaks is to add sawdust to the water, hoping that as the water leaks out, the seam will plug up with wet sawdust. This method only works one time out of ten and is a very temporary fix. I have never heard of such repairs lasting for more than a few months. A similar concept is a product called Tub O'Gold, which is a chemical compound that is added to the water and allowed to leak through the seams. The compound is supposed to clog the leak, filling it with a white paste-like material. Again, I have found this product to work once in awhile, but the results are not long-term. Moreover, this product requires extensive brushing, vacuuming, and filter cleaning to remove the excess dough that is created, which never seems to completely clear from the tub. After many experiences with such quick or easy fixes, I would recommend disassembly and reassembly—realignment of the staves—as a way to repair leaks before I would mess with sawdust or Gold.

Leaks in the croze

The seams between the floorboards rarely leak, although if they do you can caulk them as described previously. Leaks are more likely to be found in the croze. Leaking in the croze can be caused by pressure or wood decomposition. Figure 11-11 shows how a deck can apply pressure to the top of the staves, kicking them out-

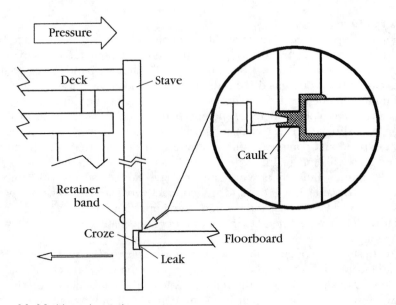

11-11 Hot tub repair.

ward at the base near the croze, and causing separation. The solution is to relieve the pressure and tap the base of the staves back in with a rubber mallet. A similar pressure leak can be caused if a band is tightened too much just above the level of the croze. In each of these cases, loosen and reposition the lowest band on the tub so that it wraps around the floor and croze joint. Tighten the band until the staves are replaced securely to the floor.

Pressure leaks are also caused if the bottom of the staves are resting on the deck, supporting the weight of the tub. Remember, the joists are designed to support the weight of the tub through the floor, not the staves. If the tub has settled for some reason, you can raise it and reposition the joists or piers by emptying the tub and using a car jack to raise it off of the deck. To raise it, apply the jack to a band, not a stave, or you will create more pressure damage. Use extreme caution when jacking a hot tub off the ground. The forces are extremely great and the jack head is in contact with a small band of metal (which can snap if it is rusted or cracked), and if the tub shifts suddenly the jack can snap out of place with deadly force. I have jacked many tubs to examine the underside or reset the supports, and have always succeeded by working slowly and with helpers to steady the tub. You also don't want to take any chances when the tub is raised on a jack by reaching under the tub. Use boards or a telepole to push materials out from under the tub or maneuver new support pieces into place. A $20 telepole is easier to replace than an arm.

If the wood has deteriorated, the croze can be caulked with silicone sealant. Because it is virtually impossible to insert organic materials into the croze, this is the only time when using sealant is recommended. As shown in Fig. 11-11, drill a hole in the stave directly behind the floorboard in the vicinity of the leak. Fill the voids with caulk. You might need to drill holes in several staves to stop all the leaks. The hole should be no larger than necessary to accommodate the tip of the caulk tube. You can plug the hole with a short length of wooden dowel of the same diameter as the hole or with more caulk.

Leaks in fittings

As with other types of spas, hot tubs can leak around jet, drain, or skimmer fittings. Wood is more likely to leak in these areas because the wood expands and contracts with temperature extremes, while the plastic does not.

Always try to tighten fittings first. A loose fitting is usually the source of the leak. If that fails, follow the techniques already described. As wood ages and deteriorates, the hole around the fitting might enlarge or rot. In this case you might want to fiberglass (as described later) over the face of the fitting where it meets the wood.

Leaks from wood decomposition

As the tub ages, the wood will begin to decompose. This can be accelerated by poor water chemistry maintenance, but the process is inevitable.

I have added another two or three years to the life of a decaying wooden hot tub by fiberglassing the interior. The entire interior of the tub can be glassed, or you can do just the croze joint or specific staves that are rotting or leaking. The steps are as follows.

1. Drain the tub. Allow it to dry for several days so that there is no surface moisture. If the wood is still moist deep inside, it can dry out by breathing through the exterior, because only the interior surface will be glassed.

2. Sanding. Use a rough grain sandpaper (40 grit) to strip off any decomposing wood and expose the best of the hard wood below. The roughness will also aid in creating a mechanical bond for the fiberglass. Wipe out the sanding dust completely, or wash the tub and allow it to dry again.
3. Fiberglass patch kits are sold in most pool supply houses, all boat supply stores, and in many hardware stores. Fiberglass is actually a collective term applied to two distinct parts of a process. Fiberglass itself is a cloth woven from yarns of various twist and ply construction into a variety of sheets, strips, tapes, or chopped loose fibers. The fiberglass provides the strength, but by itself is not stiff and will not adhere to a surface. Liquid resin is the other half of the fiberglass process. Resin is a clear or slightly golden liquid that is poured or brushed over the fiberglass fabric. The resin dries hard, providing the bond and stiffener, while the fabric provides the strength.

   Unless large areas of the hot tub are deteriorated and the structural integrity of an area is in doubt, fiberglass cloth is not needed for this repair. The wood takes the place of the fiberglass cloth, providing the framework onto which the resin is applied.

   Follow the product preparation directions on the patch kit.
   Generally, after sanding and cleaning the surface to be glassed you will mix a two-part resin to paint the surface to be repaired. One part of the resin mixture is resin, while the other is a catalyst that starts a chemical reaction that leads to hardening. The two cannot be premixed or the product will set up inside the can.

   After mixing the resin components, liberally brush the mixture on the area to be patched. As the mixture thickens it will adhere better to vertical surfaces, so you might want to wait a few minutes before application. In any case, most mixtures will set up within ten to fifteen minutes, so be ready to work quickly.

   If there is structural support needed, apply a thin coat of resin, then lay up fiberglass sheets or strips. Allow 24 hours drying time (or as recommended on the particular product you are using), then apply a second coat of resin. Sand lightly between coats to ensure good bonding of each coat. Always mix more resin than you think you will need to avoid running out in the middle of the job. The first coat will require about twice as much as successive coats because the bare wood will absorb a great deal of resin.

   When fiberglassing an exposed product such as a surfboard or a piece of furniture, a clear topcoat of finishing resin is also added. But because the patch is inside a hot tub and will not be seen, this clear coat is unnecessary.
4. After allowing adequate drying time for the entire project, refill the hot tub. It will likely leak at the seams that you have not glassed, but the normal swelling should close those leaks. If you have glassed the entire interior, the hot tub is now virtually a plastic spa.

Fiberglass spas can be patched and repaired using the same techniques. As with other repair methods, the best way to learn the nuances of preparation and application is to test the products on scrap wood or other materials before applying them at a job site. Working with fiberglass and resins is very easy and a great solution to otherwise impossible leak repairs.

Cleaning and maintenance

Routine service for a wooden hot tub is similar to any other body of water, especially concerning chemical balance, vacuuming, skimming, and brushing. There are, however, some important differences about maintaining wood and prolonging its life with proper service methods.

Wood decomposition

Overchlorination will strip lignin out of the wood. Lignin is the white, pulpy cellulose material that binds the organic material of the wood together. The tub will appear to be growing a white fur that will brush off and clog the filter and strainer basket with what appears to be wet newspaper.

The obvious prevention of this decomposition is to lower the doses of chlorine. Because of this problem, I don't use floaters or chlorine tablets in wooden tubs. Keep an eye on the filter and pump strainer because they will clog frequently once this decomposition has started.

The solution to this problem is to drain the tub, allow the wood to dry, and sand it down to hard, good wood. Use a coarse grit paper (40) to take off the worst of the material, then go back over the surface with a finer grit (120) to smooth and seal the grain. Refill and reswell the tub.

Chemical balance

As the previous problem indicates, chemical extremes can cause severe problems with wooden tubs. Since the wood itself is slightly acidic and the addition of bathers adds more acid, the problem with hot tubs is usually low pH requiring a regular addition of soda ash. The acidic water will not harm the wood, but remember that it will strip metals from the components of the circulation equipment. Metal stains are not visible on the wood, so there is no early warning signs of such problems. In other words, there is no substitute for regular chemical testing and balancing of hot tub water.

Since heat and high bather activity can deplete chlorine residual between service calls, I usually premeasure a small amount of liquid chlorine and ask the customer to add it in the middle of the week. I don't use granular products in hot tubs because they can sink to the bottom before they dissolve, and at full strength will begin wood decomposition. Under no circumstances should you throw a chlorine tablet into the bottom of the tub.

Algae and wood

If you discover an algae bloom in a hot tub, superchlorinate the water and brush all of the affected surfaces. This process forces sanitizer into the grain of the wood where the algae is growing. Drain the tub and rinse with fresh water to remove excess chlorine, brushing and rinsing thoroughly. Refill the tub and treat the water normally. If you try to treat an algae bloom in a wooden tub without draining it, you will either not kill all the algae or the excess chlorine will begin wood decomposition.

WATER FEATURES

Fountains, koi (large goldfish) ponds, waterfalls, wading pools, and other decorative systems are grouped together under the collective term water features. Water features are essentially small swimming pools, requiring circulation, filtration, and chemical balance like any other water vessel. The differences between water features and swimming pools are based on the intended use of the water feature.

Fountains

Fountains are usually powered by small submersible pump and motor combinations and are circulated without filtration. The only way to clean a fountain is to remove as much debris as possible with a leaf rake and spa vacuum or to drain the vessel completely, scrub it out, and refill it.

Larger fountains or waterfalls are powered by larger pool-style pump and motor systems and are filtered. They include skimmers for vacuuming and often include water fill lines with automatic fillers. Since most fountains and waterfalls are shallow with extensive water surface area, evaporation is a problem. First, keeping enough water circulating is important. Second, the high rate of evaporation can leave extensive scale deposits on decorative tile or rock at the waterline.

The other hazard of fountain maintenance is wind. Not only will wind accelerate evaporation, it can blow water out of the fountain. Again, the answer is to keep the water supply adequate to prevent the pump from running dry. In extreme conditions, shut down the unit.

Building fountains and waterfalls is a specialty that has been elevated to new levels in recent years with artificial rock and unique lighting. Designers are realizing that water provides a unique living enhancement to indoor and outdoor settings and the limits are only the imagination. Shopping centers and office buildings, as well as numerous Las Vegas hotels, are among the many new users of waterfalls and fountains in architectural design (Fig. 11-12). Because of the unique engineering requirements of each design, however, there are no simple or standard outlines for designing and building waterfalls or fountains. If your customer wants to add a water feature to a yard or existing pool, I recommend an expert and work with the designer or builder to complete the job. The hydraulic requirements of lifting and moving water for fountains and waterfalls are unique and demanding.

Koi ponds

Koi, large goldfish-like carp, have become more popular in recent years, and are kept in decorative ponds in residential and commercial settings. A colleague who sold koi to a Hollywood celebrity tells me some of them cost as much as $20,000 each.

A koi pond has special needs unlike a typical pool or spa. First, the circulation must operate 24 hours a day to sufficiently oxygenate the water. Preventive maintenance and being aware of problems before they become critical are essential. Second, because chlorine will kill the fish, it cannot be used to keep the pond clean. Lastly, koi, their food, and decomposing decorative plants generate more decaying debris than a DE or cartridge filter can handle. Therefore, koi ponds use only sand or biological filters. Sand filters were described in the chapter on filters. A biological filter is composed of several layers of gravel through which the water is circulated to create a natural filtration process.

11-12 Water feature.

Any cleaning procedure that shuts down the circulation system must be performed promptly to avoid suffocating the fish. You must also avoid disturbing settled debris which will cloud the water to the point of suffocating the fish. To clean a biological filter, the fish must be caught (use your leaf rake and telepole, but be careful not to scrape or bruise the fish) and placed in a clean barrel or drum. The drum can be filled with water from the pond before draining it so that the fish are transferred to a similar environment to lessen the stress.

Drain the pond with a submersible pump and shovel the gravel of the biological filter out of the pond. Lay the gravel out on a concrete surface or put it in barrels for cleaning. Use a hose to wash and stir the gravel until the rinse water is completely clean. Do not use any cleansing chemicals. Scrub out the pond itself, again without chemicals or detergents. Replace the gravel in the biological bed over the pond suction drains. Refill the pond.

Because municipal water supplies include sanitizers that will harm or kill the fish, you must neutralize chloramines in the water before returning the fish to the pond. These can be obtained at most garden and aquarium stores. Follow the directions on the package for the amount and procedures. Carefully return the fish to the pond, with the water from the barrel if it is clean. Using as much of the old water as possible will help the fish acclimate. Restart the circulation as soon as possible.

An aquarium or garden store will be able to suggest various water additives that will help medicate the fish to prevent infection in scrapes or injuries the fish might have sustained during the procedure. They will also provide specialty algicides that are not toxic to the fish but which will keep algae to a minimum. Of course, the fish will eat much of the algae and a properly operating biological or sand filter will keep the water clear, so don't be obsessive about algae removal from koi ponds.

The biological filter not only filters particles out of the water, but it breaks down organic materials by creating an enzyme-rich bed of bacteria that hastens decomposition of debris. It takes several days for the bacteria to reproduce and up to a month for sufficient decomposition activity to keep up with the amount of excrement, excess food, and plant material found in the typical koi pond. To aid this process, the aquarium or garden store that supplies other koi food and supplies can provide a biological filter enhancer that begins and accelerates the bacteria bloom. During the first week or two after you have cleaned the pond, it might become cloudy and debris might accumulate on the bottom, after which the biological filter will catch up with the decomposition and cleaning.

You can always help the filter by skimming and vacuuming as much debris out of the pond as possible, just as you would any other pool or spa. As noted earlier, continuous circulation is also necessary because koi absorb large amounts of oxygen from the water which must constantly be replenished. Keep a sharp eye on the circulation, quickly removing blockages or cleaning dirty sand filters before the flow slows to critical levels.

Finally, don't make the same mistake I once did. I was called to the home of actor Lou Gossett, Jr., to clean the indoor pool and as I passed through the courtyard to the house I noticed an extremely dirty fountain. On my way out, I thought I would do Lou a favor and clear up the water in the fountain for him, so I poured a gallon of liquid chlorine in the murky water as the start of a cleanup. Unfortunately, after I had poured most of the chlorine I noticed little mouths sucking air through the surface of the murky water. It turns out that deep in the brown water were some very expensive koi. I quickly tried to rescue them, clean out the fountain, and replace the water with something less toxic, but unfortunately my actions were too late. The moral of this story has value in all aspects of the water industry. Look before you pour.

Wading pools

A word about wading pools. Shallow pools, usually designed for children, are sometimes individual units with their own equipment or are combined with a pool and spa. Many designs today have a wading pool, spa, and pool on different levels with water circulating through one to the other. Like attached spas, three-port valves, often automated, are built into the equipment so that with the touch of a button or the twist of a knob the system will heat and filter the wading pool only.

There are no special requirements for wading pool maintenance. Because the wading pool is usually used by children, extra care should be given to provide a clean environment and balanced chemicals. Excess chlorine can irritate adult skin, but can be particularly harmful to young skin and eyes. Repairs usually involve extra cleaning (kids throw a lot of things in pools) and leaking through three-port valves as described previously.

WATERFALLS AND THE VANISHING EDGE

Finally, a few words about waterfalls. Adding a fall to an existing pool requires a decision to add the fountain head to the circulation system, treating it essentially like any other return line, or adding a separate suction line, pump, and return to the waterfall. Many technicians are intimidated by the engineering required for installing a fountain or waterfall to a pool or water feature and miss out on valuable jobs. The truth is that the manufacturer of the display has done all this work for us!

Manufacturers of each fountain head, spray design, or waterfall weir provide data sheets explaining flow requirements, resistance (dynamic head expressed in feet as previously described in this book), plumbing size requirements, and spray characteristics (such as diameter or distance of fall curve). Of course, flow and head change as the height, volume, and diameter of the discharge changes.

Actual installation of a fountain head or spray display is simple plumbing as described in other chapters, with the addition of a gate or ball valve to regulate the flow, thereby adjusting the height or diameter of the discharge. The actual fountain head can be plumbed onto standard plumbing and fittings, distributed directly off of a return line of the pool.

In the case of a dedicated system, with its own plumbing and a pump separate from the circulation system, the plumbing process requires a separate suction and return line added where it is most practical or aesthetic. The only additional concerns with a separate system are sizing the plumbing and pump.

Once again, the answer lies in the specification sheet provided by each manufacturer. For waterfalls, a chart details the required flow rate (expressed in gallons per minute) for every foot of desired projection. If, for example, you wanted the fall to curve out over the pool (curve projection) 2 feet before falling into the pool, you might require a flow rate of 35 gpm. A curve projection of 4 feet might require double that flow rate. The specifications also recommend piping size, usually suggested by the size of the adapter fitting on the fountain head or weir itself.

Pump sizing is then a matter of following the calculations of total dynamic head in the system and using a pump that can overcome that resistance and still deliver the desired flow rate. In this case, a slightly larger pump and related plumbing is better, since you can always reduce the flow with a valve, but it is impossible to increase the flow once you have completed the installation.

Not all wholesale pool-supply warehouses stock fountain heads or waterfall weirs, because the variety and sizes are endless. The warehouse staff should be able to recommend a special supply house or manufacturer that does. These supply houses and manufacturers can also supply suggestions and technical advice. You can request copies of the manufacturer's specification charts in advance to plan the work and discuss it with your customer.

One unique use of falling water is the vanishing edge (Fig. 11-13). Also called the negative edge, this waterfall employs one or more sides where the top of the wall is slightly below the water level of the rest of the pool. The water then evenly cascades over this "low wall", pouring into a catch basin. From the basin, it is pumped to the equipment and returned to the pool in a normal fashion.

Vanishing edges are a feature that must be designed into a pool in the first place or added during a major remodeling. The engineering is demanding. You must ensure that the basin can hold enough water to allow return to the pool before the

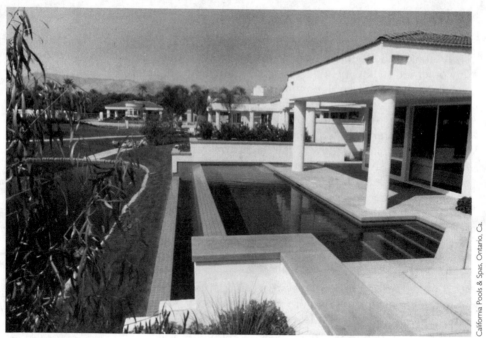

11-13 Vanishing edge pool and catchment basin.

pool becomes empty, without overflowing when the pump is turned off. For the technician, the vanishing edge is the ultimate skimmer, since the overflow acts as a natural removal system for leaves and dirt. The basin must be cleared of leaves or other debris regularly to avoid clogging, much like a huge skimmer. The basin must also be vacuumed, or dirt can be brushed toward the suction drain, where it will be taken to the filter. The suction drain is normally designed to accommodate a vacuum hose and the basin itself is shallow enough to reach in, remove the screen or drain cover, attach your hose and vacuum, just like the skimmer.

12
Commercial pools

The major difference between residential and commercial pools is that equipment, design, and maintenance standards are recommendations for residential pools but are mandatory for commercial installations. A commercial pool or spa might be defined as one that is used by the public, even if the users are members of a private club or a residential establishment. In fact, many building codes and health departments mandate rules and standards for residential pools and spas as well, but such regulation is impractical to enforce. Rules about minimum chemistry standards, signage, covers, safety equipment, and turnover rates are becoming more common in an increasing number of state and local jurisdictions.

Commercial pools and spas are generally dealt with in a fairly serious manner. Health inspectors visit the pool at least once a year in most areas and more frequently if they uncover problems. Because of liability issues, health inspectors are not shy about closing a commercial pool until violations are corrected and imposing fines if they feel compliance is not forthcoming. As a professional technician, you can lose your license (or be prevented from obtaining one) if you repeatedly fail to maintain a commercial pool or spa properly.

These factors should be considered before accepting a maintenance contract or repair at a commercial installation. You need to obtain a copy of all applicable building and health codes for your area to know what standards you must meet in your work. The apartment manager or property owner will probably not be familiar with the regulations, which are constantly being updated and changed, so it is your professional responsibility to know the rules and comply. In the event of an injury or other problem related to the pool, you will be held liable along with the property manager, so it is your job to understand the rules and advise them accordingly.

Since commercial work requires a great deal of time and attention to detail, be sure to include those factors in your job estimates. Where you might be able to provide satisfactory service to a residential customer with one or two calls per week, a commercial pool or spa might require daily visits. Especially with commercial spas, you might find that heavy bather loads create exceptional demands on the equipment, chemicals, and your time.

TYPES OF COMMERCIAL POOLS

At small apartment buildings or health clubs, the commercial installation might be a typical residential pool or spa being used by members of the public. In this case, the equipment and operational requirements will be familiar to you. The only differences in such cases will be added attention to mandated signage and some specialized equipment such as extra pressure gauges or sight glasses (detailed later).

At larger multifamily dwellings, schools, or public installations, the pool and spa might be ten times the size and volume of its residential counterpart. Competitive swimming pools, for example, are usually 50 meters in length (164.05 feet) by 23 meters (75 feet) in width. The depth is 4 to 6 feet for swimming, 6 feet for water polo, and 12 feet minimum for diving competition. Therefore, a typical Olympic-style competition pool might contain over 500,000 gallons of water; about 25 times the volume of a typical 18-by-30-foot residential pool. Commercial spas and wading pools are designed in similar multiples.

Large public pools have a skimming gutter that runs the length of the pool on each side. The surface water flows over the edge of the gutter at all times, skimming surface oils and debris, and drains into suction lines in the bottom of the gutter. The gutter is no different than the residential pool skimmer, it is simply larger.

Another type of commercial pool gaining in popularity is the free-form pool with multiple rockscapes and waterfalls. Because of the decoration and landscaping, some of these pools might seem larger than they actually are, and volume calculations become important and challenging. As with free-form or kidney-shaped residential pools, the solution to tricky volume calculations is to diagram the pool and divide it into sections, estimating the volume for each and adding the results together for a total.

VOLUME CALCULATIONS

Other than complex free-form volume calculations that actually become rather simple if the pool is divided into sections, the volume calculations for commercial pools is not different from residential pools. The methods described in chapter 1 work just as well for larger pools.

There are two important differences between calculating volumes of small pools and large ones. First, the accuracy of the estimate becomes even more critical with a commercial pool, because liability and public health are involved. Because you are dealing with large numbers, even a small percentage error results in mistakes of thousands of gallons and therefore significant errors in chemical applications, turnover rates, and other maintenance calculations.

Slope calculations

The second important difference in dealing with commercial volume calculations is that errors in estimating the slopes of the sides and floor can result in volume estimates that are off by 20 percent. Remember, 20 percent of a 500,000-gallon pool is 100,000 gallons, or about five times the entire size of a typical residential pool.

As noted, the basic formulas for volume calculations remain the same. Figure 12-1 depicts the profile of a typical public pool. Calculate the volume of each sec-

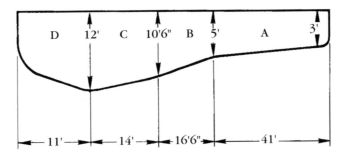

12-1 Volume calculations in commercial pools. Sta-Rite Industries, Inc., Delevan, Wisc.

tion (A through D) and add them together. The additional step in commercial calculations is to adjust the estimate for the slope of the pool walls to the floor by deducting a percentage of the total.

Figure 12-2 shows a chart for a typical public pool slope. This chart will work for most installations, but it is best to determine the exact calculation by consulting the pool plans or builder. In this example, let's assume that the pool in Fig. 12-1 measured 450,000 gallons total. By reading across the bottom portion of the chart in Fig. 12-2, you will find a line for 450,000 gallons. Follow the line up to the point where the curve intersects it. Now look horizontally to the scale on the left side of the chart to determine the percentage of reduction. In this example, the number on the left side of the chart is 2, meaning 2 percent. You would therefore lower your volume calculation for this pool by 2 percent to arrive at an accurate estimate: $450,000 \times 0.02 = 9000$ gallons. The actual total volume therefore is about 441,000 gallons. As you can see, a difference of 9000 gallons would make a significant difference in calculating the amount of chemicals needed to balance the water chemistry.

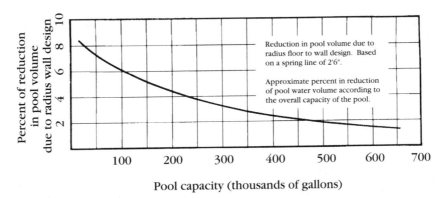

12-2 Slope calculations. Sta-Rite Industries, Inc., Delevan, Wisc.

Bather loads

Signage around commercial pools and spas requires maximum bather loads. Usually, the technician is the one who needs to fill in the blank on such signs. Although specifics might vary in your location, the following general rules will serve if local codes do not specify.

- One bather per 15 square feet of surface area in portions of the pool that are 5 feet deep or less.
- One bather per 20 square feet of surface area in portions of the pool that are more than 5 feet deep.
- Subtract 300 square feet from the total surface calculation for every diving board.
- For example, the shallow end of a pool (less than 5 feet deep) is 40 feet × 20 feet = 800 square feet; 800 ÷ 15 = 53 bathers. The deep end (more than 5 feet deep) is 40 feet × 80 feet = 3200 square feet. There is one diving board, so you deduct 300 square feet, arriving at 2900 square feet; 2900 ÷ 20 =145 bathers. The total occupancy of this pool is 53 + 145 = 198.

Bather displacement

In a public pool that might have several hundred bathers at one time, how much water is displaced by the volume of people? This question will determine flow rates of replacement water, surge chamber requirements, and other technical calculations. Even in a small pool at a public location, such as a motel or apartment building, this calculation is important. If the pool is fitted with an overflow drain to waste, the bather load will displace the water out of the pool or spa. When the bathers leave, the water level might drop below the skimmer and the pump could run dry. Obviously this would be even more likely to occur in a commercial spa.

To calculate the volume of water displaced by bathers, you must also factor in the likely amount each bather is submerged. A typical bather will displace 2 cubic feet of water (an average for adults and children). But bathers in the shallow end are usually only partially submerged, while bathers in the deep end are almost completely submerged. Therefore, shallow bathers displace 75 percent of 2 cubic feet each (0.75 × 2 = 1.5 cubic feet each), while in the deep end each bather displaces 90 percent of 2 cubic feet (0.9 × 2 = 1.8 cubic feet each). Obviously this is an inexact science, because on any given day you might have more or fewer bathers in each end and there might be greater or lesser percentages of the total as children who displace less water. In any event, this calculation will help determine a likely volume.

In the example pool, 53 shallow bathers × 1.5 = 79.5 cubic feet plus 145 deep bathers × 1.8 = 261 cubic feet. The total of 79.5 + 261 = 340.5 cubic feet of water displaced. At 7.5 gallons per cubic foot, you arrive at 2554 (rounded) gallons displaced (340.5 × 7.5). The surge chamber (Fig. 12-3) needs to accommodate at least this volume of overflow.

COMMERCIAL EQUIPMENT

Most equipment for the commercial pool and spa will resemble the smaller versions used by residential installations. Pumps, air blowers, heaters, cleaning, and other equipment are often built of heavy-duty materials and increased capacities. The following notes detail equipment and supplies that are unique to commercial water technology.

Surge chamber

As noted earlier, heavy bather loads displace a great deal of water. In large commercial pools, this water is not drained off to a waste drain but flows into a holding tank called a *surge chamber* (Fig. 12-3). When bathers leave the pool and the water level drops, water from the surge tank is pumped back into the pool. Since each bather leaving the pool (or splashing) takes water out of the pool on their body, replacement water is also directed into the surge chamber, where the level of the pool is monitored. Each surge chamber is designed differently, but understanding this basic concept will make the various components understandable when you examine one.

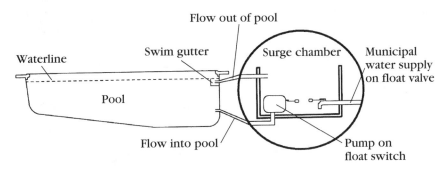

12-3 Surge chamber components.

Slurry feeder

Large pools require a great deal of filter grid square footage and require a great deal of DE. The answer is a slurry tank or feeder where DE is mixed with water (it doesn't actually dissolve, but becomes suspended in the water) and is pumped at an even rate into the filter to precoat the grids.

Some slurry feeders are simply an open concrete pit in the pool equipment room where this mixing takes place. A motorized paddle mixer (called an agitator) is used to keep the DE suspended in the water, which is then pumped into the suction flow of the pump and sent to the filter grids. Other systems use a slurry tank, from which the mixture is vacuum fed to the pump, or a pressure system where the mixture is forced into the system. Modern systems use injectors and various pumps for measuring, mixing, and moving the mixture, but the objective and components are the same on each type. As with any DE precoating, the filter must first be clean (either from a teardown or backwashing), then the appropriate amount of DE is applied, followed by normal filtration.

Filter

Filters themselves will vary for commercial applications, but fall into the same categories detailed previously. DE, cartridge, and sand filters are all made for large commercial pools and spas. When one large filter is inadequate, several filters are plumbed together in series with a manifold, as in Fig. 12-4.

12-4 Commercial filters in series with slurry feeder and series manifold plumbing. A: U.S. Government Printing Office, B: Jacuzzi, Inc., Little Rock, Ark.

Cartridge filters are also manufactured for commercial use (Fig. 12-5), but as discussed in the chapter on filters, they have not been widely adopted because of the large amount of cartridge surface area required to filter large bodies of water. Cartridges can also be difficult to clean, the pleating creating crevices in which oils and dirt can become lodged.

12-5
Commercial cartridge filter.
Harmsco, Inc., North Palm Beach, Fla.

1 Retaining wing nut
2 Stainless steel lid
3 Rim gasket
4 Stainless steel spacer disk
5 Standpipe O-ring
6 Pipe cap
7 Holding pipe
8 Standpipe
9 Pleated cartridge
10 Rod O-ring
11 Filter tank

Commercial filtration generally requires pressure gauges on the incoming and outgoing plumbing of the pump and filter. Sight glasses are often required on waste lines and the outflow plumbing of the filter. Flow meters might be required on the heater and filter plumbing as well. Make sure these are always clean, free of leaks, readable, and reflecting accurate pressures and flows. In other words, keep the filter clean and the other equipment well-maintained and there will be no problems.

Gas chlorinator

Even distribution of sanitizer is a problem in the large volumes of water found in commercial pools. Also, because bathers are likely to use the pool or spa immediately after servicing, it is important to be sure chemicals are not lingering in harmful concentrations at any point in the water.

Chlorine gas is the most efficient and least expensive sanitizer. Injecting it into the circulating water is an effective way of gradually adding chemical to a pool or

spa. The cost, availability, and dangers of working with chlorine gas generally prohibit residential users from sanitizing with it, but for commercial pools it is the preferred method.

Figure 12-6 diagrams a typical chlorine gas feeding system. The gas is supplied
in cylinders that discharge the chlorine gas through a pressure regulator, volume
control (to control the ultimate strength of the chlorine mix), and visible volume meter to a water reservoir. When the gas is injected, it mixes with the water to form a
solution that is then pumped into the circulation system for delivery to the pool or
spa. By regulating the time of operation and volume of gas delivered, the sanitizing

12-6 Chlorine gas feeder. U.S. Government Printing Office of the water can be managed. The incoming water supply might be fill water for the pool or part of the circulation water diverted for this purpose.

Modern electronic control technology is being applied to chlorine gas delivery systems, so if you plan to service a pool with a complex unit, familiarize yourself with its operation by reading the manufacturer's literature or calling them for technical support. Since chlorine gas is deadly, be sure you know what you are doing before adjusting or repairing any system. Better yet, call a factory-trained technician and watch him work before attempting repairs yourself.

Chlorine generators

Considering how expensive it is to sanitize a large pool or spa and considering how relatively inexpensive salt, electricity, and water are, some installations make their own chlorine. Chlorine generators produce chlorine gas by electrolysis. The generator consists of a plastic tank with two compartments separated by a filter-type membrane. The larger side, which includes a positively charged electrode (called the *anode*), is filled with a salt (sodium chloride) and water solution. The smaller compartment, which includes a negatively charged electrode (called a *cathode*), is filled with distilled water or municipal water (If it is not too hard). The membrane allows electricity to pass through but restricts the chemicals produced to one side.

Chlorine gas is generated in the salt solution compartment when the electricity separates the chlorine from the sodium. The gas rises above the water and some passes through the membrane, mixing in the smaller chamber with the fresh water to form sodium hydroxide (which is drained off periodically). Hydrogen gas formed in the process is vented out of the generator. Salt is added periodically to recharge the system. The chlorine gas is then injected through plastic plumbing into the circulation plumbing. The electrical current is supplied through a transformer that converts 110- or 220-volt ac into 12-volt dc. Regulating the current regulates the volume of chlorine produced. Some variations of the chlorine generator include systems that produce the chlorine right in the circulation water flow for direct application to the pool or spa water.

The commercial equipment room

All pool or spa equipment rooms should be kept clean and neat, but this is critical in the commercial equipment room. Chemicals should never be stored near equipment, where heat or water leaks could mix with the chemicals, with toxic results.

A sump pump should be installed in a pit in the floor of the equipment room. Since leaks in commercial plumbing can be very substantial, a submersible pump with an automatic float switch should be provided that will drain water from the room to a waste line. Commercial equipment is expensive and water rising from a major leak can cost much more than a sump pump.

Safety equipment

Most building codes require that the deck be clear and unobstructed all around the pool or spa, with any furniture set back at least 5 feet. This provides a margin of safety so people don't trip and fall in the pool and so that bathers can easily exit the pool or spa at any point.

Part of the technician's responsibility at commercial pools is safety. Periodically check the ladders, diving boards, slides, and rails, tightening any loose fittings. Be sure all pool rules and mandated signage are in plain view and completed with all required information (with maximum bather loads, emergency numbers, etc.).

Special safety equipment is usually required by health departments. These might include a shepherd's crook, toss ring, and first-aid kit. The toss ring must be attached to a rope that can reach to the greatest distance of the pool. The shepherd's crook (Fig. 12-7), attached to a sturdy pole of at least 16 feet, must have at least an 18-inch opening and be able to support up to 150 pounds. Many jurisdictions also require a floating safety rope across the pool at the point where the major break in the slope of the floor begins, usually about the 4- to 6-foot depth.

12-7 Shepherd's crook specifications. U.S. Government Printing Office

Before performing any remodeling procedure on a commercial pool, check the building codes for safety requirements. For example, if you are replastering the pool, the area might require contrasting colored stripes at the edge of each step leading into a pool or spa. It might require depth markers of a certain size or other general safety features (Fig. 12-8).

Another example is if you are adding equipment. If you install a diving board, be sure you know the requirements of water depth and area clearances. Many jurisdictions require a minimum of 10 feet of water depth, 15 feet of clearance above any part of the diving board surface, 10 feet clear on each side of the board, and at least

12-8 Commercial pool and spa. California Pools & Spas, Ontario, Ca.

16 feet of horizontal water surface in front of the furthest reach of the board. If you do not update the pool or spa to include these requirements, you can be held liable for accidents that might result.

OPERATIONAL REQUIREMENTS

As with any commercial installation, you must obtain a copy of the codes and regulations applicable in your area. The following are a few common standards to look for.

- For health and safety reasons, most jurisdictions require that the circulation equipment be operating whenever the pool is open to bathers.
- Service logs must be maintained and kept available in the equipment room for inspection so that a health department inspector can verify compliance with regulations and compare servicing to water testing results.
- **Always** close the pool or spa to users whenever you are performing cleaning or chemical service. Swimmers have been severely injured or killed when caught in vacuum hoses in the water, have tripped and fallen over service equipment on the deck, or have been harmed by chemical concentrations not yet distributed through the water. Believe me, it's easier to listen to a little complaining about the pool or spa being closed than to deal with an injury or lawsuit. Also, don't be intimidated by demands to use the pool when you feel the pool is unsafe for any reason. I have closed pools and spas down for days in the middle of summer when I thought there was an unsafe condition such as loose coping stones, algae blooms,

superchlorination, rusted ladders or rails, etc. You will be held responsible, so take the initiative and never hesitate to close the pool down. After servicing, look around the deck for wet areas, spills, or tools left behind.

- Many regulations require deck and pool lighting of certain specifications, whether the pool is used at night or not. It is your job to make sure that the bulbs are working and the timers are operating correctly.
- Make sure access to the heater temperature controls is locked. I have never serviced a commercial pool or spa (especially at apartment buildings or condos) where some users didn't insist that the water be warmer, while others insisted it be cooler. If you leave controls exposed, someone will turn the temperature up, burning out chemicals prematurely and (in spas) perhaps creating dangerous conditions for users. Make equally sure that time clocks and chemical supplies are locked away.
- Drains, gutter suction ports, deck skimmers, and other lines must be covered or fitted with a grate. Make sure these are always in place.
- Some jurisdictions require minimum visibility so that any injured person at the bottom will be seen. Some standards call for visibility at a certain level (such as 6 feet); others call for the main drain cover to be clearly visible. Of course, if the water is cloudy it is a sign that there is some problem and you might want to close the pool and take corrective action.
- If you are going to service many commercial accounts, take a course in first aid and cardiopulmonary resuscitation (CPR). Accidents do happen around pools and spas, so it pays to be prepared.
- The biggest mistake made in servicing commercial accounts is hasty chemical testing or using old reagents. In either case, you obtain false results and take the wrong actions with chemical applications. Since a large pool might have very different chemistry in one area than another, it pays to take your time and check several areas before making decisions about balancing. Also, be sure the water has circulated for a few hours before testing so that the results are more likely to be the same everywhere in the pool. Remember that the health inspector can close down the pool for any reason, so be cooperative, do a thorough job, and don't give him a reason to close the pool.

13
Winterizing

If you work in a cold climate, I probably don't need to tell you that a body of water needs to be prepared for the winter months. But even in warmer climates, you might discover that some of the following information is useful.

In the late 1970s I can recall a cold February morning in Ft. Lauderdale, Florida, when I found ice forming on the surface of my pools. My sister taught kindergarten and called to tell me how delighted her class was to experience snow for the first time in their young South Florida lives. Sure enough, snow was actually falling, albeit for only an hour and it melted within minutes.

Fast-forward a decade to another February in Malibu, California. I drove along Pacific Coast Highway for my service route and saw snow on the rails of the Malibu Pier. Reaching into the skimmer baskets to clean them and grasping the metal telepole was numbing. Later that morning I went to my accounts in the adjacent mountain areas and found frozen pipes, cracked pump strainer pots, and ice over the standing water. To my colleagues in truly cold climates, I apologize for making this bout of moderately cold weather sound like a New England winter, but the point here is that no matter where you are, unusual weather can create the same short-term problems that are regular seasonal conditions in other climates. In short, be prepared by understanding how cold weather might effect your pools and spas.

The most important single concern from cold temperatures is freezing. Water expands when it freezes, meaning that if it is trapped inside pipes and equipment it will expand and crack them. PVC pipes, fittings, and ABS plastic pump parts are most susceptible, but soft copper heat exchangers and galvanized fill-line pipes are not exempt either. Expanding frozen water will also crack tiles and plastic skimmers.

The second problem, which relates to the seasonal winter closure of a pool or spa, is potential damage from algae and debris. Although algae does not grow well in temperatures below 55°F, especially if the water is shielded from sunlight by a cover, prolonged periods of stagnation will permit and promote algae development.

No matter where you are, winterizing and reopening pools and spas can be among the most profitable jobs of the year. The procedures don't require difficult work and customers will be pleased to pay reasonable fees rather than pay for major repairs. Keep notes of what works in your area and specifically at each pool, making the job easier next year.

TEMPERATE CLIMATES

In unusual conditions, such as those described previously, for normally temperate climates, the best protection during winter is to keep an eye on weather forecasts. When temperatures are predicted to drop near freezing, set the circulation to run 24 hours per day, preferably with the heater on at the lowest thermostat setting. This is especially important for pools and spas in valley or mountain regions where temperatures might be slightly colder than those predicted in metropolitan areas.

I find prevention to be easier than reaction. With 200 customers, it is difficult to visit every location to turn on the circulation for 24-hour operation, and go back when the weather warms up to return the settings to normal. Instead, during winter months, I set my customer's time clocks to run two hours during the day, two hours in the late evening, and two hours in the early morning. This way, the water circulates and the heater comes on every few hours and prevents any freezing from setting in. Be sure to consult with the customer before setting the heater to come on, because they might object to the added gas expense. I usually give customers an idea of the cost of replacing equipment and digging up broken pipes versus a few dollars over the winter for the heater. I haven't had a single refusal.

Continue normal service during the winter. Wind and rain means that pools will flood, and stain-causing leaves and debris can harm the surfaces or clog the system. In short, the work in winter doesn't lessen, it simply changes slightly. You will use less chlorine to maintain the same residual and you can reduce filtration times to a total of four hours per day (staggered as described earlier). Some specialty control systems, especially for spas, include freeze protection. Temperature sensors detect freezing conditions and operate the circulation system and/or heater automatically. Of course, if the pool or spa is kept heated, it's business as usual.

COLDER CLIMATES

In climates where winter means actually wearing long pants and jackets, most pools and spas are closed for the season. The shutdown process requires some planning to avoid damage to the installation and to make reopening in the spring a simple task. The following steps will help you make these preparations for your customers and provide some options to achieve the goals regardless of your skills or the availability of specialized equipment. Aboveground plastic spas can be drained completely, although in-ground gunite and plastic spas should be treated just like pools. In either case, the plumbing and equipment must be prepared for winter.

1. Balance the chemistry. Etching or scaling conditions of water will harm the pool or spa even when the circulation is off, so before closing make sure the chemistry of the water is balanced.
2. Cleaning. Dirt and debris left in the water during long periods of stagnation will leave stains and/or be much harder to remove several months later. Therefore, thoroughly service and clean the pool before closure.
3. Algae and stain prevention. To prevent algae growth and staining during the closure, superchlorinate the water. A simple formula is to triple the normal superchlorination that you use for the particular installation. A more precise approach is to raise the residual to 30 ppm for plaster surfaces, 10 ppm for vinyl or plastic. Some technicians recommend lower residuals for vessels

that will be covered or those that will be closed for less than four months, however I would stay with these recommendations regardless. Also, add a chelating agent to prevent metals from dropping out of solution and staining the surfaces. Finally, add an algicide that will inhibit black algae growth (I prefer silver-based products).

Do not leave tablets or floaters in a pool or spa during closure. Since the water isn't circulating, the dissolving chemical isn't either, and the extreme concentrations in one area can do structural or cosmetic damage, particularly to vinyl or plastic surfaces. A vinyl pool owner told me he left a floater in his pool during one winter and found it had sunk when he reopened the pool in the spring. The concentration of chlorine on that spot not only bleached a large area of the blue vinyl, it weakened the area and caused leaks later in the summer that required extensive patching.

Make sure that you have adequately circulated the chemicals before shutting off the system for the winter. Finally, there are several winterizing kits on the market with chemicals, antifreeze, and instructions. If you are new to winterizing, try one of these the first time. Even for old hands, you might discover that a package deal is less costly than buying supplies individually, because there are often specials offered by chemical companies.

4. Shut down the equipment. Turn off the circulation equipment at the breakers and tape over them to prevent someone from turning them back on. Turn off all manual switches and time clocks and remove the trippers as an added precaution in case someone does turn the breakers back on. Pay special attention to lights, which might be on household circuits rather than wired through the equipment breakers. Some technicians don't disconnect underwater lights if the water level will remain above the fixture all winter. The light provides a little warmth to the water periodically and draws the owner's attention to the pool or spa for regular inspection. I have seen light lenses crack, however, as the extremely cold water and extremely hot light fixtures contact, so I don't recommend this practice.

5. Lower the water level. Pump out 24 to 36 inches of water, or at least enough to drop the level 18 inches below the tile and skimmer line. When water freezes and expands, it can crack the tile and plastic skimmer components, so the goal is to lower the level to a point where winter rain or snow will not raise it back up to those delicate areas. If the equipment at the installation permits, lower the level by vacuuming to the waste line on your final cleanup. This will save you time by accomplishing two goals at once.

The vessel should not be drained completely for the season, because hydrostatic pressures might cause cracks or make the entire vessel pop out of the ground. Even vinyl-lined pools will ripple when empty and will not smooth out completely when refilled. These folds in the vinyl can then stretch and weaken the material. Complete draining is not advisable unless you are also prepared to spend a great deal of time carefully smoothing and refilling in the spring.

6. Clear the lines. Perhaps the most important objective of winterizing is to

protect the plumbing from freeze damage. There are several methods, some easier than others. The method you choose might depend on your equipment, skills, and the availability of water in your area.

The most effective method of protection is to drain the pool, completely evacuating the lines. Of course, you then refill it to the level discussed in step 5. If the water needs draining anyway for chemistry reasons, as discussed in the chemistry chapter, this is an opportunity to accomplish two tasks at once. The cost or availability of water in your area might prohibit annual draining, but if you can, it is the most effective way to be sure all water has been removed from the system. Even small amounts of frozen water can crack pipes and fittings.

With the pool empty, remove the collar/nozzle fittings from the return lines and plug them with tapered expandable rubber winterizing plugs (Fig. 13-1). Be sure to use tapered plugs, because they will pop out if freezing water in the pipe expands. You're better off with a popped plug than a cracked pipe. The main drain might still have water in the bottom of the plumbing, so try to suck it out with a wet/dry shop vacuum or mop it out with a sponge. Just to be sure, pour a cup of antifreeze into the main drain before plugging it. For antifreeze, use a mixture of one part propylene glycol to two parts water. Your supply house will have propylene glycol or premixed products. Never use automotive antifreeze, which is corrosive. With all of the lines plugged, refill the pool to the level described previously.

13-1
Test plug and tapered winterizing plug.

If you are unable to drain the pool, you can blow the lines out with air (described later) and plug them underwater. Some references suggest you can plug the lines first and then use a vacuum to suck water from the lines, but it is unlikely that all of the water (or even most of it) will be removed. If you have no way to blow the lines, you might try using a wet/dry vacuum, preferably a high-powered model available at tool rental shops. After you have vacuumed as much water out of the lines as possible, pour the antifreeze solution into the pipes by pouring it into the strainer pot and allowing it to flow back toward the pool. If there is a check valve in the line, the mixture will not get any closer to the pool than the valve. If that part of the plumbing is exposed or you can access the suction lines by disassembly of the three-port valves or gate valves, pour the mixture in at those points.

The return lines can also be filled with antifreeze by starting the pump with the strainer pot lid open. With the pump running, pour the mixture into the pump. It will be distributed throughout the equipment, plumbing, and return lines. When you see antifreeze discharging from all of the return lines, turn the system off and plug the return outlets. You might want to add some food dye to the mix to make it easier to see when it is discharging to the pool. Some water will remain in the lines, so as you can see, without draining the pool, clearing the lines is guesswork at best.

The other method of clearing lines underwater is to blow them with air. You can buy or rent a tank of compressed air (useful for clearing obstructions from lines also) from a plumbing supply house or a company that provides compressed gasses (look in the phone book under Gas). Attach a reinforced rubber line with a garden hose attachment on the end and add your blow bag to the end. Insert the blow bag into the suction side of the pump (be sure to get on the pool side of any check valve and remember to open any gate valves) and force compressed air through the lines. Go to the pool while the air is bubbling out of the lines and plug them. The lines will be filled mostly with air, but some water remaining in the lines is inevitable with this method. Do the same on the return lines, inserting the blow bag at the heater outlet pipe or disassembling a valve on the return side of the equipment system.

The blow-out method requires diving on the main drain to plug it. If you are unable to do that, you can try to flood the area with antifreeze by using the acid spotter described previously. You can also insert a long 2-inch PVC pipe into the pool, placing one end on top of the main drain. Pour the antifreeze mixture into the pipe. Since the mixture is heavier than water, it will sink to the bottom and flow into the main drain plumbing.

7. Remove equipment. Any equipment in the pool or circulation system that might be damaged by exposure to the elements (or which might be stolen) should be removed and stored. Especially at commercial pools and spas, remove ladders, rails, safety equipment, deck furniture, and anything else that might be subject to weathering over the winter or which might be stolen.

The circulation equipment should be disassembled and important components stored. If the pump and motor is plumbed with unions, you can easily remove the entire unit and disconnect the electrical connections. If there are no unions, you might want to cut the plumbing and add unions when you reinstall the unit in the spring to make the process easier next season. If the plumbing makes removal of the pump difficult, unbolt the wet end and motor from the volute and remove those.

The filter should be disassembled, cleaned, and drained thoroughly and the grids or cartridges put in storage. Freezing water can cause fabric deterioration. Sand filters should be cleaned and drained. Since rain might refill an open filter tank, close the tank and leave the drain plug out.

Close the gas valve to the heater and any supply valve (if it is a dedicated line for the pool heater) at the meter. The heater has drain plugs on both sides. I recommend removing the heat exchanger and burner tray and storing them after draining. Drain the water from the pressure switch tube as well.

Drain any solar panels and leave the plumbing fittings or gate valves open to the atmosphere. Even in winter the heat in a solar panel can be intense and expanding air can be as hazardous to the panels as freezing water.

Be sure the lines to the equipment are empty or are filled with antifreeze. Remove any spa jet or automatic cleaner booster pumps and store them. Be sure you have drained the automatic cleaner plumbing or have filled it with antifreeze. Usually the automatic cleaner plumbing outlet is at or near the tile line, so the water level will probably be lower than this making draining easy and plugging unnecessary. Leave all valves open and disassemble any three-port valves. Finally, if the pool has a fill line, turn off the water supply and drain the line.

8. Extra precautions. Put a little dirt or gravel in an empty plastic bottle and leave it in the skimmer. The weight will help it stay upright in the skimmer if rain or snow refill it with water. Should the water subsequently freeze, the ice will compress the bottle, not crack the skimmer. Similarly, leave some plastic milk or chlorine jugs floating on the surface of the pool. Fill them about one-quarter with water so they will grip if the water freezes and not just pop out onto the surface of the ice. Again, the goal is to create an expansion joint so that the ice will crush the jugs, not the pool or spa walls.

If there are exposed pipes that you suspect might still contain water, such as in the equipment area, under decks, or in self-contained spa cabinets, wrap them with insulation tape (a good idea anyway to reduce heat loss when operating the system). In extremely cold areas, wrap them with electric insulation tape, available at most hardware stores on a seasonal basis. These tapes actually warm the pipe with a low-level electric current.

On spas, the air blower ring must be drained. When the spa is empty, use the blower to evacuate the line. If you need to refill the spa, pour anti-freeze over the air holes so it penetrates the openings. Then dry the surface and apply duct tape over the holes before refilling the spa. When you reopen the spa, you will probably need to clean the tape adhesive off with acetone, but the extra effort is better than finding a leaking spa.

If metal ladders, rails, and light fixtures cannot be removed, protect them against corrosion with a coat of petroleum jelly. Cover the diving board.

9. Cover the pool or spa. Some cover is better than none, because it will inhibit algae growth and keep heavy debris out of the pool. Sheet vinyl covers sold for pools are very inexpensive and can be held in place with sand- or waterbags around the edge of the pool. If the pool has a mesh cover, put a sheet vinyl cover over that, because the mesh cover will not keep out sunlight and dirt.

10. Close off access to the pool. Whether it is a commercial or residential pool or spa, blocking access to a closed installation is an important safety precaution. Yellow caution tape strung around the pool, locked gates and fences, and extra signage will help keep the pool or spa from becoming an inadvertent hazard and will limit your liability. At the same time, remove any deck furniture or other loose items that might be stolen, thrown, or

blown into the pool. Store the automatic cleaner and water supply hose indoors. The hose should be laid out as straight as possible or coiled in large loops to avoid permanent bends.

11. Take this opportunity to properly dispose of any extra chemicals or test kit reagents that won't be used during the winter and will not be potent in the spring. Soda ash and acids are about the only water chemicals that will still be good after prolonged storage, but make sure they are packed in watertight containers and are stored in well-ventilated areas away from water or heat sources.

During the winter you still need to check on the pool. Snow or rain might raise the water level or sink the cover. Animals or heavy debris might fall in the pool and are better removed now than in the spring. Reopening the pool or spa is essentially the reverse of the shutdown procedure, with emphasis on balancing the water before restarting the circulation system.

14
Basic electricity

It is electricity that powers pumps and motors, ignites gas heaters, and operates controls. When major renovation or installation of electrical circuits is required, call a professional electrician and subcontract the job. When troubleshooting short circuits or other specialized electrical problems, an electrician will solve and repair it faster than you can, so again, call a professional and let him do his job.

Having said that however, the water technician is required to make basic electrical connections, troubleshoot underwater lights that won't work, switch appliances from 110 volt to 220 volt, and so on. Understanding the basic concepts of how electricity works, is controlled, and is conducted, will keep you both safe and profitable. In preceding chapters that included electrical appliances, I explained the particulars of wiring and controls. This chapter explains some of the pertinent facts of how electricity gets to the appliance, the wiring, how it is controlled, and related safety issues.

ELECTRICAL TERMS

A comprehensive glossary of pool and spa terms appears at the back of this book, but to more easily understand the concepts of electricity, a few definitions are presented here.

- *Amperage* (*amps*) is the term used to describe the actual strength of the electric current. It represents the volume of current passing through a conductor in a given time. Amps = watts ÷ volts.
- *Arc* or *arcing* is the passage of electric current between two points without benefit of a conductor. For example, when a wire with current is located near a metal object, the electricity might arc (pass) between the two.
- *Circuit* is the path through which electricity flows.
- *Conductor* is any substance that carries electric current, such as a wire, metal, or the human body.
- *Current* refers to the rate of flow between two points.
- *Cycle* is a complete turn of alternating current (ac) from negative to positive and back again.
- *Gauge* refers to the size of an electric wire. Heavier loads can be carried on

heavier gauge wires, however the numbering system of wire gauges works in reverse. A 10-gauge wire, for example, is thicker than a 14-gauge wire.

- *Line* refers to a wire conducting electricity.
- *Load* is an appliance that uses electricity.
- *Volts* is a basic unit of electric current measurement expressing the potential or pressure of the current. Volts = watts ÷ amps.
- *Watt* is a measurement of the power consumption of an appliance. One watt is equal to the volume of one amp delivered at the pressure of one volt. Watts = amps × volts.

Other terms are explained throughout the following text.

ELECTRICAL THEORY

Until I started in the water business and started working with electricity, I only knew that when I flipped a switch I wanted the light to go on. I learned some basic motor wiring and then got shocked a few times and decided to consult some books and friends who are professional electricians for more detail. A detailed description of electrical theory would take a separate volume, so this section only outlines the basic concepts that make the specific applications a bit easier to understand.

Electricity is much easier to comprehend if you picture it as water flowing through pipes. For electric current to flow from one place to another there must be a difference in pressure, called *potential*. A wire, for example, with no current flowing through it has zero potential—there is no current there yet. When the wire comes in contact with something that has electrical potential, it will accept the current until the potentials are equal in both places.

Figure 14-1 shows two tanks of water. Tank A is full and the weight of the water creates pressure at the valve. Tank B is empty, so there is no water pressure in it. If you open the valve, the pressure of tank A forces water into tank B until each tank contains the same amount of water, equalizing the water pressure in each. Electricity works the same way, flowing from the place of greater electrical charge to the place of lesser or no charge.

14-1 Electrical potential illustrated. U.S. Government Printing Office

The earth has no electrical charge, it has zero potential. Therefore, electrical panels and appliances are grounded, so that if current takes an unexpected path it is attracted to the earth. Electric service to a home, for example, includes a copper stake or rebar driven into the soil that is attached to a connection bar in the electrical panel with a heavy gauge wire. Similarly, an appliance that is grounded includes a wire attached to the appliance, run back to the electrical panel, and attached to this grounding bar. In this way, if a wire inside a motor were to touch the metal case of the motor, for example, the current would be conducted harmlessly through the ground wire to the grounding bar and into the earth.

Electricity, like water, takes the path of least resistance. Therefore, in the preceding example if you were to touch that motor case, the electricity would travel through you to the earth, unless that path were broken in some way, by rubber-soled shoes for example. More on that in a moment.

To take advantage of this natural desire of electricity to flow, we create circuits and power appliances. Figure 14-2 shows a simple circuit created with a battery and a small appliance—the light bulb. Without getting into more scientific detail than is needed, be aware that current flows from the negatively charged side of a battery to the positive, as shown. This diagram represents a *closed circuit*, because there is no interruption in the flow of electricity from the negative terminal, along the wire, through the appliance, and back along the wire to the positive terminal of the battery. If you were to cut the wire at any point and interrupt the flow of current, you would have what is called an *open circuit*.

14-2
A simple electrical circuit (closed). U.S. Government Printing Office

In a simple 110-volt circuit, current travels along a wire, called the *hot* line, from the electrical panel to the appliance (called the *load*) and back to the panel. The current is then sent through a *neutral* line combined with other such return lines to ground. As long as the wires are insulated (covered with plastic or other nonconductive material) and the appliance is not damaged, the electricity will stay in this path. As noted previously however, electricity will take the line of least resistance, so if it finds a break in the insulation or can flow to a metal casing, it will find the ground by the shortest route, usually through whoever touches that spot.

This unintentional route to the ground without first returning to the electrical panel is called a *ground fault*. As noted, if you are part of the path to the ground, you will be shocked. Another way to be shocked is to become part of the circuit. If you touch a metal casing that is electrified, then touch another conductive surface, the electricity will pass through you as if you were a wire. In other words, you become part of the circuit. This can also occur if you touch a hot line and a neutral or ground line, again becoming part of the circuit. These examples are called *short circuits*.

The practical application of these concepts is discussed more fully later; however, the following summary is helpful. To protect against ground faults and short circuits, grounding of appliances keeps you from being grounded, while bonding wires (heavy gauge wires that connect together all appliances and metal surfaces in an area) keep you from being part of the circuit.

The human body is operated by electricity—very small electrical impulses that stimulate muscles or transmit information as energy in the brain. Therefore, the body is designed to be an excellent conductor of electricity. As with any appliance, however, too much electric current delivered to the appliance can destroy it, especially the heart and brain.

An electric current of 30 milliamps (1 milliamp = 0.001 amp) will cause muscles to contract uncontrollably, meaning if you grab a hot line of more than that you probably won't be able to let go. In a short time, the current disrupts the normal heartbeat and breathing, causing death. Children can be killed in this manner with as little as 10 milliamps. To put that in perspective, a 40-watt lightbulb at 110 volts uses about 0.3 amp, or ten times the 30 milliamps that can kill you. Taking that further, a typical pool motor drawing 10 amps is using enough current to kill you 300 times.

As noted above, conductors are any substance that allows the free movement of electric current. *Insulators*, on the other hand, are substances that do not conduct. Examples of each are

Conductors	Insulators
Silver	Dry air
Copper	Glass
Aluminum	Rubber
Brass or bronze	Plastic
Iron or steel	Ceramic

Every conductor offers some resistance to the movement of electric current, like friction inside a water pipe. Some conductors conduct better than others because they offer less resistance. Although I won't refer much to these variations in resistance, it is important to know the concept and terminology. *Ohms* are the units used in measuring resistance. The better the conductor, the lower the ohms reading. The shorter the length of a conductor, the lower the ohms.

Before leaving general electrical theory and applying it to the world of pool and spa appliances, I need to briefly describe the type of current used. The battery in Fig. 14-2 provides direct current (dc). Touch the wire to it and it delivers its rated voltage without question. A battery is designed to deliver a certain voltage at all times until it is exhausted. For example, a radio battery of 9 volts will always deliver 9 volts. If the appliance uses very little power, it might draw that 9 volts slowly, say at 1 amp per hour. Another appliance might draw the 9 volts at 2 amps per hour, meaning the battery will be exhausted twice as fast. In these dc battery circuits, all current flows in one direction, from the negative side of the battery toward the positive. Therefore, the polarity of the appliance and battery must agree—the positive terminal of the battery must be connected to the positive side of the appliance. You see an example of this in the fact that batteries can only be inserted into a radio in one direction for the radio to function.

Alternating current (ac) travels in one direction then the other (alternating), so the appliance does not have to be connected to the power source in any special order. Unlike dc voltage, ac can be stepped up or down with a transformer, permitting the transmission of high voltage along municipal power lines that is transformed to lower voltages at each home or business. Because of this inherent versatility, ac is used in virtually all residential and commercial applications.

Alternating current is delivered to the home for consumption by appliances designed to accept it at either 110 volts or 220 volts (there are larger voltages in heavy-duty commercial applications, but those are best left to the electricians). Both designations are averages, since current supply varies slightly and operates most appliances in a range of 108 to 127 volts and 215 to 250 volts. Thus, you will sometimes see voltages expressed for appliances as 110, 115, 120 or 220, 230, 240.

Alternating current is also delivered at a certain rate. As noted, the alternating of the current one way, then the other, creates one complete cycle each time it reverses direction. The speed of that reversal can be controlled and makes a difference to appliances such as CD players or tape recorders that depend on a certain rate. In the United States, power is delivered at 60 cycles per second (60 hertz). In Europe and much of the rest of the world, it is delivered at 50 hertz. That is why you can take a voltage converter on vacation to step the voltage down from 220 to 110, but you can't operate appliances that require a certain cycle timing.

ELECTRICAL PANEL

The municipal power supply enters the home or business as two (or three if there is heavy equipment use) lines (*phases*) of 110-volts ac and one neutral line in a protected metal box called the electrical panel. Figure 14-3 diagrams the concept of the electrical panel.

The power supply enters the panel and is connected to bars. Circuit breakers are attached to the bars. If the breaker is attached to one phase, it delivers 110 volts (single phase) to anything that is connected to it. If the breaker is designed to be connected across both phases, it delivers 220 volts (dual phase). All neutral lines returning to the panel are connected to the neutral bar, which is in turn connected to a ground. In this way, both 110- and 220-volt ac breakers are found in the same panel.

Circuit breakers

The supply lines are generally designed to carry 100 amps for the typical residential user. Each circuit breaker is designed to carry a specific load and break the circuit open when the load exceeds that value. Typical circuit breakers are 15, 20, 25, 30, and 50 amps, depending on the requirements of the appliances (or probable total of appliances on the same breaker). Wiring attached to the breaker leading to the appliances is sized in accordance with the amperage of the breaker.

When electrical volume exceeds the rating of the breaker, it opens the circuit and disconnects the power supply to the appliance or circuit in question. Such overload might occur as the result of an unintentional ground or short circuit at the appliance (or wiring to it).

Depending on the design of the breaker, resetting is accomplished in one of several ways. Sometimes it is not obvious which breaker has tripped. One style of breaker looks as if it is still on. You need to push the switch fully to off, then back to on to reset it. Another style pops halfway between on and off, again requiring a hard push to off before going back to on. Another has a small window displaying a red flag when the breaker is off. Some of these require waiting up to 30 seconds before the breaker can be reset. Another type is off when a tab pops out and is reset by pushing the tab back in. In short, be aware that a tripped breaker might require some detective work.

It is also important to know that each manufacturer makes electrical panels to accommodate their breakers only. None that I know of are interchangeable, although some generic brands copy the major manufacturers.

Troubleshooting and replacement

When a breaker will not reset, it might mean that the breaker is faulty or the circuit is overloaded (demanding too much current). An overloaded circuit can be the result of an appliance that is faulty, an unintentional ground, or a short circuit in the wiring, or it might be that there are too many appliances on the same circuit (or one that is too large for the circuit.)

Troubleshooting is simple. First, check the appliances on the circuit. Does their total amperage exceed the rating of the breaker? If so, remove the extra appliances or wire them to a circuit that can handle the load.

If that is not the problem, disconnect each appliance from the circuit one at a time, resetting the breaker after each disconnection. Be sure the disconnected wires are taped off and no bare wires are touching each other. When you have removed the faulty appliance, the breaker will stay on. You now know which appliance to repair.

If the breaker is still tripping, the problem might be in the wiring between the breaker and the appliance. Make a visual inspection (with the breaker off) of all the wiring that is accessible. If you don't find a frayed or broken wire or two bare wires touching each other, disconnect the wiring from the breaker. To do that, turn off the main service breaker that feeds the entire panel. Remove the faceplate from the breaker panel. Make sure the breaker in question is off (an added safety in case the main breaker is still on for any reason). Unscrew the wire lug screw at the base of the breaker (Fig. 14-3) and pull the load wires from the breaker. Turn the main service back on and reset the breaker in question. If it still pops off under this no-load condition, then the breaker itself is faulty and must be replaced.

Never try to repair a breaker. If you are unable to locate a replacement and need to get the equipment operating again, look at the remainder of the breakers in the panel. Often there are spare breakers in the panel that can be used for replacement. Sometimes a breaker of a comparable amperage is servicing a circuit that is not needed as much as the pool equipment and you can make a temporary switch. Al-

Neutral main wire

Double-pole breaker

Red

Black

Single-pole breaker

Ground

Double-pole breaker

Grounding bar

Neutral bus bar

Creative Homeowner Press, Upper Saddle River, NJ

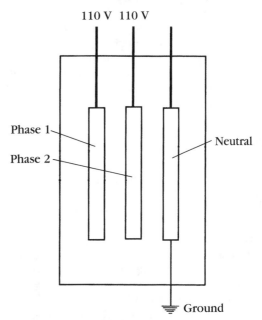

110 V   110 V

Phase 1

Phase 2

Neutral

Ground

14-3 The electrical panel and circuit breakers.

ways replace a breaker with one of the same amperage. You can use a lower amp breaker on wires designed to carry higher amps, but only if you know the appliance uses less amperage than the rating of the smaller breaker.

To replace a breaker, turn off the main service breaker. Place your flat-blade screwdriver on the front, top edge of the breaker and pry it out of the panel. Some breakers fit tightly, so apply firm, even pressure. If you have not disconnected the load wires, do so as described earlier. Look at the back of the breaker and the design of the hook connection that fits into the electric bar of the panel. When you have your replacement, reconnect the load wires to the new breaker, and return it to the panel reversing the steps taken to remove it. Put the panel faceplate back on and turn on the main service breaker.

If the breaker did not trip when you disconnected the load, the reason for the breaker tripping off must be in the wiring between the breaker and the appliance. Since you were unable to find a problem with the wiring during your visual inspection, you might need to replace the wiring. At this point, I recommend calling an electrician.

Sometimes electrical problems at the appliance or the tripping of a breaker is caused by a loose breaker. If you find that the breaker is loose when you first try to remove it, try pushing it back into the panel, and try your appliance again. If it won't seat firmly, replace the breaker.

You might think that it is easier to disconnect the load and check the breaker first, prior to following all of the other checks, but I present the steps in this order, beginning at the appliance, because it is usually here that you will find the problem.

Older homes might still have fuses. Fuses perform the same function as circuit breakers, but fuses must be replaced each time the overload breaks the circuit (blows the fuse). Fuses either clip or screw in place. As with breakers, always replace a fuse with one of the same amperage.

Whenever you approach a breaker panel, do so with great respect. Water, frayed wiring, or a poor previous service work might have created problems at the panel that you cannot anticipate. If you are planning to work on a panel, it's best to have a helper around to get help in case of electric shock. Other safety measures include wearing rubber gloves and boots, standing on a piece of dry wood to further insulate you from the ground, and leaving one hand in your pocket, so you can't inadvertently touch one hand to a live wire or panel and the other to a ground.

Wiring

Pulling new wires in a circuit or adding a circuit is a job best left to a professional electrician. But you might be called upon to replace the wires from an appliance to a J-box in the equipment area or the wiring of a heater connection, so it is valuable to know a few things about requirements.

Gauge and type

The gauge of the wire refers to its thickness and, therefore, its ability to handle volume and pressure of current (amps and volts). AWG refers to American Wire Gauge and is the system described in Fig. A-3. AWG wire is designed to operate under temperatures as high as 140°F.

Whenever you run wire for any reason, consult the chart in Fig. A-3 to be sure you use the correct type. Remember, you can always use wire that is heavier (lower

AWG number) than the breaker and appliance require, but never use wire that is thinner (higher AWG number) than required.

Wire is stranded or solid. There is less resistance in solid wire than stranded, so this should be your first choice. Wire is generally available in copper. During times of copper shortages and high prices, aluminum was used for wiring homes, but it has been the cause of overheating and fires and should be replaced whenever possible.

Wires are sold in various colors. Green wire is always ground. Black and red are used for hot lines, white for neutral. When you need many hot lines to many circuits, they might be colored in any of the many other colors available. If you must use a wire color not in keeping with this code, tape the correct color tape over the wire or clearly label it. Never assume that the previous technician used the correct colored wire. Check everything as you go and try to leave wiring better than you found it.

If you must run a circuit yourself, be sure to encase the wires in conduit. Use flexible, waterproof conduit and connectors for outdoor installations, such as wires from heaters or motors to J-boxes or time clocks. Often a subpanel (a small breaker panel supplied from the main household panel with one large circuit) is located in the pool equipment area, so you might be making short runs between appliances and the subpanel. Lastly, be aware that wires of different voltage should not be run in the same conduit. You can run numerous circuits of the same voltage in a conduit, but you need to run a separate conduit for every group of circuits of different voltage. Never run anything else in an electrical conduit, such as air hoses or water lines.

When terminating wires to be attached to connections in appliances or at other terminal posts, use crimp connectors rather than simply wrapping the bare end of the wire around the post. Wrapping can come loose or be squeezed off the post. Figure 14-4 shows the correct way to make wrap connections if you have no other choice. Bend the wire in the same direction as you will tighten the screw, so when you tighten the screw it also tightens the wrap.

Figure 14-4 also shows a typical crimp connector. The connectors are available in various sizes and with various connection ends (called the *tongue*). The insulation is stripped off to accommodate the barrel of the connector. Using a crimping tool, secure the wire to the connector. Crimping sets are available for a few dollars at any hardware store.

Since most pool and spa applications are wired directly between appliances and switches, you won't be dealing with too many outlets. With portable spas, however, you might encounter a few. It is important to recognize the appearance of outlets so you don't try to plug 110-volt appliances into 220-volt outlets. Figure 14-5 shows three electrical outlets typically found in and around pool and spa installations. These outlets are designed so that the plug can be inserted only one way to prevent reverse polarity. With ac, polarity is important with motors, which run in a certain direction. Reversing the polarity can reverse the direction of the motor. This might be significant with blowers or pump motors in a skid pack installation, for example.

Bonding and grounding

A bonding wire is an important (and under some building codes, required) safety component of any pool or spa equipment area. Since electricity will always flow

14-4
Wire wrapping and crimp-on
connectors. U.S. Government Printing Office

along the path of least resistance, a grounding wire (back to the panel and then into the ground) might offer more resistance to the current than a person touching the source of electricity and standing in a puddle of water. The bonding wire is a path of less resistance than the human body, so stray current is conducted along it instead of you in the event you become part of the circuit.

14-5 Various electrical outlets.

A bonding wire is a solid 8-gauge wire that is connected to a lug on the exterior of each appliance in the equipment package. No conduit is needed because current is not normally carried by this wire. Some installers run one end of the bonding wire into the ground (or connected to a metal stake in the ground at the equipment area) as an additional safety measure.

The gauge of the ground wire of any appliance must be as large as the hot wire(s) so it is capable of efficiently conducting electricity away from the appliance in the event of a short circuit or stray current. Similarly, at the breaker panel, the main ground wire must be as large as the largest hot wire in use. All pool and spa equipment must be grounded.

Ground fault interrupter (GFI)

When equipment or wiring fails it might draw more current than the appliance can use, burning out the appliance. The circuit breaker is designed to break the circuit when demand exceeds the rating of the breaker. As noted earlier, it takes so little current to kill a human that the typical breaker will deliver a lethal dose before breaking the circuit. In other words, circuit breakers are designed to protect equipment, not humans.

The GFI is designed to protect humans. It is a circuit breaker that detects problems at a low enough level to protect you before lethal doses are delivered. It breaks a circuit when it detects a ground fault. Figure 14-6 diagrams the GFI concept. The GFI constantly measures the current going out of it (to the appliance) and coming back into it. If an inadvertent grounding takes place, such as if the metal case of an appliance were electrified, and you touch it, completing a pathway for current to the ground, the GFI detects the drop in the current it is receiving and breaks the circuit. The GFI detects variations as low as 0.005 amp, which is about half the lethal charge to a child and about one-sixth a lethal dose to you. The GFI cuts the circuit within one-fortieth of one second, so it is not only sensitive, it's quick.

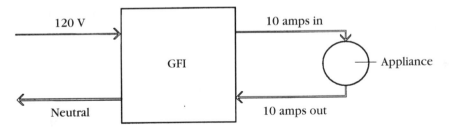

14-6 Ground fault interrupter (GFI).

There are three basic styles of GFI that you will likely encounter in pool and spa work. The first looks like a standard circuit breaker in the electrical panel, but it has a test button in the face of the breaker in addition to the on/off breaker switch. By pressing the test button, you are simulating an unbalanced current condition inside the breaker and thereby testing the efficiency of the GFI. The GFI breaker resets the same way a normal panel breaker does.

The second type of GFI is built into a wall outlet, such as the type you might install for plugging in a portable spa. It also contains a test button and a switch to reset the GFI.

The third type is a portable GFI, a unit that plugs into a wall outlet. The appliance is then plugged into the GFI, making the outlet a GFI outlet.

All types of GFIs, like any other mechanical device, are subject to failure and should be tested at least every month.

If they provide so much safety, why aren't all breakers and outlets GFIs? The first answer is probably cost, since the GFI breaker or outlet costs two to four times as much as a standard one. The practical reason is that some appliances or circuits operate normally with slight variations in current, so the GFI would constantly be breaking circuits for the wrong reason.

In fact, slight variations might occur in the pool or spa equipment, causing the GFI to trip even though everything is functioning properly. Unfortunately, many technicians disconnect or bypass GFIs for that nuisance reason. Panel GFIs are somewhat prone to this problem, because often there is slight current leakage inside the panel, and long wire runs to the equipment location can create slight variations in current that cause the GFI to trip. For this reason, it is best to locate the GFI as close to the appliance(s) as possible.

If a GFI keeps breaking the circuit, you troubleshoot the problem in the same manner as any other breaker. As described earlier, start by disconnecting the appliance and resetting the breaker, check the wiring, disconnect the load at the breaker, etc. If you have a GFI serving a skid pack, the problem can exist in any single piece of equipment, so these must be disconnected one at a time and the GFI reset after each to detect the appliance with the problem. Start with the light, then proceed to the blower, electric heater, pump motor, control devices (yes, the problem might exist in the control panel). I recommend this sequence because electrical problems in portable spas are most often in lights and blowers. Since the heater will only operate with the pump running, you want to test the heater next, then complete the list.

The National Electric Code (NEC) specifies that electrical outlets located within 15 feet of the water's edge must be protected by a GFI and that circuits for all underwater lighting be so equipped. Underwriter's Laboratory (UL) requires all portable spas be wired with a GFI.

Switches

Pool and spa equipment is not wired directly to circuit breakers in the electrical panel. Instead, these circuits are interrupted at some point by switches to control the operation of each appliance. A breaker should never be used as the on/off switch for an appliance because repeated switching will weaken the breaker.

I discussed air switches, time clocks, and other control devices earlier, but all these are switches, turning appliances on or off by completing or breaking an electrical circuit. An understanding of the basic concept of switches will help you troubleshoot electrical problems, especially when you encounter an appliance that isn't coming on because it isn't getting current.

Figure 14-7 depicts a basic switch, which is a break in the hot line of a circuit. This is the most basic on/off switch, called a single pole, single throw (SPST) switch. This switch handles one circuit (single pole) each time the switch is thrown. The

SPST

14-7
Switch diagrams.

SPDT

second drawing depicts a single pole, double throw (SPDT) switch. In this case, there is still only one circuit of electric current, but when this switch is thrown one direction, it electrifies one appliance, and when it is thrown the other way, it electrifies another appliance. Depending on the appliance(s), you might use several variations of poles (circuits) and throws (destinations for the current). By understanding these basic concepts, you will recognize whatever type of switch you encounter.

The other concept regarding switches that you will encounter with pools and spas is multiple switching. There is often more than one control or switch on each appliance. For example, there might be a wall switch and a time clock, either of which can turn on a pool light. There are often air switches and time clocks controlling spa equipment and a simple on/off switch attached to each appliance so you can shut it off easily for service work.

There are two kinds of circuits and, therefore, two kinds of switching. First, switches wired in *series* are those which operate together. Figure 14-8 depicts a se-

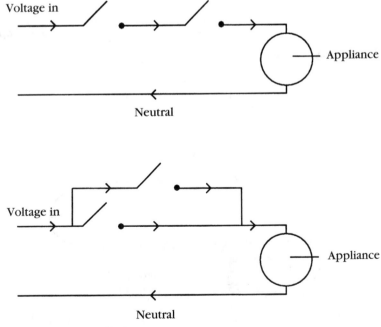

14-8 Series and parallel circuits and switches.

ries circuit, where the electric current cannot pass along the line unless each switch is closed. An example of a series circuit and series switches is the control circuit in a heater, where each control switch must be closed before the entire circuit is completed and the ignition of the heater is fired. Another example is a string of Christmas tree lights, where if one bulb burns out, the circuit is broken and all the lights go off.

The other type of circuit and switching is *parallel*. Figure 14-8 also depicts a parallel circuit, where there is more than one way for the circuit to be completed, each independent of the other. The drawing shows that the current will reach the appliance if either switch is closed. But closing both is not necessary, and if they are it does not deliver any more or less current to the appliance because both switches depend on the same source of current. An example of parallel circuits and switches is the pool light that is controlled by a switch in the home and by a time clock in the equipment area. By understanding these concepts, you will be able to detect why an appliance is not operating or why it is operating after you turned the switch off.

Relays

A *relay* is a switching device on a circuit that controls current flow in another circuit. Figure 14-9 depicts a typical relay circuit. When the relay circuit is electrified, it energizes an electromagnet that pulls the two halves of the relay together. In doing so, the contacts of the controlled circuit are brought in contact, completing the circuit. Relays are normally used as safety devices. The purpose of this type of control

14-9 Relay.

is to use a low-voltage circuit (the relay circuit) to turn on or off a higher voltage circuit (controlled circuit). For example, a safe 12-volt circuit can be used near a pool or spa to control a dangerous 220-volt circuit that operates a pump motor or blower.

Relays can also be used to control low-voltage situations. For example, a millivolt control on a heater cannot be located more than 20 feet from the heater (see the chapter on heaters). If you want a heater switch inside your home, for example, you might run a standard 110-volt switch that activates a relay in the equipment area, which in turn controls the low-voltage millivolt circuit of the heater.

Relays allow you to run thinner, less expensive wires over long distances. For example, a high-amperage blower or pump motor would require extremely heavy wires to be controlled from 200 feet away. Since the small electromagnet of the relay uses a small amount of current, you can run much smaller, cheaper wires along that 200 feet to control the relay, which can be located near the appliance. The heavier wiring for the appliance only needs to travel to the nearby relay and back.

Some control and time clock makers include relays in their designs. Understanding how they work will make your troubleshooting easier. The most important concept to remember is that relays are just switches. Instead of requiring you to flip a switch however, the relay turns on or off when powered by electric current, thus turning on or off the appliance.

Safety

This section should be first in considering basic electricity, however I have left it toward the end because by now you should have an appreciation for the many ways that pool and spa equipment and wiring can deliver lethal power. With this in mind, I hope you will take these few words about safety more seriously.

I will never forget grabbing an old light fixture that had been taken from the water and was lying on the deck and feeling a surge of 220 volts run through my body. Every muscle in my body contracted violently. I could see, but everything was reddish and hazy. I could hear perfectly, but could only force a gurgle out of my mouth. No matter what I tried or thought, I couldn't release the grip my hand had on that light fixture. If my helper had not been near the light switch, I wouldn't be writing this book today.

I have also been shocked on numerous occasions by metal time clock boxes (fortunately they're now mostly made of plastic, but most existing installations are still metal). In most cases, I sensed the current as if I had been bitten and pulled my hand away. But as my story illustrates, you might not always be able to remove yourself from harm.

I have already discussed how little current is actually needed to harm you. It should also be noted that it takes about five times as much dc to kill you as ac, and you will almost always be dealing with ac. Moreover, 60-cycle American ac is the most dangerous frequency. It isn't important to understand why, just to know that in the United States, we deal with a dangerous charge.

Safety with electricity then is twofold. First, be aware. Never go into an equipment area and start grabbing wires, components, or disassembling things without turning off the power supply. Even once you have disconnected appliances, there might be current from another source, such as a parallel circuit or switch or a short circuit in another appliance that is connected by a bonding wire or water on the ground. In other words, always treat electrical circuits and appliances as if they were an animal that could bite you at any time.

Second, take precautions. Wear rubber-soled shoes and rubber gloves, and don't stand in water when you work on equipment. Have a helper around when you plan to do installations or electrical work. Take the extra time to walk over to the electrical panel and turn off the supply breaker in addition to any on/off switches that are closer.

Finally, never take shortcuts with wiring or installations, such as leaving off the bonding or ground wires, running circuits of different voltages in the same conduit, using wire of the wrong color because you are too lazy to make another trip to the hardware store, etc. There is a law in most states that refers to the *last expert* on a job. If you were the last technician to work on a job site, you can be held liable for all of the faults with the work, even those not your own. The law assumes that as an expert, you will upgrade the job to code as needed, shut down the job site, or refuse to work on the installation until everything is made right. Of course, your liability is even greater if you made the faulty installation.

Water appliances work outdoors under severe conditions and are subject to electrical failures. When a motor overheats, for example, for any of the many reasons discussed previously, the windings can melt and make contact with the metal casing, making the case hot. Check valves on blowers sometimes fail and water enters the appliance, bringing water and electricity together and electrifying any area that is in contact with that water.

Water and electricity are dangerous allies. Water is a very effective conductor, and an electrical fault in the equipment room, contacting a leak in an appliance, can charge the entire pool or spa and make it lethal. For your safety and that of your customers, not to mention the very existence of your business, be careful and take no shortcuts when working with electricity.

TESTING

The last general area of basic electrical knowledge that is useful to the water technician is how to test for the presence and parameters of electric current. A dear friend of mine is well into his eighties and continues to work as a professional electrician. When he goes on jobs with me and needs to know if a circuit is hot, he licks his fingers and sticks them into the connection. He tells me 110 volts is good for you now and then, but that 220. . . !

As just discussed, you will probably get away with inadvertent contact with electric current if the situation permits you to quickly withdraw your hand or finger. But I strongly advise against using my friend's test methods, especially when a completely adequate, modern multimeter costs under $20.

As the name implies, the multimeter has multiple functions, testing circuit voltages, continuity, and resistance. Figure 14-10 shows a typical multimeter. It has a positive and a negative test lead and a switching device to set the meter for reading dc or ac (reading various ranges of each), resistance, or continuity. The meter is battery powered for continuity and resistance testing because you must send current into a line to test if it is continuous (unbroken) or broken and to test the amount of resistance in a conductor.

14-10
Multimeter.

Testing for the presence of current at a connection or appliance is simple. It does require guessing what you expect to find because you need to set the tester for ac or dc and for the range of voltage you expect to find. For example, when testing the control circuit of a millivolt-controlled heater, you would set the meter for dc current in a voltage range of 0 to 1 volt (since you will be testing a circuit with up to 750 millivolts, which is equal to 0.75 volt). Similarly, if you are looking for the presence of current at your motor, you set the meter for ac in the voltage range of either 0 to 110 or 0 to 220 volts. Electronically controlled heater circuits operate on 25-volts ac, so you would set the meter for ac in a range of 0 to 50 volts. Generally you can't harm the meter by feeding it less current than the range you have chosen, but you can destroy it by feeding it more. Therefore, if you are uncertain about the voltage being tested, start with the 220-volt range and work down.

When testing dc circuits, remember that polarity (positive and negative) makes a difference. You must touch the positive meter lead to the positive contact of the appliance or switch and the negative lead to the negative contact. If you reverse these, you will see the meter register negative voltage. When testing ac voltage, the polarity doesn't matter, and you can touch either lead to either side of the circuit.

Figure 14-10 shows a meter testing for the presence of current entering a circuit breaker. The easiest way to understand testing is to remember that the meter is an appliance. It needs a hot and a neutral line to operate.

When testing 110-volts ac, touch one lead to the suspected hot line and one to a neutral line or to ground. When testing 220-volts ac, perform the same test on each of the two hot lines, then touch one lead to each hot line at the same time. If each line individually reads 110 volts, but when tested together it does not read 220 volts, it means the two hot lines are being supplied by the same phase of the power supply and therefore will not deliver 220 volts. This usually denotes a faulty breaker.

When buying a multimeter, make sure it can test millivoltage for working on millivolt heaters. Some meters won't accurately read less than 10 volts, and therefore are useless with millivoltage. Most electronic meters are pocket-size and can self-range, which is to say you need only dial in ac or dc and the meter will detect the voltage and adjust accordingly.

Continuity testing is useful when you suspect a broken connection. To test continuity of a line, set your meter for continuity testing and touch one lead to each end of the suspect circuit. If the meter reads positive or beeps, it means there is continuity. Before conducting such tests, make sure your meter is working properly, and that the battery which electrifies the circuit is working, by touching the two test leads together. This should represent a good connection and a complete circuit, and you should get a positive reading.

Since most of the wiring and installation you do uses good conductors, you will probably not use the resistance measuring function of the multimeter much. If the continuity tester on your meter is not working for some reason, you can use the resistance test to check for continuity. Resistance is measured in units called ohms. The higher the ohms, the more resistance exists in the circuit. If there is no resistance, however, it means there is not a complete circuit, thereby also verifying continuity.

When appliances are operating poorly, there might be a drop in voltage between the panel and the appliance. Check the voltage at the appliance, then at the breaker, while the appliance is operating. There will be a slight difference because of some voltage drop as a result of heat loss and resistance along the length of wire, but it should be no more than 2 to 5 volts. If it is greater, follow the troubleshooting procedures outlined previously to determine where the loss is being created. Like water that is not flowing in a pipe (pressure is the same everywhere in the pipe), when the appliance is not operating, the voltage (pressure) should be the same everywhere along the line.

To test amperage you need an amp probe. The amp probe is a meter with a large, open clip on the end. The clip is looped over the wire and the amperage in the wire is detected by the probe without actual contact with the current.

As with other procedures in pool and spa work, practice in the workshop makes perfect in the field. When you buy your multimeter, ask the seller to demonstrate it for you and be sure it comes with an owner's manual with complete instructions. Try it out on your electrical panel and on small appliances in your workshop. Always be careful to touch only the connection you intend to test. For example, if you touch the hot line in a time clock and inadvertently lay the same probe against the metal face of the clock, it will create a short circuit, sparking and burning the points of contact.

15
The toolbox

Throughout this book I have detailed procedures requiring average hand tools and some specialized equipment. Nothing is more frustrating than getting a job near completion, only to discover you are missing a particular allen wrench or nut driver. A trip back to the workshop or to the hardware store can ruin a carefully planned schedule, not to mention costing you time and money.

The following list represents what I carry in my truck at all times. It represents not only basic tools, but also basic plumbing and equipment parts. This inventory has helped me in literally thousands of jobs. For ease of reference, I will list the entire inventory, then follow it with some notes about why you should carry certain items.

Every tool you can buy is available in a quality (expensive) version and several economy (cheap) models. Experience has taught me that cheap tools provide cheap results. Buy the best tools you can afford because they are the way you will make a living and you will abuse them beyond the endurance of cheaply made products. All metal tools should have rubber or plastic insulation on the handles.

I don't list here any of the specialty supplies or tools mentioned in the book for infrequent jobs such as plaster patching, acid washing, etc. This list is intended to be only those items you should always have in your truck for day-to-day operational convenience. I also don't list any chemicals, because those are not carried as spares, but rather should be loaded into the truck fresh daily as needed.

GENERAL HAND TOOLS

You should carry the following hand tools.
- Leatherman tool (on your person)
- Wrench set (¼ to 1 inch)—open box on one end, closed on the other
- Flat-blade screwdrivers (one long with small blade, two average length with medium blade)
- Flat-blade screwdriver or breaker bar (very large)
- Phillips-head screwdrivers (one small, one medium)
- Allen-head wrench set
- Needle-nose pliers (one long and narrow, one stubby with wire cutter)
- Steel hammer
- Channel-lock pliers

- Vise grips
- Wire stripper and cutter
- Standard loose-jointed pliers (one medium, one large)
- Nut driver set (¼ to ⁹⁄₁₆ inch) with extender
- Calipers
- Ratchet socket driver (adaptable to sockets on nut driver)
- Cold chisel
- Rubber mallet
- Scissors
- Small bolt cutter (dikes)
- Set of wood chisels
- 18-inch pipe wrench (2)
- Pump seal setter
- Hacksaw with four 12-teeth per inch and four 36-teeth per inch blades
- Small brass (or stainless steel) wire brush
- Small steel rattail file
- Medium flat file
- Impeller wrench
- 25-foot steel tape measure
- Wire coat hanger
- Razor knife
- 2-inch drain flush (blowbag)
- Multimeter (that tests millivolts to 240 volts)
- Four jumper wires (about 1 foot long with alligator clips on each end)
- Electrician's snake (100 foot)
- Indelible felt-tipped marking pen
- Carpenter's pencils
- Rubber gloves
- Cloth or leather gloves
- Fire extinguisher
- First-aid kit
- Floating thermometer
- Flashlight

SPECIAL TOOL SETS

The following special tools are also needed.
- ⅜-inch electric drill with various bits
- Hole saw kit for electric drill (1- to 2½-inch diameters)
- Copper soldering kit
- 100-foot extension cord
- Submersible pump
- Seal driver (1-inch slip coupling with 1-inch plug on one end)
- Sand filter vac
- Polaroid camera

SUPPLIES

You should carry the following basic supplies.
- Silicone-based grease lubricant
- Quart of PVC cement
- Quart of PVC primer
- Pipe dope or joint stick
- Teflon tape (three rolls)
- Paper or rubber pump gaskets (carry the four most common styles)
- Pump seals (the four most common styles)
- Pump shaft extender and three spare allen-head set screws
- 4 feet of ⅛-inch clear plastic hose
- Electrician's tape
- Plumber's strap tape
- Fine sandpaper
- Duct tape
- WD-40 or similar spray lube
- Cleanup rags
- Assortment of sheet metal screws, nuts, machine screws, and bolts
- Hose clamps (stainless steel, ¾ inch to 4 inches)
- Wire nuts (various sizes)
- Three-port valve repair kit (variety of O-rings and valve body gaskets, screws, etc.)
- Business cards, notepads, and pens

CLEANING TOOLS

The following tools are required for pool and spa cleanup.
- 8-by-16-foot telepole with magnet tip on the end
- 4-by-8-foot telepole
- 16-inch stainless steel mesh leaf rake
- 18-inch wall brush
- 9-inch stainless steel brush
- Vacuum head
- Vacuum adaptor plate
- 50 feet of vacuum hose
- 20 feet of vacuum hose
- Vacuum hose connector
- Bronze skimmer diverter valve
- Spa vacuum and spare socks
- Acid spotter
- 100 feet of backwash hose
- Leafmaster
- 50 feet of garden hose
- Garden hose ends (one male, one female)
- Tennis balls (2)

- Tile soap (in quart squirt bottle)
- Tile brush (clip on)
- Barbecue scrub pad (also for tiles)
- Pumice block and clip-on pumice stone
- Plastic garbage bags
- Garden hose adapter (NPT threaded ¾ inch)
- Test kit
- Chemicals
  ~ DE
  ~ Sanitizers (liquid for pools, granular for spas)
  ~ Acid
  ~ Soda ash
  ~ Algicide

HEATER REPAIR KIT

The following supplies are needed for heater repair.
- Scripto lighter (BBQ-style throw-away)
- Honeywell IID
- Millivolt pilot assemblies (including generator)
  ~ Clip-in type (Teledyne Laars)
  ~ Screw-in type (Raypak)
- Electronic pilot assemblies
  ~ Teledyne Laars style with bolt-on electrode
  ~ Raypak style with clip-on electrode
- Orange high-tension cords
  ~ Teledyne Laars style with bare extension wire and ceramic
  ~ Raypak style (longer) with clip on end
- Pressure switch with adapter for different models
- Set of high-limit switches (Teledyne Laars and Raypak)
- Electronic thermistor
- Transformer
- Wire of various lengths and colors (available with connectors in kits called harnesses)
- Flange gaskets (1½- and 2-inch pairs) (Teledyne Laars and Raypak)
- Small dental-type mirror for examining pilot flames

PLUMBING PARTS

The following plumbing parts should be carried in your toolbox.
- 1½- and 2-inch PVC fittings (same list of each):
  ~ Straight slip couplings (4)
  ~ 90-degree slip couplings (4)
  ~ 45-degree slip couplings (4)
  ~ T fitting
  ~ 10 feet of pipe
  ~ MIP × slip (2)
  ~ FIP × slip (2)

~ Mission clamps (2)
~ Plugs (2)
~ Caps (2)
~ Reducer bushings—2 inch to 1½ inch (4)

NOTES ON TOOLS AND SUPPLIES

The Leatherman tool is a pocketknife extraordinaire. There are few products that I am willing to name by brand, but this is one of them (along with Channel Lock pliers and Vise Grips). A standard pocketknife might have a cutting blade and screwdrivers that will save you a few trips back to the truck for tools, but Leatherman is solid stainless steel (important around pools), includes a truly useful needle-nose pliers with wire stripper/cutter, files, ruler, and several sizes of screwdrivers. It stays on my belt 24 hours per day and it has saved me countless hours of wasted time running back to the truck for pliers to loosen time clock trippers, a screwdriver to pry open a skimmer cover, etc., etc., etc.

The hammer should have a steel or graphite handle rather than wood because it will get wet.

The long, thin flat-blade screwdriver is helpful when adjusting pressure switches and other hard-to-reach, delicate components. The medium screwdrivers are for general duty. Carry two because you will be using this tool frequently and for many procedures, meaning when one chips or breaks, you have another.

The very large (about 12-inch long), sturdy flat-blade screwdriver is to use in tight quarters as a wedge or breaker bar. I can't tell you how many times I use this to persuade components to come apart, such as filter lids. It also makes a good small chisel for hard-to-reach areas.

The most common nut driver you will use is $\frac{5}{16}$ inch. That is the standard nut size on the terminal connection board inside most motors, where you will be connecting wires. It is also useful in other applications, such as for the square-head sheet metal screws used to assemble some heater cabinets. The sockets should be adaptable to a ratchet driver for ease in removing pump bolts and cabinet screws when performing major disassembly of equipment. The bolts holding many pumps together are ½ inch or $\frac{9}{16}$ inch.

You might be surprised to see wood chisels on the tool list of a water technician, but I am frequently cutting open a deck or wooden tub to accommodate pipes or conduit. They are also useful for chiseling PVC fittings.

You will need two 18-inch pipe wrenches. One holds one side of a plumbing fitting or the appliance, while the other twists another fitting or pipe on or off.

The small wire brush is useful for cleaning parts before reassembly.

Every time you make a repair, save the extra screws, bolts, and nuts, etc., or save the ones you replace if they are still in good shape. I carry a plastic box of these general-duty fasteners and am never caught short on a job when a particular nut gets lost (most often filter lid retainer rings and terminal posts inside motors). Pump bolts ($\frac{5}{16}$ inch by varying lengths) are often rusted and need to be replaced, so it's good to carry a supply of those.

The wire coat hanger is useful for poking, clearing, hooking . . . well you get the idea—a stiff, narrow wire has a lot of uses.

The 10-foot lengths of pipe can be cut to any lengths that will fit in your truck, because you can always reconnect them to a longer length as needed with straight slip couplings. The number of fittings is enough to back you up on a job and is not meant to replace careful planning of a plumbing job.

A
Facts and formulas summary

The following facts and formulas are found throughout the text and are compiled here for easy reference. Rules quoted represent industry standards and might vary in different health department or building codes.

VOLUME CALCULATIONS

Length × width × average depth × 7.5 = volume (in gallons) of square or rectangular pool

3.14 (pi) × radius squared × average depth × 7.5 = volume (in gallons) of a round pool

0.45 × (A + B) × length × average depth × 7.5 = volume (in gallons) of kidney or irregular-shaped pool

MEASUREMENTS

7.5 gallons in each cubic foot of water
1 inch = 0.0833 feet.
radius = diameter ÷ 2
1 gallon of water weighs 8.3 pounds
1 gallon of chlorine weighs 10 pounds
16 ounces dry weight = 1 pound dry weight
64 ounces liquid = 1 gallon
8 ounces liquid = 1 cup liquid
1000 watts = 1 kilowatt
1 milliamp = 0.001 amp (30 milliamps will cause muscle contractions)
1 psi = 2.31 feet of head (resistance)
1 psi = 28 inches water column pressure (WCP)
1 inch of mercury = 1.13 feet of head
1 micron = one-millionth of 1 meter = 0.0000394 inches (the human eye can detect objects as small as 35 microns)
1 grain of table salt = 100 microns
¾ of 1 volt = 750 millivolts = 0.75 volts
1 Btu (British thermal unit) is the amount of heat required to raise 1 pound of water 1°F

1 therm (measurement of gas supply) = 100,000 Btu per hour
1 gallon of fuel oil = 140,000 Btu
1 kilowatt = 3412 Btu
1 ton (heat pump rating) = 15,000 Btu
A typical bather displaces 2 cubic feet of water
1 hp = 746 watts

Conversion table

Gallons to liters	gallons × 3.785 = liters
Liters to gallons	liters × 0.2642 = gallons
Ounces to milliliters	ounces × 29.57 = milliliters
Milliliters to ounces	milliliters × 0.0338 = ounces
Inches to millimeters	inches × 25.4 = millimeters

DATA

Normal rotation speed of a pump motor is 3450 rpm (revolutions per minute). On two-speed pumps, the low speed rotates at 1750 rpm.

Alternating current (ac) in the United States runs on 60 hertz (cycles). In Europe and many other foreign countries it runs at 50 hertz.

PVC plumbing is designed to carry water at temperatures of less than 100°F.
CPVC plumbing is designed to carry water at temperatures above 100°F.

Head created by various plumbing and equipment components:
- A 1½-inch × 90-degree elbow = 7.5 feet of straight 1½-inch pipe.
- A 2-inch × 90-degree elbow = 8.6 feet of straight 2-inch pipe.
- A 1½-inch × 45-degree elbow = 2.2 feet of straight 1½-inch pipe.
- A 2-inch × 45-degree elbow = 2.8 feet of straight 2-inch pipe.
- Filters create 5 to 7 feet of head.
- Heaters create 8 to 15 feet of head.

Sand filters strain particles as small as 60 microns.
Cartridge filters strain particles as small as 20 microns.
DE filters strain particles as small as 7 microns.
A 1-pound coffee can holds ½ pound of DE.
1 pound of DE will coat 10 square feet of filter grid area.
Sand filters are designed to use #20 silica sand.
A heater pilot light creates 1100°F of heat.
Electricity supplied to an electronic ignition pilot is 10,000 to 20,000 volts.
The chemical processes when chlorine (in any form) is added to water:

Cl_2 (chlorine) + H_2O (water) = HCl (muriatic acid) + HOCl (hypochlorous acid)
HOCl = H^+ + OCl^- (a positive ion of hydrogen and a negative ion of hypochlorite)
at pH 7.2 = 80 percent killing HOCl + 20 percent useless OCl^-
at pH 8.0 = 20 percent killing HOCl + 80 percent useless OCl^-

Urine and perspiration = ammonia
One active swimmer produces approximately 1 quart of ammonia/hour.
Liquid chlorine is sold in a solution of 10 to 15 percent strength.

Chlorine comparison chart:

Product	pH	Available chlorine	Common form	Stability
Cl_2 gas	Low	100%	gas	Very unstable
Sodium hypochlorite	13+	12.5%	Liquid	Unstable
Calcium hypochlorite	11.5	65%	Dry granular	Stable
Dichlor	6.8	60%	Dry granular	Very stable
Trichlor	3.0	90%	Tablet (or granular)	Very, very stable

Effectiveness comparison

1 pound Cl_2 gas = 1 gallon sodium hypochlorite = 1.5 pounds calcium hypochlorite = 1.75 pounds dichlor = 1.15 pounds trichlor = 2.8 pounds lithium hypochlorite

Cost comparison

Sodium hypochlorite = 2 times the cost of chlorine gas
Calcium hypochlorite = 3 times the cost of chlorine gas
Dichlor = 5 times the cost of chlorine gas
Trichlor = 3 times the cost of chlorine gas
Lithium = 5 times the cost of chlorine gas

Ninety-five percent of chlorine in water can be lost on a sunny day in two hours (without stabilizer).

Amount of product required to obtain a change in 1000 gallons of water:

Item	Raise or lower 10 ppm	Product	Product amount required (dry weight unless otherwise noted)
Calcium hardness	raise	Calcium chloride	2 ounces
Total alkalinity	Raise	Bicarb of soda	2.5 ounces
Total alkalinity	Lower	Sodium bisulfate (dry acid)	2.5 ounces
Total alkalinity	Lower	Muriatic acid	¼ cup (liquid)
Stabilizer	Raise	Cyanuric acid	1.5 ounces

Amount of chlorine needed to raise residual in 1000 gallons of water by 1 ppm (multiply results by 30 for superchlorination procedures)

Availability of product (percent)	Product amount
12	⅛ cup (liquid)
50	¼ ounce (dry weight)
80	⅙ ounce (dry weight)

STANDARDS

Water may not move at more than 8 feet per second through copper pipe.
Water may not move at more than 10 feet per second through PVC.
Water may not move at more than 8 feet per second through any type of suction pipe.

Turnover rates:

    Pools must turn over in 6 hours.

    Spas must turn over in ½ hour.

    Wading pools must turn over 1 hour.

Filtration standards:

Filter style	Maximum flow (gpm/sq. ft.)	Minimum backwash flow (gpm/sq. ft.)
High-rate sand	15	15
DE	2	2
Cartridge	0.375*	No backwash

*Note that cartridge filters are rated at 1 gpm/square foot maximum flow rate on most residential applications and the more stringent 0.375 gpm/square foot on commercial installations.

Rapid sand filters are designed for flow rates of less than 3 gpm.

High-rate sand filters are designed for flow rates of 5 to 20 gpm.

The normal filter operating pressure differential (between incoming pipe and outgoing) is between 2 and 4 psi. When it reaches 10 psi, the filter must be cleaned. Expressed another way, if you know the normal operating pressure of your filter (as observed on the pressure gauge), you must clean the filter when the pressure is more than 10 psi higher.

Water flow must not exceed 100 gpm through 1½-inch heater plumbing.

Water flow must not exceed 125 gpm through 2-inch heater plumbing.

Water flowing out of a heater should not be more than 10°F warmer than the water flowing into the heater.

Pool water should not exceed 82°F.

Spa water should not exceed 104°F.

The on/off switch of a millivolt heater must not be located more than 20 feet from the heater.

Heater installations:

- The rear and nonplumbed sides must have at least 6 inches of clearance.
- The plumbed side must have at least 18 inches of clearance.
- The front must have at least 24 inches of clearance.
- The top must be at least 5 feet below (and 4 feet horizontal distance) any window, door, or vent.
- The top must be at least 3 feet below any overhang.
- PVC gas lines must be buried at least 18 inches below ground (12 inches below concrete).
- Metal gas lines must be buried at least 12 inches below ground (6 inches below concrete).

To accommodate a diving board, a pool must be at least 7½ feet deep, 15 feet wide, and 28 feet long.

Recreational water chemistry parameters:

- Chlorine residual   1.0 to 3.0 ppm
- Total alkalinity   80 to 150 ppm

- pH   7.4 to 7.6
- Hardness   200 to 400 ppm
- Total dissolved solids   Less than 2000 ppm
- Cyanuric acid   30 to 80 ppm

Competition or public swimming pools

Dimensions:

50 meters long (164.05 feet) by 75 feet wide by 4 to 6 feet deep for swimming; 12 feet minimum for diving.

15 feet minimum clearance required above any diving board; 10 feet on each side; 16 feet of horizontal water surface in front of the furthest reach of the board.

Swimming lanes are 8 feet wide.

Occupancy:

1 bather per 15 square feet of surface (in portions of the pool that are 5 feet deep or less).

1 bather per 20 square feet of surface (in portions of the pool that are 5 feet deep or more).

Subtract 300 square feet from the total surface calculation for each diving board installed.

Alternate calculation method:

1 bather for every 36 square feet of surface (regardless of depth)

1 diver for every 100 square feet of surface

Electrical outlets or switches located within 15 feet of the water's edge must be protected by a ground fault interrupter (GFI).

All portable spas must be protected with a GFI.

FORMULAS

Seventy-five percent of the surface area of a pool is the required surface area of solar panels to adequately heat that pool.

Each spa jet requires flow of 15 gpm.

Each spa jet requires ¼ hp from the pump and motor.

Hole size requirements for air blower rings:

If the air ring will have ⅛-inch holes, drill 30 of them.

If the air ring will have $\frac{5}{32}$-inch holes, drill 73 of them.

If the air ring will have $\frac{3}{16}$-inch holes, drill 50 of them.

If the air ring will have ¼-inch holes, drill 30 of them.

To estimate the horsepower of a blower:

1 hp = 6.5 amps at 110 volts = 3.25 amps at 220 volts

2 hp = 13 amps at 110 volts = 6.5 amps at 220 volts

Volts × amps = watts

Heater Btu output rating ÷ (gallons × 8.33) = degrees temperature rise/hour

Breakpoint chlorination method (for every 20,000 gallons of water):

1 gallon of anhydrous ammonia + 8 gallons liquid chlorine

CENTURY MOTOR SERIAL NUMBER DATE CODES

1960	AS	1970	K	1980	X	1990	BH	2000	BU
1961	A	1971	L	1981	Y	1991	BJ	2001	BW
1962	B	1972	M	1982	Z	1992	BK	2002	BX
1963	C	1973	N	1983	BA	1993	BL	2003	BY
1964	D	1974	P	1984	BB	1994	BM	2004	BZ
1965	E	1975	R	1985	BC	1995	BN		
1966	F	1976	S	1986	BD	1996	BP		
1967	G	1977	T	1987	BE	1997	BR		
1968	H	1978	U	1988	BF	1998	BS		
1969	J	1979	W	1989	BG	1999	BT		

Plant of manufacture (assigned serial numbers):

St. Louis   10000 thru 29999
Corinth   30000 thru 39999
McMinnville   40000 thru 49999
Lexington   50000 thru 99999

Month of manufacture is designated as 1 for January, 2 for February, and so on.

A-1 Century Motor serial number date codes.

FRICTION LOSS IN PIPE AND FITTINGS

The following table is designed for calculations with PVC pipe; however, for most applications it also applies to copper (which actually flows 5 to 10 percent slower).

Friction flow chart for Schedule 40 rigid PVC pipe
Friction loss of water in feet per 100 feet length of pipe.
Based on Williams & Hazen formula using constant 150. Sizes of standard pipe in inches.

U.S. Gals. per Min.	¾" Pipe Vel. Ft. per Sec.	Loss in Feet	1" Pipe Vel. Ft. per Sec.	Loss in Feet	1¼" Pipe Vel. Ft. per Sec.	Loss in Feet	1½" Pipe Vel. Ft. per Sec.	Loss in Feet	2" Pipe Vel. Ft. per Sec.	Loss in Feet	2½" Pipe Vel. Ft. per Sec.	Loss in Feet	3" Pipe Vel. Ft. per Sec.	Loss in Feet	4" Pipe Vel. Ft. per Sec.	Loss in Feet	6" Pipe Vel. Ft. per Sec.	Loss in Feet	U.S. Gals. per Min.
1	.60	.25	.37	.07															1
2	1.20	.90	.74	.28															2
3	1.80	1.92	1.11	.60	.43	.07													3
4	2.41	3.28	1.48	1.02	.64	.16	.47	.07											4
5	3.01	5.8	1.86	1.52	.86	.25	.63	.12											5
					1.07	.39	.79	.18											
6	3.61	7.0	2.33	2.15	1.29	.55	.95	.25	.57	.07									6
8	4.81	11.8	2.97	3.6	1.72	.97	1.25	.46	.76	.14	.54	.05							8
10	6.02	17.9	3.71	5.5	2.15	1.46	1.58	.69	.96	.21	.67	.09							10
15	9.02	37.8	5.57	11.7	3.22	3.07	2.36	1.45	1.43	.44	1.01	.18	.65	.07					15
20			7.42	19.9	4.29	4.2	3.15	2.47	1.91	.74	1.34	.30	.87	.12					20
25			9.28	30.0	5.36	7.9	3.94	3.8	2.39	1.11	1.67	.46	1.08	.16					25
30			11.14	42.0	6.43	11.1	4.73	5.2	2.87	1.55	2.01	.65	1.30	.23					30
35					7.51	14.7	5.52	7.0	3.35	2.06	2.35	.88	1.52	.30	.88	.07			35
40					8.58	18.9	6.30	8.9	3.82	2.63	2.64	1.11	1.73	.39	1.01	.09			40
45			8" PIPE		9.65	23.5	7.09	11.1	4.30	3.28	3.01	1.39	1.95	.48	1.13	.12			45
50			700		4.37	.66	10.72	28.5	7.88	13.5	4.78	4.0	3.35	1.69	2.17	.58	1.26	.16	50
60			750		4.70	.75			9.46	18.9	5.74	5.6	4.02	2.36	2.60	.81	1.51	.21	60
70			800		4.99	.82			11.03	25.1	6.69	7.4	4.69	3.14	3.04	1.09	1.76	.28	70
80			850		5.37	.95					7.65	9.5	5.35	4.0	3.47	1.39	2.02	.37	80
90			900		5.64	1.06					8.60	11.8	6.03	5.0	3.91	1.73	2.27	.46	90
100			950		5.94	1.23					9.56	14.4	6.70	6.1	4.34	2.10	2.52	.55	100
125			1000		6.25	1.28					11.95	21.8	8.38	9.2	5.42	3.19	3.15	.85	125
150			1050		6.57	1.40	U.S.						10.05	12.8	6.51	4.5	3.78	1.18	150
175			1100		6.89	1.51	GAL.								7.59	5.9	4.41	1.57	175
200			1150		7.20	1.65	PER MIN.		10" PIPE						8.68	8.8	5.04	2.08	200
225			1200		7.51	1.79	1500		5.99	.92			9.76	9.4	5.67	2.52	2.50	.35	225
250			1250		7.82	1.94	1600		6.40	1.03			10.85	11.5	6.30	3.05	2.78	.42	250
275			1300		8.13	2.09	1700		6.80	1.15					6.93	3.6	3.05	.48	275
300			1350		8.45	2.23	1800		7.20	1.27					7.56	4.3	3.33	.58	300
325			1400		8.77	2.37	1900		7.60	1.41					8.19	5.0	3.61	.67	325
350			1450		9.08	2.53	2000		7.99	1.54					8.82	5.7	3.89	.79	350
375			1500		9.39	2.68	2100		8.39	1.67					9.45	6.5	4.17	.88	375
400			1550		9.70	2.85	2200		8.80	1.82					10.08	7.3	4.44	.99	400
425			1600		10.00	3.01	2300		9.20	1.99							4.72	1.11	425
450			1650		10.32	3.21	2400		9.60	2.17							5.00	1.22	450
475			1700		10.64	3.40	2500		10.00	2.34							5.28	1.36	475
500			1750		10.96	3.59	2600		10.40	2.50							5.55	1.50	500
550			1800		11.27	3.78	2700		10.79	2.69							6.11	1.80	550
600			1850		11.58	3.98	2800		11.18	2.88							6.67	2.10	600
650			1900		11.89	4.19	2900		11.59	3.08							7.22	2.44	650
700			1950		12.21	4.40	3000		11.99	3.28							7.78	2.79	700
750			2000		12.52	4.61	3100		12.39	3.32							8.33	3.19	750
800							3200		12.79	3.35							8.89	3.6	800

A-2 Friction loss in pipe.

GENERAL WIRE SIZING CHART

Run (ft)	30	40	50	75	100	125	150	175	200
Amps				Minimum AWG size wire recommended					
5	14	14	14	14	12	12	12	12	12
10	14	14	14	12	12	12	10	10	10
15	12	12	12	10	10	10	8	8	8
20	10	10	10	10	10	8	8	8	8
25	10	10	10	10	8	8	6	6	6
30	8	8	8	8	6	6	6	6	4

Interpolate as needed between values given. Values given are based on 110/120 volts and are heavier than actually required with 220/240 volt equipment. However using these values for either voltage will ensure safe, adequate installations.

A-3 General wire sizing chart.

B
Labor reference guide

What is a fair price to charge a customer for a repair or maintenance procedure? The answer to this question will vary significantly from one part of the country to another, depending on the cost of doing business in your area. The one basic consideration in setting fees for a job is labor.

Labor rates for a qualified pool and spa technician vary in the United States from $30 to $60 per hour. Overhead, insurance, complexity of the job, and the experience of the technician all might be factors in determining the rate. Moreover, when I do a simple cleanup job of a pool or spa, I charge $15 per hour. My time is worth more than that, but the experience required to do that type of work doesn't justify more. When I do repair or installation work however, I charge $50 per hour, based on my skills and expertise. Regardless of the rate, a fair price for a job is determined by establishing a reasonable amount of time the job should require, then multiplying by the appropriate hourly labor rate in your area.

What is the *reasonable* amount of time for a job? In the automotive industry, guidebooks establish that it might take four hours for a brake job, two hours for a tune-up, six hours to rebuild a carburetor, and so on. The mechanic can then charge according to these industry standards, regardless of how long the procedure actually took him to complete. Using this method, both mechanic and customer can be assured of a fair price for the job.

In setting a standard for the amount of time a job should require, I include several factors in the calculation and leave other factors out. For example, if special parts are required, I must consider the time spent going to get those parts that aren't normally in my workshop. The time it takes to unload tools (and pack up after the job as well), set up, and prepare a job are just as much a part of the work as the repair itself. The job might require returning to make adjustments or other follow-up work, and this must be included in the initial estimate of time required.

Warranty is another factor. Heater manufacturers have factory technicians that will make repairs after the initial installation while the product is under warranty (or after the warranty expires for a fee). If I install a heater and something goes wrong, my customer doesn't want to hear that the factory technician will be along in a week or so. Therefore, even if the problem is in the equipment and not my installation, I must be prepared to make the repair or remove the equipment, exchange it, and install a replacement. This type of extra warranty must be provided for in the original labor charge. Of course, you can't charge a customer for your mistakes. I don't charge four hours for a job that takes two, just to cover the possibility that I might have goofed and will need to come back to do the work a second time.

Travel time is another consideration. In Malibu, where many of my customers live 10 miles up a winding canyon road, I have to consider the time required to stop by and make the initial estimate, the time to go back and do the work, and the time to go back and check on the results. Keep in mind also that recommended hours are total work hours for the job, including helpers (when necessary) for work such as carrying a heavy heater for a replacement procedure (and removing the old one).

Finally, consider the working conditions. If you have to remove several pieces of equipment to access the one you need to work on, or if the repair requires hanging off the edge of a steep hill working with one hand and holding on for dear life with the other, then add an appropriate number of hours to the estimate. Also, some jobs require considerable time to determine what work is needed, especially when troubleshooting heaters. That time invested must be included in the total hours for the job. In other words, when setting your rates and hours, give careful consideration to *all* aspects of the work you do or might have to do.

The following guidelines are based on many years of personal experience and supervision of employees. They are based on residential installations, not commercial. I provide them here as a reference only. They should be adjusted as needed in your circumstances. Remember, they include reasonable setup and cleanup time, as well as the actual work time I believe each job should require when performed by a technician of average skills working under average conditions. I have not listed items requiring individual estimates that might vary widely depending on size, difficulty, or condition of the pool, such as painting or breaking in plaster on a pool, leak repair, or replacing a skimmer. I have also not included estimates for repair or replacement of chlorinators or automatic pool cleaners because there are so many different makes and models, each of which has unique demands.

Procedure	Hours
Filters	
Tear down/clean filter (cartridge or DE)	1
Open sand filter, breakup/clean sand	2
Tear down sand filter, change sand	4
Install new filter (remove and dispose of old unit)	4
Pumps and motors	
Replace seal, seal plate, gasket, or other nonplumbed pump components	1.5
Replace volute, strainer pot, or other plumbed component	2
Replace motor (including electrical)	2
Heaters	
Replace heater	8
Replace flange gaskets or thermostatic mixing valve	1
Troubleshoot and replace defective control circuit component(s)	1.5
Replace pilot assembly	2
Replace main gas valve or burner tray component(s)	2.5
Replace manifold headers, heat exchanger, or firebrick	3
Ream heat exchanger	4

General procedures

Drain, acid wash, refill, and rebalance chemistry (pool)	10
Drain, acid wash, refill, and rebalance chemistry (spa)	4
Replace diving board, rail, ladder, or slide (on existing identical mount)	3
Cut floating pool cover to size, install on roller straps	2
Rebuild three-port valve	1
Replace (or replace a component in) a time clock	1
Replace an underwater light bulb	1.5
Replace an underwater light fixture	2
Replace an air blower	1.5

Appendix C

Typical pool and spa health, safety, and building codes

The county, state, and possibly the city in which you work govern the construction, equipment installation, and maintenance procedures for pools and spas.

Reading through the codes for your area in detail can help you understand what is required of a professional pool and spa technician. Among other important facts, the codes tell you what the inspector (health or building) might be looking for. Reading these codes also adds important specifics to the guidelines presented throughout the book.

Some codes deal with construction requirements, some deal with operational requirements, and others deal with health and safety issues. You need to know them all, especially if you are maintaining commercial accounts.

You can obtain copies of codes for your area through contractor's book stores (often expensive, because you have to buy the entire code just to get the parts relating to pools and spas) or the governing agency itself (very often it can send you copies at little or no cost, covering only those topics of interest).

So what information should you look for in your local codes? Since space is limited and since many code examples have been given throughout the book already, the following is designed to give you just a flavor of the information found in these valuable references. The prevailing code regulations for Los Angeles County and the state of California have been cited as examples in this book and are identical or similar to the codes in jurisdictions around the country. Using Los Angeles as an example, here's a brief list of codes and topics with which you should become familiar:

California Health and Safety Code, Division 20, Chapter I, Sections 24100–24109.
- Definitions of "public swimming pool" and "lifeguard service"
- "Sanitary, healthful and safe" operation of pools
- Health inspector responsibilities
- Penalties for violating health code regulation

California Administrative Code, Environmental Health, Chapter 20, "Public Swimming Pools"

- Definitions of "swimming pool," "wading pool," "special use pools," and "spa pools"
- Who is required to comply with these codes? (It's a long list!)
- Who is exempt?
- Construction plan approval
- Maintenance recordkeeping
- Circulation and filtration requirements
- Standards for water clarity, pH, cyanuric acid, sanitizer, allowable bacteria counts, and cleanliness
- Life-saving equipment and signage requirements
  - Bather rules (caps, showering, etc.)
  - Pool closure rules

California State Building Code, Chapters 2–90, "Public Swimming Pools"

- "The provisions of this Chapter shall apply to the construction, installation, alteration, addition, relocation, replacement or use of any public swimming pool and to its appurtenant auxiliary areas and facilities and to its mechanical equipment relating to piping."
  - Extensive definitions of the components of pools and spas
  - Pool marking requirements (depth and lane markers)
- Stair, tread, ladder, and rail requirements
- Safety fencing and signage
- Gas chlorination and other chemical feeder equipment requirements
- Minimum filtration requirements and turnover rates
- Waste water disposal laws

County of Los Angeles, Department of Health Services, Swimming Pool Plan Approval Requirements

- Information needed to submit a plan
- Layout, slope, and measurement requirements for a pool or spa
- Treads, risers, coping, decking, rain, and ladder placement and minimum requirements
- Local filtration, turnover rate, and water-quality requirements
- Plumbing sizing, flow rates and gauging
- Signage, covers and safety equipment required
- Bather loading
- Special regulation variations for spas

D

Meter reading—gas, electric, and water

As noted previously in this book, customers often ask how much it costs to operate a heater or pump motor. They also want to know the amount of water used to refill a pool after complete or partial draining for a maintenance or repair procedure.

In the case of water usage, you can easily measure the vessel and calculate the volume; however, you might still want to verify your estimate. Your customer might also have an automatic water-fill system, and if you suspect a leak in the pool, you might want to monitor water consumption during a certain period.

Estimating consumption of water, gas, or electricity is a matter of reading the related meter at the supply outside the home. Since all three of these utility meters are similar, one description is applicable for reading the others.

Two types of utility meter are in common use. The newer style uses a simple digital readout, like the odometer in your car, to record a running total of the volume of the commodity that has passed through the meter.

The second type, more widely in use, uses five or six dials to record this information and, although easy to read, is more likely to be misread than a simple digital readout. Figure D-1 shows five dials, let's say of an electrical meter measuring kilowatt hours. When the needle points between two numbers, always read the lower number. These dials read "16064".

D-1 Typical meter recording dials.

After one hour, if these dials read "16066," then 2 kilowatt hours have been used. Refer to your customer's electric bill to learn the charge for each kilowatt hour to calculate the cost of this electricity used.

When using this technique to analyze utility use, make sure the appliance you are measuring is the only thing running during the test. For example, if you wanted to assess the power consumption of a pump motor, go through the house (with the customer's permission!) to shut off all other electric appliances. Read the electric meter and run the motor for 15 minutes. Read the meter again and multiply the kilowatt hours used by four to determine the hourly use (4×15 minutes = 1 hour). Then multiply the hourly consumption by the number of hours per day you expect the motor to run to arrive at a daily usage.

Water or gas consumption and cost estimates are performed the same way. Again, make sure that the appliance being tested is the only one in operation during the test. If measuring water consumption of an automatic filler system, you might need to test for an entire 24-hour period, because the filler might not be running at all times. In that case, you would see no consumption in the 15 minute test and have no data for drawing conclusions. Similarly, if you found the water running during the 15 minutes, you could not conclude that it would run all of the time. You might have tested during the only 15 minutes in every 24 hours that the filler operates. Conducting a 24-hour test is difficult, because it requires shutting down all other water use. You might ask your customer if they are planning to be gone over a 24-hour period when other water would not be in use. In that case, be sure no automatic sprinklers come on during that period either.

Another simple way to test water use by an automatic filler is to shut off the water supply to the filler for 24 hours. When you turn it back on, the filler will start providing water to the pool. Record the amount of water it takes to bring the pool level back up to a point where the filler shuts itself off. This quantity should be the volume used every 24 hours. Again, during this fill-up time, be sure no other water is being used in the house that might invalidate your conclusion.

One last word of caution about meter reading. On some meters, the last dial records tenths. If that were the case in Fig. D-1, the meter actually reads "1606.4" and obviously the use was $\frac{2}{10}$ of a kilowatt hour, not two kilowatt hours. The dials clearly indicate what is being recorded, so take note as you write down the results.

Glossary

Not all terms in this glossary are used in the text, but because you will encounter them in other aspects of the industry or in regional variances they are included for your reference. Common words or measurement definitions not specific to the water industry are not included, but can be found in any standard dictionary. Ordinary words with colloquial industry meaning are defined only by those references.

ABS Acrylonitrile butadiene styrene. Rigid plastic pipe similar to PVC, usually manufactured black, used for drainage systems.

ac Alternating current. Electrical charge that flows from negative to positive, then reverses direction. The basic form of electricity used in most homes and businesses and, therefore, in most pool and spa equipment.

acetone A highly flammable solvent used to clean plastic surfaces and tools.

acid A liquid or dry chemical that lowers pH when added to water, such as muriatic acid.

acid demand The amount of acid required (demanded) by a body of water to raise the pH to neutral (7).

acidity The quality, state, or degree of being acid.

acid spotter A siphon device that attaches to a telepole for insertion into a body of water to siphon acid from a bottle (connected by means of a plastic hose). The acid can then be directed and concentrated on a particular location for stain removal.

acid wash The procedure of cleaning plaster with a solution of muriatic acid and water.

adapter bracket The part of a pump that supports the motor and connects the motor to the pump.

aerator A pipe vented to the atmosphere, sometimes with an adjustable volume control, added to a waterline to mix air and water prior to discharge.

aggregate The major component of plaster, composed of sand, marble dust, pebbles, or other solid matter.

air blower *See* blower.

air relief valve A valve on a filter that permits air to be discharged from the freeboard .

air switch A pneumatic-mechanical control device used to operate spa equipment safely. A button, located in or near the water, is depressed sending air pressure along a plastic hose to an on/off switch.

algae Airborne, microscopic plant life of many forms that grow in water and on underwater surfaces.

algicides A group of chemical substances that kill algae or inhibits their growth in water.

algistat A chemical that inhibits the growth of algae.

alkalinity The characteristic of water that registers a pH above neutral (7).

aluminum sulfate Alum. An additive for sand filters that prevents the sand from combining, hardening, and thus not filtering impurities from water.

ambient temperature The average, prevailing surrounding temperature.

American Gas Association (AGA) The national institute that sets standards for installations using natural gas as a fuel.

American National Standards Institute (ANSI) The national institute that sets standards for construction.

American Public Health Association (APHA) The national institute that sets standards for issues relating to public health and safety.

American Wire Gauge (AWG) The standard used to specify wiring of certain thickness and capacity.

ammonia Natural substance composed of nitrogen and hydrogen that readily combines with free chlorine in water forming chloramines (weak sanitizers).

amperage (amps) The term used to describe the strength of an electric current. It represents the volume of current passing through a conductor in a given time. Amps = watts ÷ volts.

anhydrous Lacking water. A characteristic of a substance that makes it readily combine with water.

antisiphon valve A control device added to the domestic water supply line to prevent contaminated water from flowing backward into the pipe.

antisurge valve A check valve used in air blower plumbing to prevent water from entering the blower mechanism.

antivortex The property of a plumbing fitting that prevents a whirlpool effect when water is sucked through it. Used on main drain covers.

arc (arcing) The passage of electrical current between two points without benefit of a conductor. For example, when a wire with current is located near a metal object, the electricity might arc (pass) between the two.

automatic gas valve The valve that controls the release of natural or propane gas to a heater. Also called the combination gas valve.

available chlorine *See* chlorine, free available.

backfill Dirt, sand, or other material used to fill the gaps between a pool wall and the surrounding excavation.

backwash The process of running water through a filter opposite the normal direction of flow to flush out contaminants.

bacteria Any of a class of microscopic plants living in soil, water, organic matter, or in living beings and affecting humans as chemical reactions or viruses.

balance The term used in water chemistry to indicate that when measuring all components together (pH, total alkalinity, hardness, and temperature) the water is neither scaling or etching. The Langlier Index value of balanced water is zero.

balloon fitting A plumbing connector made of rubber or flexible plastic that readily adapts to pipe of varying sizes.

bank A chemical residual in water that is inactive but can be released by adding a catalyst.

barb fitting A plumbing fitting with exterior ribs, connected by insertion into a pipe (usually flexible pipe).

base An alkaline substance.

bather Any person using a pool or spa.

bather load The number of bathers in a pool or spa at one time.

bayonet Refers to a light bulb using two nipples to engage corresponding slots in the socket.

bicarbonate of soda A chemical used to raise pH and total alkalinity in water. Also called bicarb or baking soda.

biological filter A method of filtration using bacteria to expedite decomposition of organic debris, usually in conjunction with sand as a filter media.

bleach Colloquial term for liquid chlorine.

bleed To remove the air from a pipe or device, allowing water to fill the space.

blister Refers to an air pocket in a plaster surface.

blow bag Also called a drain flush or balloon bag. A device attached to a garden hose that expands under water or air pressure to seal an opening, forcing the water or air into that opening.

blower An electromechanical device that generates air pressure to provide spa jets and rings with bubbles.

bluestone *See* copper sulfate.

blue vitriol *See* copper sulfate.

boiler *See* heater.

bond beam The top of a pool or spa wall, built stronger than the wall itself to support coping stones and decking.

bonding system The wiring between electrical appliances and the ground to prevent electric shock in case of a faulty circuit. All appliances are grounded to the same wire.

booster pump A pump added to a spa system to add pressure to the jets.

break down *See* tear down.

break-in The procedure of curing and preparing new plaster for service in a pool.

break point chlorination The application of enough chlorine to water to combine with all ammonia (creating chloramines), then to destroy all chloramines, leaving a residual of free chlorine. The break point is, therefore, the point at which chlorine added to the water is no longer demanded to sanitize and is available to become residual chlorine. Some references define break point as ten parts of chlorine for every one part of ammonia present in the water when the pH is between 6.8 and 7.6.

bridging The condition existing when DE and dirt closes the intended gaps between the filter grids in a DE filter, reducing the flow rate through the filter and reducing the square footage of filter area.

broadcast Dispersal of chemicals over the widest possible surface area of a body of water.

bromine (Br_2) A water sanitizing agent. A member of the halogen family of compounds.

brown coat The undercoat of plaster applied over old plaster or as a base coat over gunite.

Btu British thermal unit. The measurement of heat generated by a fuel. The amount of heat required to raise 1 pound of water 1°F (when at or near 39.2°F).

buffer A substance that tends to resist change in the pH of a solution.

burner tray The component in the bottom of a heater that controls the burning of gas.

bushing An internal plumbing fitting that fits (threaded or slip) into a connector fitting to reduce the internal diameter so as to accommodate a smaller pipe size than the connector fitting was designed to accept. *See also* reducer.

calcium A mineral element typically found in water.

calcium bleed The condition in plaster where calcium leeches from the mixture.

calcium carbonate (CaCO₃) The mineral precipitated out of water, deposited on pool and spa surfaces, commonly called scale.

calcium hypochlorite (Ca(OCl)₂) A granular form of chlorine (widely produced under the brand name HTH), generally produced in a compound of 70 percent chlorine and 30 percent inert materials.

cam lock The device that holds or releases two halves of a telepole.

cap A plumbing fitting attached to the end of a pipe to close it completely.

capacitor A device used to store an electrical charge to be released in one short burst.

capacitor start-capacitor run motor (CSR) An ac single-phase electric motor controlled by a capacitor for both starting and running. The motor does not require an internal start switch or governor. *See also* energy efficient.

capacitor start-induction run motor (CSI) An ac split-phase electric motor with a capacitor connected in series to a separate starter winding that shuts off after the motor has reached operating speed. The capacitor winding jump starts the motor, then an induction winding takes over to run the motor. The motor requires an internal start switch and governor.

cartridge The element in a filter covered with pleats of fabric to strain impurities from water that passes through it. Generally strains out particles larger than 20 microns.

caulking Material used in a joint to create a waterproof seal.

caustic Acidic or etching in nature.

cavitation Failure of a pump to move water when a vacuum is created because the discharge capacity of a pump exceeds the suction ability.

centrifugal force The outward force created by an object in circular motion. The force that is used by water pumps to move water.

C frame A type of motor housing resembling a C. Adapts to a particular style of pump.

channeling Creation of a tube or channel in a filter media through which water will flow unfiltered. Channeling is caused by calcification of the media (hardening of the sand in a sand filter, for example) and is resolved by breaking up the sand and treating it with alum.

check valve A valve that permits flow of water or air in only one direction through a pipe.

chelating agent Chemical compounds that prevent minerals in solution in a body of water from precipitating out of solution and depositing on the surfaces of the container.

chine The portion of a stave (in a wooden hot tub) that extends below the croze.

chine joist The sturdy plank that supports the floor of a hot tub.

chloramine A compound of chlorine when combined with inorganic ammonia or nitrogen. Chloramines are stable and slow to release their chlorine for oxidizing (sanitizing) purposes.

chlorinator A device that delivers chlorine to a body of water.

chlorine (Cl_2) A substance made from salt that is used to sanitize water by killing bacteria. A member of the halogen family, chlorine is produced in gas, liquid, and granular form.

chlorine demand The amount of chlorine required (demanded) by a body of water to raise the chlorine residual to a predetermined level.

chlorine, free available That portion of chlorine in a body of water that is immediately capable (available) of oxidizing contaminants.

chlorine lock The term applied to chloramine formation, when ammonia is present in the water in sufficient quantity to combine with all available chlorine.

chlorine residual The amount of chlorine remaining in a body of water after all organic material (including bacteria) has been oxidized, expressed in parts per million. The total chlorine residual is the sum of all free available chlorine plus any combined chlorine (chloramine).

chlorophyta Green algae.

circuit The path through which electricity flows.

circulation system The combination of pipes, pump, and any other components through which water flows in a closed loop (from the body of water, through the components, and back to the same body of water).

clamping ring The metal band that applies pressure to and holds the lid on a filter tank or holds together two halves of a pump.

closed loop Any system that recirculates the same water, rather than constantly taking in a new supply and discharging it after only one pass through the system.

close nipple *See* nipple.

colloidal silver A compound of silver, used as an algicide.

colorimetric The name given to a chemical test procedure where reagents are added to water and change color to reflect the presence and strength of a substance. The color is compared to a color chart to evaluate the volume of that substance. Colorimetric tests are used to detect the presence of chlorine in water.

comparator The color chart or other device used to compare the color of a treated water sample with known values. Used in chemistry test kits.

compression fitting A plumbing connection that joins two lengths of pipe by sliding over each pipe and applying pressure to a gasket that seals the connection.

condensation The process by which a substance changes from a gas to a liquid (called the condensate).

conditioner A chemical that slows the decomposition of chlorine from ultraviolet light. Conditioner, usually cyanuric acid, also helps prevent spiking of pH between high and low extremes.

conductor Any substance that will carry electric current, such as a wire, metal, or the human body.

control circuit A series of safety and switching devices in a heater, all of that must be closed before electric current flows to the pilot light and automatic gas valve to ignite the heater.

coping The cap on the edge of a pool or spa mounted on the bond beam.

copper sulfate A blue granular substance added to a body of water to kill algae. Also called bluestone or blue vitriol.

corona discharge A method of producing ozone by passing electricity through oxygen and water.

coupling A plumbing fitting used to connect two lengths of pipe.

cove The curved radius between the wall of the pool and the floor.

CPVC Chlorinated polyvinyl chloride. The designation of plastic pipe that can be used with extremely hot water.

croze The milled groove in the stave of a wooden hot tub into which the floorboards are inserted.

CSI *See* capacitor start-induction run motor (CSI).

current Refers to the rate of electrical flow between two points.

cyanophyta Blue-green algae or black algae.

cyanurates Chlorine sanitizers combined with buffers, such as dichlor and trichlor.

cyanuric acid *See* conditioner and stabilizer.

cycle A complete turn of alternating current (ac) from negative to positive and back again (*see* hertz). Also refers to a filtration period (*see* filter run).

dedicated circuit An electrical circuit used for only one specific appliance.

delamination Separation or failure of the bond in layered materials, such as plaster to gunite or fiberglass to resin.

diameter The distance across the widest part of a circle.

diatomaceous earth (DE) A white, powdery substance composed of tiny prehistoric skeletal remains of algae (diatoms), used as a water filtration media in DE filters. DE filters can remove particles larger than 5 to 8 microns.

dichlor *See* sodium dichloro-s-triazinetrione.

dielectric An insulating material that prevents electrolysis or corrosion when applied between two surfaces of dissimilar metal.

diffuser A housing inside a pump covering an impeller that reduces speed of the water but increases pressure in the system.

discharge The flow of water out of a pipe or port.

diverter A 6-inch long by 1½-inch-diameter plastic or bronze adapter pipe that fits into a skimmer port to facilitate connection of a vacuum hose. The diverter can divert all suction to the skimmer, closing off the main drain or vice versa.

dog *See* tripper.

DPD Diethy phenylene diamene. The chemical reagent used to detect the presence of free available chlorine in a body of water.

drain flush *See* blow bag.

dry fit To assemble PVC plumbing without glue to check that you have prepared the components correctly.

duty rating An evaluation parameter of pool and spa motors that are designed for continuous duty, meaning they could run 24 hours a day for their entire service life without stopping. The nameplate on the motor shows this by the rating, continuous duty.

dynamic head *See* head.

Dynell A type of material used to cover filter grids.

effluent The water discharging from a pipe or equipment.

elbow The term for a plumbing connector fitting with a 90-degree bend.

electrolysis The production of chemical changes (usually corrosion) in metal by passing an electric current through an electrolyte (usually water with mineral content).

electrometric A chemical test involving an electronic analysis meter.

element A filter grid. Also the electric heat-generating rod of an electric heater.

end bell The metal housing or cap at the end of an electric motor.

energy efficient A specific design of pool and spa motors that have heavier wire in the windings to lower the electricity wasted from heat loss. *See also* capacitor start, capacitor run.

equalizer line A pipe that balances the flow of water between two locations. Usually found in a pool skimmer, the equalizer line allows the suction line to draw water directly from the pool in the event that the water level drops below the point where suction would normally occur from the surface.

erosion system A type of chemical feeder in which granular or tablet sanitizer is slowly dissolved by constant flow of water through the device.

escutcheon plate The decorative metal ring that slips over the end of a rail or ladder improving the finished appearance of the installation.

etching Corrosion of a surface by water that is acidic or low in total alkalinity and/or hardness.

ethylenediamine tetra-acetic acid (EDTA) A reagent used for testing calcium hardness, added a drop at a time until the solution turns blue. The number of drops is compared to a chart to evaluate the hardness of the sample.

expansion joint The gap between the coping and the deck that allows for normal expansion and contraction of the materials, preventing damage that would otherwise result from pressure of the two against each other.

feathering The process of tapering the edge of a coating material, such as paint, resin, or plaster, to blend invisibly with the existing materials.

female Plumbing fittings or pipe with internal threads or connectors.

fiberoptics Underwater lighting devices that illuminate by sending light along thin plastic cable from a remote source.

fill water Water added to a pool or spa to replace water lost to evaporation or other reasons. Also called make-up water.

filter A device for straining impurities from the water that flows through it.

filter cycle *See* filter run.

filter filling The technique used to prime a pump by filling the filter with water and allowing it to flow backwards into the pump.

filter run The time between cleanings, expressed as the total running time of the system, also called the filter cycle. Care must be taken when using the terms *run* or *cycle*, because some technicians mean the number of hours the system operates each day rather than the total time between cleanings. Regionally, one term might be used for one definition and the other for the second. I use *run* for the daily operation, *cycle* for the time between cleanings.

FIP A female threaded plumbing fitting.

fireman's switch An on/off control device, mounted in a time clock, that turns off a heater 20 minutes before the time clock turns off the circulating pump and motor. This allows the heat inside the heater to dissipate before shutdown.

flange gasket The rubber sealing ring that prevents leaks between a heater and the circulation pipes.

flapper gate The part in a check valve that swings open when water is flowing in the intended direction but swings shut when water attempts to flow backward.

flare fitting A threaded plumbing fitting that requires a widening of the pipe at one end.

flash test The method of dropping chemistry test reagents directly into pool or spa water rather than into a vial containing a test amount of that water.

flex connector A coated metal pipe with threaded fittings on each end designed to bend freely for connection in tight quarters or at odd angles. Usually used to connect a gas pipe to an appliance (such as a pool heater) indoors, allowing the appliance to be moved without breaking the connection.

floater A chemical feeder system whereby a sanitizer tablet is placed in the device and is allowed to float around the body of water. The tablet dissolves and the sanitizer is released into the water.

float valve A plumbing device that restricts or shuts off the flow of water based on a lever that is attached to a float and that thereby rises and falls with the water level.

flocculate The process of adding a chemical to a body of water which combines with the suspended particulate matter in the water, creating larger particles that are more easily seen and removed from the water. Also called a clarifying agent or coagulant.

flow meter A device for measuring the rate of water passing through a given pipe, expressed in gallons per minute (gpm).

flow rate The volume of a liquid passing a given point in a given time, expressed in gallons per minute (gpm).

fluorine A little used (very costly) halogen sanitizer.

flux The chemical compound applied to copper before soldering to enhance the even flow of the solder and to prevent oxidation of the metal when heating.

forty-eight (48) frame A housing design for a specific type of pool or spa pump motor.

four-pass unit A type of heat exchanger found in pool and spa heaters that directs the flow of water to be heated through the unit four times before returning it to the body of water.

freeboard The vacant vertical area between the top of the filter media and the underside of the top of the filter.

free chlorine Also called available chlorine. It is chlorine in its elemental form, not combined with other elements, available for sanitizing the water.

free-flow filter A type of sand and gravel filter that allows unpressurized passage of water through the filter media (sand).

free joint A colloquial expression in plumbing. When connecting several joints in a plumbing job, the free joints are the ones you can complete without committing to the end result; the ones that are still reversible.

freon A gas used in refrigeration and ozone production (*see* ozonator).

full-rated A characteristic of motor design meaning that the motor operates at its designated horsepower (1, 2, etc.). *See also* up-rated and service factor.

fusible link (fuse link) A safety device located near the burner tray of a heater that is part of the control circuit. If the fuse link detects heat in excess of a preset limit, it melts and breaks the circuit to turn off the heater.

galvanized Iron pipe coated with zinc or another alloy to prevent corrosion.

gasket Any material (usually paper or rubber, but sometimes caulk or other pastes) inserted between two connected objects to prevent leakage of water.

gate valve A valve that restricts water flow by raising and lowering a disc across the diameter of the pipe by means of a worm drive. *See also* slide valve.

gauge Refers to the size of an electrical wire. Heavier loads can be carried on heavier gauge wires. The numbering system of wire gauges works in reverse. A 10-gauge wire, for example, is thicker than a 14-gauge wire. Also refers to a measuring device, as in pressure gauge.

gelcoat A thin, surface, finishing coat of resin sprayed over fiberglass.

geophone An acoustic listening device used for leak detection.

glazed Refers to surfaces that have been covered with a clear, protective coating, applied under heat, such as glazed tile.

gpm Gallons per minute.

grains per gallon (gpg) A unit of measurement used in old-style test kits rather than the more common parts per million (ppm). To convert grains per gallon to parts per million, multiply the gpg by 17.1. For example, if your old kit resulted in a reading of 10 gpg, that is equal to 171 ppm.

grid Frame covered with fabric used as a filter media; also called septa or element.

ground fault interrupter (GFI) A type of circuit breaker. A sensing device that determines when electricity in a circuit is flowing through an unintended path, usually to earth, creating a hazard of electrocution. The GFI detects current variations as low as 0.005 amp and breaks the circuit within one-fortieth of a second.

gunite A dry mixture of cement and sand that is mixed with water at the job site and sprayed onto contoured and supported surfaces to build a pool or spa, creating the shell.

halogens A family of oxidizing agents including chlorine, bromine, iodine, and fluorine.

hardness Also called calcium hardness. The amount of dissolved minerals (mostly calcium and magnesium) in a body of water.

Hartford loop Used in air blower installations. A method of plumbing the air delivery pipe above the water level of the spa, then back below the water level. The loop prevents water from siphoning into the blower.

head The measurement of pressure (expressed in feet) in a water circulation system (or parts thereof) created by friction, resistance, distance, or lift. Dynamic head + static head = total dynamic head (TDH).

heater A device that raises the temperature of water using natural gas, electricity, propane, solar, or mechanical energy for fuel. To be called a heater, the device must covert at least 70 percent of its fuel into heat (no more than 30 percent lost in venting). Over 80 percent efficiency, the device is called a boiler.

heat exchanger The copper tubing in a heater through which water flows. The water absorbs rising heat that is generated from the burner tray below.

heat pump A type of pool and spa heater. Like conventional heaters, the heat pump circulates water through the unit and transfers heat to the water from a fuel. Instead of using gas, electricity, or solar heat as a fuel, the heat pump takes warmth out of the air that is created by compressing a gas.

heat riser A metal pipe plumbed directly to the heater to facilitate dissipation of heat before plumbing with PVC. Also called a heat sink.

hertz Unit of measurement in ac indicating the number of cycles per second in the current as generated. In the United States, ac is generated at 60 hertz.

high-limit switch A safety device used in the control circuit of heaters. When the high-limit switch detects temperatures in excess of its preset maximum, it breaks the control circuit to shut down the heater.

high-rate sand filter A filter using sand for the filtration media designed for flows in excess of 5 gpm but less than 20 gpm (less than 15 gpm in some codes) per square foot. Strains impurities larger than 50 to 80 microns.

hopper The low bowl portion of the pool around the main drain.

horsepower (hp) The standard unit of measurement that denotes the relative strength of a mechanical device. One horsepower equals 746 watts or the power required to move 550 pounds one foot in one second.

hose bibb (also bib) The faucet to which a garden hose is attached.

hydration The process of adding moisture to a dry substance.

hydraulics The science of water movement.

hydrochloric acid (HCl) Muriatic acid.

hydrostatic pressure The force created by water that tends to push pools up out of the ground.

hydrostatic valve A check valve located in the main drain of a pool to relieve hydrostatic pressure created by rising groundwater. The valve allows groundwater into the pool, but does not let water out.

hypochlorous acid (HOCl) A form of free available chlorine resulting from the solution of an active chlorine compound added to water.

impeller Rotating part of a pump that creates centrifugal force to create suction. The impeller is said to be closed if it is shrouded (covered) on both sides of the vanes, or semiopen if it is shrouded on one side, while the interior surface of the volute creates a partial shroud on the other side.

inlet *See* return.

intermittent ignition device (IID) The electronic control and switching device used in electronic ignition heaters to operate the control circuit and automatic gas valve. Often called the brainbox.

iodine (I_2) An element used to sanitize water. A member of the halogen family like chlorine and bromine.

IPSSA Independent Pool and Spa Servicemen's Association.

J-box Short for junction box. The metal container in which wires are connected along a circuit or conduit.

jet (hydrojet) The discharge fitting through which water is returned to a spa at a high rate of speed and/or pressure.

jetted tub A bathtub fitted with hydrotherapy spa-type jets.

joint stick A paste used on threaded plumbing to prevent leaks that is available in the form of a crayon-like stick for easy application.

joist The support beam under the floor of a wooden hot tub or similar structure.

keyed shaft The shaft of a motor that has a groove for securing setscrews of the shaft extender. Used with specific designs of pumps.

kilowatt One thousand watts of electrical power. Electricity is sold by the kilowatt-hour, meaning a certain fee is charged for every 1000 watts delivered per hour.

Langlier index An indicator (like pH) that measures the scaling or etching ability of a body of water. A Langlier Index of zero indicates perfect, balanced, neutral water (neither scaling nor etching).

laterals The horizontal filter grids at the bottom of a sand filter, installed in the underdrain.

latiance The mineral and cement dust formed on the surface of a pool shell when the concrete dries. It must be removed prior to plastering.

lazy flame A natural or propane gas flame in the burner tray of a pool or spa heater that burns in slow, wavering licks or stuttering, rather than the normal strong, clear blue flame burning straight upward.

leafmaster A brand name; leafmaster is a term applied to any device that vacuums large debris from a pool by means of water pressure created with a garden hose.

leaf rake A large open net secured to a frame that attaches to a telepole that is used to skim debris from the surface of the water.

lignin The white, pulpy cellulose material that binds the organic material of wood together. When a wooden hot tub is overchlorinated, stripping out the lignin, the tub will appear to be growing a white fur that will brush off and clog the filter.

line A wire conducting electricity. Also refers to a run of pipe carrying water.

lithium hypochlorite (LiOCl) Organic, granular, highly soluble form of chlorine, produced at around 35 percent available chlorine.

load An appliance that uses electricity.

macintosh bag filter An old style of filter (no longer in use) that uses a canvas bag as the filter media. Also called a coffin filter. Not manufactured in the past 30 years and never were legal for commercial installation.

main drain The suction fitting located in the lowest portion of a body of water. The principal intake for the circulation system.

main valve The flow control device in the combination gas valve of a pool or spa heater that regulates the flow of gas to the burner tray.

male Any plumbing fitting or pipe with external threads or connectors.

manifold An assembly or component that combines several other components together. A pipe fitting with several lateral outlets for connecting one pipe with others.

manometer A device used for measuring the pressure of gas, expressed in inches of water column.

marcite A term used regionally (in Florida, for example) for plaster, generally to denote that there is marble dust content in the plaster mix.

media Any material used to strain impurities from water that passes through it. DE and the fabric covering a cartridge are both examples of filter media.

micron A unit of measurement equal to 0.000001 meter or 0.0000394 inch. For example, a grain of table salt is approximately 100 microns in diameter.

MIP A male threaded plumbing fitting.

mission clamp A trade name. A mission clamp is a rubber connector hub-type fitting, secured with a stainless steel clamping band, used in plumbing to join two lengths of pipe of similar or dissimilar diameter.

mogul base The threaded end of a light bulb that is approximately twice the diameter of the typical household bulb end. Used in old-style pool light fixtures.

mottling A difference in shades within a given color. The term is used mostly in plaster to denote cloudy patches, blotches, or streaks of uneven color.

muriatic acid Also called hydrochloric acid. This chemical is the most commonly used substance for reducing pH and total alkalinity in water.

nameplate The label on a motor that details the operating and manufacturer's specifications.

NEC National Electrical Code. The generally accepted standard for electrical installations and procedures.

neutral The pH reading at which the substance being measured is neither acidic nor alkaline. Neutral pH is 7.0.

niche The housing built into the wall of a pool or spa to accommodate a light fixture.

nipple A short length (less than 12 inches) of pipe threaded at each end. If the nipple is so short that the entire length is threaded, it is called a close nipple.

no-hub connector A rubber (or neoprene) plumbing fitting used to connect two lengths of pipe, attached with hose clamps or other pressure devices.

NPT National Pipe Thread. The generally accepted standard specifications for threaded plumbing pipe and fittings.

nut driver A tool like a screwdriver but fitted with a female socket that surrounds a given size of nut for loosening or tightening.

open loop A type of solar heating system that circulates the water being heated from the pool or spa, through the solar panels, and back to the pool or spa.

organic Any material that is naturally occurring (not manufactured), such as leaves, sweat, oil, or urine.

O-ring A thin rubber gasket used to create a waterproof seal in certain plumbing joints or between two parts of a device, such as between the lid and the strainer pot on a pump.

ORP Oxidation reduction potential. A unit of measure of sanitizer in water, measured with an electronic ORP meter.

OTO Orthotolidine. The test reagent used in detecting the presence of chlorine and, by the resulting color of the OTO, the amount of chlorine, expressed in parts per million.

ozonator A device using electricity and oxygen to create ozone and deliver it to a body of water for sanitizing purposes. Ozone is produced in two ways for pool and spa use. One method, called the corona discharge method, is to shoot an electrical charge through oxygen and water, creating ozone. The other method is to electrify freon (the gas used in air conditioners and refrigerators). In both cases, the production happens in a chamber in an ozonator built into the circulation system.

ozone (O_3) Three molecules of oxygen, creating a colorless, odorless gas used for water sanitation.

peristaltic chlorinator An automatic chlorine dispenser that applies pressure on a tube to create a vacuum, thereby creating suction of liquid chlorine from a tank.

pH The relative acidity or alkalinity of soil or water, expressed on a scale of 0 to 14, where 7 is neutral, 0 is extremely acidic, and 14 extremely alkaline.

phaeophyta Brown or yellow algae.

phenol red (phenolsulfonephthalein) The most widely used chemical reagent used to measure the pH in a sample of water.

photosynthesis The natural process by that green plants, such as algae, produce food from sunlight. Using the green chemical chlorophyll, photosynthesis converts carbon dioxide and water into carbohydrates (sugars) for nourishment of the plant, that then discharges oxygen as a by-product of the process.

pi A mathematical constant equal to 3.14.

pilot The small gas flame that ignites the burner tray of a heater.

pilot generator The device that converts heat from the pilot light into electricity to power a control circuit on a heater. Also called a power pile or thermocouple.

pipe dope A paste used to prevent leaks in threaded plumbing.

plaster A hand-applied combination of white cement, aggregates, and additives that covers the shell of a gunite pool or spa to waterproof and add beauty. Plaster can also be colored.

plug A plumbing fitting used to close a pipe completely by inserting it (slip or threaded) into a female fitting or pipe end.

port An opening, as in a discharge port being the opening through which water flows out of a pipe or system.

positive seal The type of multiport valve that directs all of the water flow to one direction, allowing no flow in the other lines. A nonpositive seal directs most of the water to one line, allowing some water to bypass into the other lines.

potassium monopersulfate An oxidizing chemical used to sanitize water. Used to supersanitize without using chlorine. Also catalyzes bromine.

potentiometer Also called a pot. A device that senses variance in temperature. Part of the thermostat assembly in a pool or spa heater.

power pile *See* pilot generator.

ppm Parts per million. The measurement of a substance within another substance; for example, 2 ounces of chlorine in 1 million ounces of water would equal 2 ppm.

precipitate An insoluble compound formed by chemical action between two or more normally soluble compounds. When water can no longer dissolve and hold in solution a compound, it is said to precipitate out of solution.

precoat The process of applying DE to grids in a DE filter after cleaning but before restarting normal circulation and filtration.

pressure gauge A device that registers the pressure in a water or air system, expressed in pounds per square inch (psi).

pressure sand filter A type of sand and gravel filter in which the water is strained through the filter media under pressure (as opposed to a free-flow filter). *See also* high-rate and rapid sand filters.

pressure switch A safety device in a heater control circuit that senses when there is inadequate water pressure (usually less than 2 psi) flowing through a heater (which might damage the heater) and breaks the control circuit, thereby shutting down the heater.

prime The process of initiating water flow in a pump to begin circulation by displacing air in the suction side of the circulation system.

psi Pounds per square inch.

pumice A natural soft (yet abrasive) stone substance (similar to lava rock) used to clean pool tiles.

pump A mechanical device driven by an electric motor that moves water.

pump curve The performance efficiency evaluation of a pump plotted on a graph comparing size, flow rate, and resistance (head).

PVC Polyvinyl chloride. The type of plastic pipe and fittings most commonly used in pool and spa plumbing.

quats Quaternary ammonium compounds. An organic ammonia compound used as an algicide in water.

radius One-half of the diameter of a circle.

rapid sand filter A filter using sand as a filter media that is designed for water flows not to exceed 3 gpm (5 gpm in commercial installations) per square foot.

reagent A liquid or dry chemical that has been formulated for water testing. A substance (agent) that reacts to another known substance, producing a predictable color.

rebar An abbreviation for reinforcement bar. Steel rods laid in concrete (such as a pool shell) to add strength.

rebound Gunite material that does not adhere when sprayed to form a pool or spa shell, falling off as pebbles and removed from the finished job.

reducer An external plumbing fitting that connects two pipes of different diameter. *See also* bushing.

residual The amount of a substance remaining in a body of water after the demand for that substance has been satisfied.

resin A liquid plastic substance (with the consistency of honey) applied to fiberglass fabric to strengthen and stiffen it.

retaining rod The metal rod in the center of certain filters on which is attached a retainer ring to hold grids in place.

retainer The plastic disc that fits over the top of a set of filter grids to hold them in place with the aid of a retainer rod.

return The line and/or fitting through which water is discharged into a body of water. Also called an inlet.

riser *See* heat riser. Also refers to any vertical run of pipe.

rotor The rotating part of an electric motor that turns the drive shaft that itself is driven by magnetism produced between it and the stator. Also the diverter in the backwash valve of certain types of filters.

run Any horizontal length of pipe. *See also* filter run.

sand filter A filtration device using sand as the filter (straining) media. *See also* rapid sand filter and high-rate sand filter.

sanitizer Any chemical compound that oxidizes organic material and bacteria to provide a clean water environment.

saturation The point at which a body of water can no longer dissolve a mineral and hold it in solution.

scale Calcium carbonate deposits that form on surfaces in contact with extremely hard water. Water in this condition is said to be scaling or precipitating.

scratch coat The first layer of plaster applied to smooth out uneven surfaces prior to application of a fine or finish coat of plaster. To ensure a good mechanical bond between this base coat and the final coat, the plaster is scratched while still wet to create furrows.

screed A large leveling tool, often made from a straight piece of wood, used to level out a wet concrete slab by drawing it over the surface.

seal A device in a pump that prevents water from leaking around the motor shaft.

seal plate The component in a pump in which the seal is situated.

sediment trap A short length of vertical pipe attached below a run of gas supply pipe to catch any contaminants prior to the gas entering the heater.

separation tank A container used in conjunction with a DE filter to trap DE and dirt when backwashing.

septum *See* grid.

sequestering agent *See* chelating agent.

service factor A measure of the reserve ability of an electric motor beyond its stated ability. For example, a 1-hp motor might actually function at 1.25 hp. *See also* up-rated and full-rated.

shaft extender A bronze fitting added to the shaft of a motor to lengthen the shaft to accommodate the design of the pump being used. The extender is held in place with three setscrews fastened tightly against the motor shaft.

shell *See* gunite.

shock treatment *See* superchlorination.

shotcrete A mixture of sand, concrete, and water that is delivered to the job site premixed to create a pool or spa shell (*see also* gunite). Shotcrete is applied through an air-powered nozzle.

shutoff head Zero gallons per minute. The point at which there is so much resistance to the free movement of water in a circulation system that the force of the pump cannot overcome the resistance and there is no movement of water.

sight glass A clear glass or plastic section of pipe that allows viewing of the water in the line. Used when backwashing filters to know when the discharge water is clean.

silica The type of sand used in filters and concrete, actually quartz. The size of grain is measured and numbered, as in #20 silica sand.

single-phase A single electrical circuit of alternating current.

single pole, single throw (SPST) switch The most basic on/off electrical switch, handling one line circuit (single pole) each time the switch is thrown. Single pole, double throw (SPDT) also operates one line circuit, but it alternately electrifies two different load circuits.

sink *See* heat riser.

skid pack A metal frame onto which is mounted the equipment needed to operate a portable spa, usually a pump and motor, filter, heater, blower, and control devices. Also called a spa pack.

skimmer A part of the circulation system that removes debris from the surface of the water by drawing surface water through it.

slide valve A guillotine-like plumbing device that restricts or shuts off the flow of water in a line. In essence, a gate valve that slides a disc up or down across the flow in the pipe. *See also* gate valve.

slip fitting A plumbing fitting that joins to a pipe without threads, but that slides into a prefitted space.

slurry A thin, watery mixture.

soda ash (Na$_2$CO$_3$) Sodium carbonate. A white powdery substance used to raise the pH of water.

sodium bicarbonate (NaHCO$_3$) *See* bicarbonate of soda.

sodium bisulfate (NaHSO$_4$) A chemical compound used to lower the pH and total alkalinity of water (dry acid).

sodium dichloro-s-triazinetrione ($C_3N_3O_3Cl_2Na$) Dichlor. A granular, stabilized form of chlorine sanitizer, generally about 60 percent available.

sodium hypochlorite (NaOCl) A liquid solution containing approximately 15 percent chlorine.

sodium thiosulfate A chemical used to neutralize chlorine in a test sample prior to testing for pH, without that a false reading might result.

soft water So called soft water is very low in calcium and magnesium (less than 100 ppm) and is therefore considered aggressive or hungry and likely to dissolve those minerals when it comes in contact with plaster.

solar panel A glass or plastic enclosure, usually 4-by-8 feet by a few inches thick, through which water flows absorbing heat from the sun. The basic component of a solar heating system.

solder A soft metal wire (usually lead or alloys) used as a bonding agent when plumbing with copper pipe and fittings (an action called sweating).

spalling The breaking away of plaster in thin layers.

spiking In water chemistry, when readings of a parameter alternately register extremely high then extremely low.

spin filter A style of DE filter in which the grids rotate to facilitate washing the dirt and DE off the grid surfaces.

squared A number multiplied by itself. For example, $4 \times 4 = 4^2 = 16$.

square flange A casing style of certain motors, designed to adapt to certain pumps.

stabilizer Any compound that tends to increase water's resistance to chemical change. *See also* conditioner.

stack Refers to the vent pipe of a heater. A heater installed indoors requires such a stack, while one installed outdoors uses a flat-top vent and is called stackless.

stanchion The vertical support pipe of a railing.

standing pilot A heater ignition device in which the gas flame (to ignite the main burner) is always burning.

static head The amount of resistance (pressure) created from nonmoving restrictions in a water circulation system.

stator The stationary part in an electric motor that contains the wound wires (winding) that generates magnetism to drive the rotor.

strainer basket A plastic mesh container that strains debris from water flowing through it inside the strainer pot.

strainer pot The housing on the intake side of a pump that contains a strainer basket and serves as a water reservoir to assist in priming.

street fitting Any plumbing fitting that has one male end and one female end.

submersible pump A pump and motor that can be submerged to pump out or recirculate a body of water. Also called a sump pump.

superchlorinate Periodic application of extremely high levels of chlorine (in excess of 3 ppm) to completely oxidize any organic material in a body of water (including bacteria) and leave a substantial chlorine residual. Procedure performed to sanitize elements in water that might resist normal chlorination. Also called shocking or shock treatment.

surge chamber A tank designed to temporarily hold the water displaced by bathers in a pool, returning the water when the bathers exit the pool. Designed into commercial pools where there are heavy bather loads.

sweating The soldering together of copper pipe and fittings. Also refers to the formation of condensation on the exterior of pipes and equipment containing water, caused by temperature differences between the water in the pipe and the surrounding air.

tear down To disassemble a piece of equipment for service or repair. Specifically used in reference to filter cleaning (as opposed to backwashing or other temporary cleaning), a filter teardown means complete disassembly and cleaning. Also called break down.

Teflon tape A thin fabric provided on a roll used to coat threaded plumbing fittings to prevent leaks.

telepole A metal or fiberglass pole that extends to twice its original length, the two sections locking together. The telepole is used with most pool and spa cleaning tools.

T fitting A plumbing fitting shaped like the letter T that connects pipes from three different sources.

therm The unit of measurement you read on a gas bill; is 100,000 Btu/hour of heat.

thermal overload protector A temperature-sensitive switch on a motor that cuts the electric current in the motor when a preset temperature is exceeded.

thermistor The sensor that sends temperature information to the thermostat.

thermocouple *See* pilot generator. Also called a thermopile.

thermostat A part of the heater control circuit. An adjustable device that senses temperature and can be set to break the circuit when a certain temperature is reached. It then closes the circuit when the temperature falls below that level.

Thoroseal A brand name of waterproof concrete commonly used in pool and spa repairs.

three-port valve A plumbing fitting used to divert flow from one direction into two other directions.

time clock An electromechanical device that automatically turns an appliance on or off at preset intervals.

titration A chemical test method to determine the amount of a substance in a sample of water. A sample is colored, then drops of the titrant are added to the sample. When the sample turns clear or changes to another predicted color, the result of the test is determined by counting the number of drops of titrant that were required to create that change. For example, an acid demand test is performed with titration. The number of drops of titrant required in that test determines the amount of acid to add to the water.

torque Application of an amount of force against an object. Usually refers to the amount of force required to tighten a fastener to a specified requirement.

total alkalinity The measurement of all alkaline substances (carbonates, bicarbonates, and hydroxides) in a body of water.

total dissolved solids (TDS) The sum of all solid substances dissolved in a body of water, including minerals, chemicals, and organics.

total dynamic head (TDH) The total amount of resistance (backpressure) created by restriction to free flow of water in a circulation system. TDH is composed of vacuum (restrictions to suction) and pressure (restrictions to returning water to the pool). TDH is expressed in feet.

transformer A device that changes electric current from one voltage to another.

Commonly used in electronic ignition heaters where the supply of 110 or 220 volts is converted (transformed) to 24 volts for use in the heater control circuit.

trichloro-s-triazenetrione ($C_3N_3O_3Cl_3$) Trichlor. A dry form of chlorine, produced in granular or tablet form at around 90 percent available chlorine.

tripper The small metal clamp that fits in the clock face of a time clock to activate the clock (on/off) at preset times. Also called a dog.

trisodium phosphate A detergent used to clean filter grids and cartridges, it breaks down oils that acid washing alone cannot. Commercially sold as TSP.

turbidity Cloudiness.

turnover rate The amount of time required for a circulation system to filter 100 percent of the water in a particular body of water.

UL Underwriters Laboratory. The agency that sets standards for and certifies the safety of electrical and mechanical devices.

underdrain The assembly in the bottom of a sand filter that connects laterals and plumbing for water filtration.

union A plumbing fitting connecting two pipes by means of threaded male and female counterparts on the end of each pipe.

uniseal A type of motor case designed to fit a specific style of pump.

up-rated The full-rated motor operates to its listed capacity. The up-rated motor has a similar horsepower rating but will function to even higher standards if called upon. *See also* service factor.

vacuum A device used to clean the underwater surfaces of a pool or spa by creating suction in a hose line.

valve A device in plumbing that controls the flow of water.

vanishing edge pool A pool or water feature with one or more sides that are below the overall surface water level. Water spills over the vanishing edge into a catch basin below for recirculation to the pool. Also known as negative edge or rim flow design.

Visoflame tube A trade name of the Teledyne Laars heater pilot lighting system.

volt The basic unit of electric current measurement expressing the potential or pressure of the current. Volts = watts ÷ amps.

volute The volute is the housing that surrounds the impeller and diffuser in a pump, channeling the water to a discharge pipe.

water column pressure The unit of measurement used in evaluating gas pressure with a manometer.

water feature A fall, rockscape, or other decorative design using moving water.

water hammer The phenomenon of variations in water pressure in a closed pipe that results from too rapid acceleration or deceleration of flow. Momentary pressures in excess of normal static pressure producing a hammering sound.

watt The way appliances are measured for power consumption. One watt is equal to the volume of one amp delivered at the pressure of one volt. Watts = amps × volts.

weir The barrier in a skimmer over which water flows. A floating weir raises and lowers its level to match the water level in a pool or spa. Another type is shaped like a barrel and floats up and down inside the skimmer basket.

winterizing The process of preparing a pool or spa to prevent damage from freezing temperatures and other harsh weather conditions.

Reference sources

Virtually every equipment and supply manufacturer produces valuable literature with repair and maintenance information, technical guidelines, and exploded diagrams of their products. Collect literature at trade shows, from your wholesale supply house, and anywhere else you can.

Many manufacturers conduct factory training seminars on the repair and maintenance of their products. Trade periodicals publish a list of these educational opportunities. Take advantage of them. Doug Steimele of California Pools and Spas says

> The service technician should feel free to call a reputable pool builder in the area for advice. Most builders would rather dispense some free advice than have uninformed technicians giving pool owners inaccurate information. Ask, ask, ask!

Excellent technical articles, tips and news regarding new products can be found in the following periodicals:

Aqua
1846 Hoffman St.
Madison, WI 53704
(608) 249-1153

Pool and Spa News
3923 W. 6th St.
Los Angeles, CA 90020
(213) 385-3926

Service Industry News
Swimming Pool and Spa Age
6151 Powers Ferry Rd.
Atlanta, GA 30339
(404) 955-2500

The following publications were used for reference in preparing this book.

Water Analysis Handbook, Loveland, CO: Hach Company, 1992.

Manas, Vincent T. *National Plumbing Code Handbook.* New York: McGraw-Hill, 1957.

Uniform Swimming Pool, Spa and Hot Tub Code. International Association of Plumbing and Mechanical Officials, 1988.

Taylor, Charlie. *Everything You Always Wanted To Know About Pool Care But Didn't Know Where To Ask.* Chino, CA: Service Industry Publications, 1989.

United States Navy. *Basic Electricity.* Washington, D.C.: U.S. Government Printing Office, 1969.

Wood, Robert W. 1993 *Home Electrical Wiring Made Easy*. 2nd ed. Blue Ridge Summit, PA: TAB Books.

The advice and resources of two industry membership organizations has also been of great value in compiling this book. There are also several excellent state and local industry organizations around the United States and Canada. I urge you to look into the advantages of joining one or more in your area.

Independent Pool and Spa Service Association (IPSSA)
P.O. Box 3396
Granada Hills, CA 91394
(818) 360-9505

National Spa and Pool Institute (NSPI)
2111 Eisenhower Ave.
Alexandria, VA 22314-4698
(703) 838-0083

To find companies and manufacturers named in this book, and get access to their technical bulletins and helplines, contact *Pool and Spa News* for its annual industry directory, published in January. It contains a wealth of information and is an excellent source reference.

Index

Illustration page numbers are in **boldface**.

Other Bestsellers of Related Interest

The McGraw-Hill 36-Hour Course Entrepreneurship
—*James W. Halloran*
Americans in record numbers are starting their own business—an exciting and challenging choice that requires top-notch guidance. Now entrepreneurial expertise for thousands of prospective small business owners is only 36 hours away. Tailor-made for busy people who want to master the fundamentals of successful entrepreneurial practice without wasting time and money, the McGraw-Hill 36-Hour Course on Entrepreneurship is the best and quickest way to learn how to start and run a successful business.
0-07-025877-5 $16.95 Paper
0-07-025876-7 $39.95 Hard

How to Run a Small Business, Seventh Edition
—*J. K. Lasser Institute*
From the J. K. Lasser Institute comes the latest edition of the most complete, authoritative, up-to-date small business consultant any budding entrepreneur could hire. Extensively revised to reflect new office technologies, late-breaking changes in the tax and employment laws, and even the effects of the North American Free Trade Agreement, this A to Z resource answers hundreds of basic and not-so-basic questions on everything from financing and choosing a location to hiring and firing, extending credit, buying insurance, and more.
0-07-036576-8 $27.95 Hard

Endless Referrals: Network Your Everyday Contacts into Sales $$$
—*Bob Burg*
In a sell-saturated world of junk faxes and telemarketing, virtually, everyone needs a tested system for building lucrative contacts...one that sets them apart from the pack. Complete with case studies of star networkers, Endless Referrals shows readers how to network to advance their career...set up win-win cross-promotions...start their own formal net-working association...even soup up customer service. Page after page, it's packed with proven approaches for turning any contact into a sky's-the-limit sales opportunity.
008942-6 $14.95 Paper

**Why Customers Don't Do
What You Want Them To Do—And What To Do About It**
—*Ferdinand F. Fournies*
Ferdinand F. Fournies, sales trainer extraordinaire, knows that the key to selling is at once basic and revolutionary: "What successful salespeople do is to help people to buy." So he's written the only how-to book that approaches selling from the customer's point of view. Reowned for his ability to get immediately to the heart of any situation, Fournies seasons his guidance with real-world dialogue taken directly from selling situations and training sessions. Salespeople will also appreciate the checklists designed to help them prepare for sales calls and analyze post-call results.
0-07-021701-7   **$12.95 Paper**
0-07-021700-9   **$29.95 Hard**

The McGraw-Hill Handbook of Business Letters, Third Edition
—*Roy W. Poe*
Bigger and better than ever, the new third edition of The McGraw-Hill Handbook of Business Letters packs more than 300 ready-to-use business letters and memos. Indispensable for taking fast, responsive action in scores of typical business situations, these carefully crafted letters help harried executives promote their products and services, mend fences with customers, make proposals, collect bills, or just "No" without ruffling feathers.
0-07-050451-2   **$19.95 Paper**
0-07-050425-3   **$59.50 Hard**

How to Order

☎ **Call 1-800-822-8158**
24 hours a day,
7 days a week
in U.S. and Canada

✉ **Mail this coupon to:**
McGraw-Hill, Inc.
Blue Ridge Summit, PA
17294-0840

 Fax your order to:
717-794-5291

 EMAIL
70007.1531@COMPUSERVE.COM
COMPUSERVE: GO MH

Thank you for your order!

Shipping and Handling Charges

Order Amount	Within U.S.	Outside U.S.
Less than $15	$3.45	$5.25
$15.00 - $24.99	$3.95	$5.95
$25.00 - $49.99	$4.95	$6.95
$50.00 - and up	$5.95	$7.95

EASY ORDER FORM—
SATISFACTION GUARANTEED

Ship to:

Name _____

Address _____

City/State/Zip _____

Daytime Telephone No. _____

ITEM NO.	QUANTITY	AMT.
Method of Payment:	Shipping & Handling charge from chart below	
☐ Check or money order enclosed (payable to McGraw-Hill)	Subtotal	
	Please add applicable state & local sales tax	
☐ Cards ☐ *VISA*	TOTAL	

☐ MasterCard   ☐ DISCOVER

Account No. ☐☐☐☐☐☐☐☐☐☐☐☐☐☐☐☐

Signature _____ Exp. Date _____
Order invalid without signature

**In a hurry? Call 1-800-822-8158 anytime,
day or night, or visit your local bookstore.**

Code = BC44ZNA

ABOUT THE AUTHOR

Terry Tamminen, 42, has spent nearly 20 years in pool/spa maintenance and property management. He is a passionate marine environmentalist and works as the Santa Monica BayKeeper to protect the coastal waters of Southern California. He is married to entertainment executive Karen Graff Borell.

ABOUT THE ILLUSTRATOR

Owen W. Smith is a stage designer and scenic artist. He is a professor of theatre at California State University, Northridge.